High Frequency Circuit Design

with

Keysight and MATLAB Design Examples

High Frequency Circuit Design

With

Keysight and MATLAB Design Examples

Ali Behagi

Techno Search

Ladera Ranch, CA 92694

High Frequency Circuit Design

with

Keysight and MATLAB Design Examples

ISBN -13: 978-0-9964466-7-9

Published in USA
Techno Search
Ladera Ranch, CA 92694

Foreword

Unlike the many traditional textbooks written mainly for the classroom teaching, the "High Frequency Circuit Design - with Keysight and MATLAB Design Examples" can be taught in a classroom or in a computer lab where students can use a hands-on step-by-step approach in designing RF and microwave circuits.

The author uses MATLAB Scripting to post-process the simulation data created in designing discrete and distributed impedance matching networks.

This book introduces not only a solid understanding of the high frequency concepts and components such as, Network Parameters, Transmission Lines, Resonant Circuits, Filters, Discrete and Distributed Impedance Matching Networks, Maximum Gain Amplifier, and Low Noise Amplifier, but more importantly it shows how to use the Keysight design tools to analyze, synthesize, tune and optimize these essential components in a design flow as practiced in industry.

Professor Behagi's book is valuable in that it marries the high frequency circuit design theory with many practical design examples. Learning the fundamentals of high frequency circuit design combined with the practical application of the Keysight Equation Editor and the MATLAB Scripting will broaden your potential career opportunities.

The investment in learning the foundational skills of high frequency circuit design taught in this book provides students and engineers with valuable knowledge that will remain relevant for a long time to come.

Joe Civello
Keysight Technologies
ADS Planning and Marketing Manager
1400 Fountaingrove Parkway
Santa Rosa, CA 95403, USA

Preface

The High Frequency Circuit Design book introduces the reader with a solid understanding of RF and microwave Concepts, Components, Transmission Lines, Network Parameters, Smith Charts, Resonant Circuits, Filters, Power Transfers, Discrete and Distributed Impedance Matching Networks, Maximum Gain Amplifiers, and Low Noise Amplifiers.

Almost all subject matter covered in the book is accompanied by practical examples that are solved using analytical or empirical equations. University students and practicing engineers will find this book both as a potent learning tool and as a reference guide to quickly setup designs using MATLAB scripts in an Equation Editor. The author thoroughly covers the basics as well as introducing Computer Aided Design techniques that may not be familiar to students.

The organization of the book is as follows:

In chapter 1, a thorough analysis of RF and microwave concepts and components are presented. Components such as, straight wire, flat ribbon, physical resistors, physical capacitors, and physical inductors are analyzed and their input impedance are determined.

In chapter 2, propagation of the plane waves in different media is introduced. Popular types of transmission lines such as coaxial, microstrip, stripline, and waveguide are defined and their parameters are analyzed. Several transmission line components are modeled and their electrical performance are discussed. Microstrip bias feed and directional couplers are also designed.

In Chapter 3, derivation of RF and microwave network parameters, development and use of the network S parameters, and the movement of the lumped and distributed elements on the Smith chart are presented.

In the first half of Chapter 4, the subject of series and parallel resonant circuits, the effect of load resistance on the bandwidth, the tuning and optimization of the circuit components, and the design of the tapped - capacitor and inductor are discussed. In the second half of chapter 4, design of the lowpass and highpass filters, generation of the physical models, and construction of the filter prototypes are presented.

In Chapter 5 the conditions for maximum power transfer and the equations for matching any two impedances are derived. Both analytical and graphical techniques are used to design narrowband and broadband matching networks. In several examples the impedance matching equations are derived to solve any complex impedance matching problem. Derivation of equations for the Q factor and the number of L-networks, designing with Q curves on the Smith chart, Fano's limit theorem, and the effect of finite Q on the matching networks are also treated in this chapter.

In Chapter 6, analytical design equations for quarter-wave transformer and single-stub matching networks are derived and narrowband and broadband distributed matching networks are designed.

In Chapter seven, the single-stage amplifiers are designed by utilizing two different impedance matching objectives. The first amplifier is designed for maximum gain where the input and output are conjugately matched to the source and load impedance. The second amplifier is a low noise amplifier where the transistor is selectively mismatched to achieve a specific Noise Figure.

Ali Behagi
March 2018

Table of
Contents

Chapter 1

RF and Microwave Concepts and Components

1.1 Introduction

An electromagnetic wave is a propagating wave that consists of electric and magnetic fields. The electric field is produced by stationary electric charges while the magnetic field is produced by moving electric charges. A time-varying magnetic field produces an electric field and a time-varying electric field produces a magnetic field. The characteristics of electromagnetic waves are frequency, wavelength, phase, impedance, and power density. In free space, the relationship between the wavelength and frequency is given by Equation (1-1).

$$\lambda = \frac{c}{f} \qquad (1\text{-}1)$$

In the MKS system, λ is the wavelength of the signal in meters, c is the velocity of light approximately equal to 300,000 kilometers per second, and f is the frequency in cycles per second, or Hz.

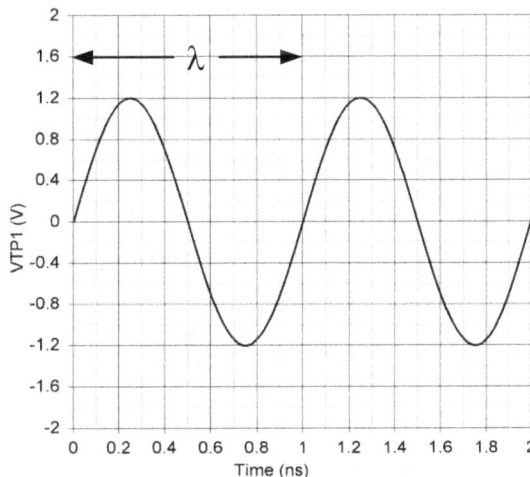

Figure 1-1 A time varying voltage waveform

The electromagnetic spectrum is the range of all possible frequencies of electromagnetic radiation. They include radio waves, microwaves, infrared radiation, visible light, ultraviolet radiation, X-rays and gamma rays. In the field of RF and microwave engineering the term RF generally refers to Radio Frequency signals with frequencies in the 3 KHz to 300 MHz range. The term Microwave refers to signals with frequencies from 300 MHz to 300 GHz having wavelengths from 1 meter to 1 millimeter. The RF and microwave frequencies form the spectrum of all radio, television, data, and satellite communications. Figure 1-2 shows a spectrum chart highlighting the RF and microwave frequencies up through the extremely high frequency, EHF, range or 300 GHz. This text will focus on the RF and microwave frequencies as the foundation for component design techniques. The application of Linear Technology software will enhance the student's understanding of the underlying principles presented throughout the text. The practicing engineer will find the text an invaluable reference to the RF and microwave theory and techniques. The numerous examples enable the setup and design of many RF and microwave circuit design problems.

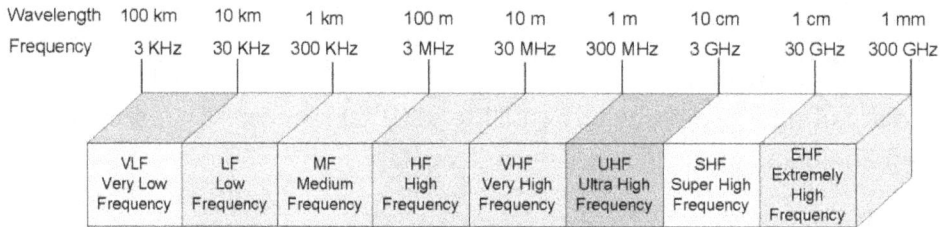

Wavelength	100 km	10 km	1 km	100 m	10 m	1 m	10 cm	1 cm	1 mm
Frequency	3 KHz	30 KHz	300 KHz	3 MHz	30 MHz	300 MHz	3 GHz	30 GHz	300 GHz

VLF Very Low Frequency	LF Low Frequency	MF Medium Frequency	HF High Frequency	VHF Very High Frequency	UHF Ultra High Frequency	SHF Super High Frequency	EHF Extremely High Frequency

Figure 1-2 Electromagnetic spectrums from VLF to EHF

The spectrum chart of Figure 1-2 is intended as a general guideline to the commercial nomenclature for various sub bands. There is typically overlap across each of the boundaries as there is no strict dividing line between the categories. The RF frequencies typically begin in the very low frequency, VLF, range through the very high frequency, VHF, range. Microwaves are typically the ultra-high frequency, UHF, super high frequency, SHF and extremely high frequency, EHF, frequency ranges. During World War II microwave engineers developed a further detailed classification of the microwave frequencies into a band-letter designation. In 1984 the Institute of Electrical and Electronics Engineers, IEEE, agreed to standardize the letter designation of the microwave frequencies. These designators and their frequency ranges are shown in Table 1-1.

Band Designator	L Band	S Band	C Band	X Band	Ku Band	K Band	Ka Band
Frequency Range GHz	1 to 2	2 to 4	4 to 8	8 to 12	12 to 18	18 to 27	27 to 40

Table 1-1 Microwave band letter designators

Engineering students spend much of their formal education learning the basics of inductors, capacitors, and resistors. Many are surprised to find that as we enter the high frequency, HF, part of the electromagnetic spectrum these components are no longer a singular (ideal) element but rather a network of circuit elements. Components at RF and microwave frequencies become a network of resistors, capacitors, and inductors. This leads to the complication that the component's characteristics become quite frequency dependent. For example, we will see in this chapter that a capacitor at one frequency may in fact be an inductor at another frequency.

1.2 Straight Wire, Skin Effects, and Flat Ribbon

In this section we will begin with a basic examination of the straight wire inductance and move into more complete characterization of inductors. Similarly we will look at resistor and capacitor design and their implementation at RF and microwave frequencies. Discrete resistors, capacitors, and inductors are often referred to as lumped elements. RF and microwave engineers use the terminology to differentiate these elements from those designed in distributed printed circuit traces. Distributed component design is introduced in Chapter 2.

Straight Wire Inductance

A conducting wire carrying an AC current produces a changing magnetic field around the wire. According to Faraday's law the changing magnetic field induces a voltage in the wire that opposes any change in the current flow. This opposition to change is called self-inductance. At high frequencies even a short piece of straight wire possesses frequency dependent resistance and inductance behaving as a circuit element.

Example 1.1: Calculate the inductance of a three inch length of AWG #28 copper wire in free space. Also calculate the reactance of the same wire at 60 Hz, 500 MHz, and 1 GHz.

Solution: The straight wire inductance can be calculated from the empirical Equation (1-2).

$$L = K\ell \left(\ln \frac{4\ell}{D} - 0.75 \right) nH \qquad (1\text{-}2)$$

Where:

ℓ = Length of the wire

D = Diameter of the wire (from Appendix A).

K = 2 for dimensions in cm and K=5.08 for dimensions in inches

Using Appendix A the diameter of the AWG#28 wire is found to be 0.0126 inches. Solving Equation (1-2) the inductance is calculated.

$$L = 5.08 \ (3) \left(\ln \frac{4 \ (3)}{0.0126} - 0.75 \right) = 93.1 \ nH$$

It is interesting to examine the reactance of the wire. We know that the reactance is a function of the frequency and is related to the inductance by the following equation.

$$X_L = 2\pi f L \quad \Omega \qquad (1\text{-}3)$$

Where:

f is the frequency in Hz

L is the inductance in Henries

Calculating the reactance at 60Hz, 500MHz, and 1GHz we can see how the reactive component of the wire increases dramatically with frequency. At 60Hz the reactance is well below 1Ω while at microwave frequencies the reactance increases to several hundred ohms.

60 Hz: $X_L = 2\pi(60)(93.1 \cdot 10^{-9}) = 35\ \mu\Omega$

500MHz: $X_L = 2\pi(10^6)(500)(93.1 \cdot 10^{-9}) = 292\ \Omega$

1 GHz: $X_L = 2\pi(10^9)(93.1 \cdot 10^{-9}) = 585\ \Omega$

It is a common practice in most commercial microwave software programs to specify resistivity in relative terms, compared to copper. Table 1-2 provides a reference of the materials used in microwave engineering.

Material	Resistivity Relative to Copper	Actual Resistivity Ω-meters	Actual Resistivity Ω-inches
Copper, annealed	1.00	$1.68 \cdot 10^{-8}$	$6.61 \cdot 10^{-7}$
Silver	0.95	$1.59 \cdot 10^{-8}$	$6.26 \cdot 10^{-7}$
Gold	1.42	$2.35 \cdot 10^{-8}$	$9.25 \cdot 10^{-7}$
Aluminum	1.64	$2.65 \cdot 10^{-8}$	$1.04 \cdot 10^{-6}$
Tungsten	3.25	$5.60 \cdot 10^{-8}$	$2.20 \cdot 10^{-6}$
Zinc	3.40	$5.90 \cdot 10^{-8}$	$2.32 \cdot 10^{-6}$
Nickel	5.05	$6.84 \cdot 10^{-8}$	$2.69 \cdot 10^{-6}$
Iron	5.45	$1.00 \cdot 10^{-7}$	$3.94 \cdot 10^{-6}$
Platinum	6.16	$1.06 \cdot 10^{-7}$	$4.17 \cdot 10^{-6}$
Tin	52.8	$1.09 \cdot 10^{-7}$	$4.29 \cdot 10^{-6}$
Nichrome	65.5	$1.10 \cdot 10^{-6}$	$4.33 \cdot 10^{-5}$
Carbon	2083.3	$3.50 \cdot 10^{-5}$	$1.38 \cdot 10^{-3}$

Table 1-2 Resistivity of common materials relative to copper

1.3 Skin Effect in Conductors

At RF and microwave frequencies, due to the larger inductive reactance caused by the increase in flux linkage toward the center of the conductor, the current in the conductor is forced to flow near the conductor surface. As a result the amplitude of the current density decays exponentially with the depth of penetration from the surface. Figure 1-3 shows the cross section of a cylindrical wire with the current density area shaded.

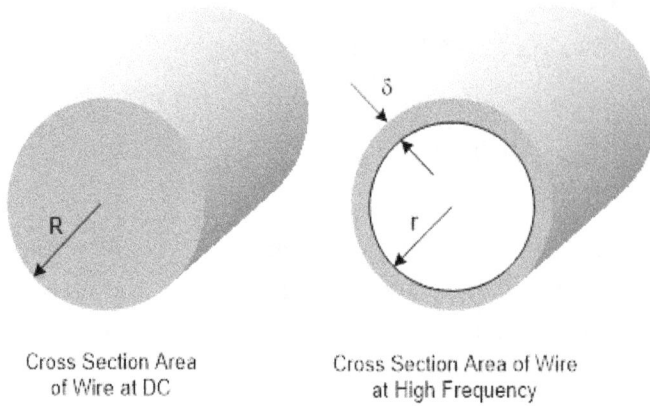

Cross Section Area
of Wire at DC

Cross Section Area of Wire
at High Frequency

Figure 1-3 Cross section of current flow in the conductor showing effect of skin depth

At low frequencies the entire cross sectional area is carrying the current. As the frequency increases to the RF and microwave region, the current flows much closer to the outside of the conductor. At the higher end of microwave frequency range, the current is essentially carried near the surface with almost no current at the central region of the conductor. The skin depth, δ, is the distance from the surface where the charge carrier density falls to 37% of its value at the surface. Therefore 63% of the RF current flows within the skin depth region. The skin depth is a function of the frequency and the properties of the conductor as defined by Equation (1-4). As the cross sectional area of the conductor effectively decreases the resistance of the conductor will increase.

$$\delta = \sqrt{\frac{\rho}{\mu \, \pi \, f}} \qquad (1\text{-}4)$$

Where:

δ = skin depth
ρ = resistivity of the conductor
f = frequency
μ = permeability of the conductor

Use caution when solving Equation (1-4) to keep the units of ρ and μ consistent. Table 1-2 contains values of resistivity in units of Ω-meters and Ω-inches. The permeability μ is the permeability of the conductor. It is a property of a material to support a magnetic flux. Some reference tables will show relative permeability. In this case the relative permeability is normalized to the permeability of free space which is: 4π (10^{-7}) Henries per meter. The relationship between relative to actual permeability is given in Equation (1-5). Most conductors have a relative very close to one. Therefore, conductor permeability is often given the same value as permeability of free space.

$$\mu = \mu_r \, \mu_o \qquad\qquad (1-5)$$

Where:

μ = actual permeability of the material

μ_r = relative permeability of material

μ_o = permeability of free space

Example 1.2: Calculate the skin depth of copper wire at a frequency of 25 MHz.

Solution: From Table 1-2, $\rho = 6.61 \bullet 10^{-7}$ Ω-inches. Using Equation (1-4), and converting the permeability from H/m to H/inch, the skin depth is:

$$\delta = \sqrt{\frac{6.61 \cdot 10^{-7}}{\left(3.19 \cdot 10^{-8}\right)\pi \left(25 \cdot 10^{6}\right)}} = 5.14 \cdot 10^{-4} \; inches$$

Although we have been considering the skin depth in a circular wire, skin depth is present in all shapes of conductors. A thick conductor is affected more by skin effect at lower frequencies than a thinner conductor. One of the reasons that engineers are concerned about skin effect in conductors is the fact that as the resistance of the conductor increases, so does the thermal heating in the wire. Heat can be a destructive force in high power RF circuits causing burn out of conductors and potentially hazardous conditions to personnel. Also the frequency dependence of the skin effect may make it difficult to maintain the impedance of a transmission line structure. This effect will be examined in Chapter 2 with the study of transmission lines.

As the frequency increases, the current is primarily flowing in the region of the skin depth. It can be visualized from Figure 1-3 that a wire would have greater resistance at higher frequencies due to the skin effect. The resistance of a length of wire is determined by the resistivity and the geometry of the wire as defined by Equation (1-6).

$$R = \frac{\rho \ell}{A} \qquad \Omega \qquad\qquad (1\text{-}6)$$

Where:

ρ = Resistivity of the wire

ℓ = Length of the wire

A = Cross sectional area

Example 1.3: Calculate the resistance of a 12 inch length of AWG #24 copper wire at DC and at 25 MHz.

Solution: The radius of the wire can be found in Appendix A. The DC resistance is then calculated using Equation (1-6).

$$R_{DC} = \frac{\left(6.61 \cdot 10^{-7}\right) \cdot 12}{\pi \left(\dfrac{0.0201}{2}\right)^2} = 0.025 \ \Omega$$

To calculate the resistance at 25 MHz the cross sectional area of the conduction region must be redefined by the skin depth of Figure 1-5. We can refer to this as the effective area, A$_{eff}$.

$$A_{eff} = \pi \ (R^2 - r^2) \qquad\qquad (1\text{-}7)$$

Where: $r = R\text{-}\delta$.

For the AWG#24 wire at 25 MHz the A$_{eff}$ is calculated as:

$$A_{eff} = \pi \left(\frac{0.0201}{2}\right)^2 - \pi \left(\left(\frac{0.0201}{2}\right) - 5.14 \cdot 10^{-4}\right)^2 = 3.14 \cdot 10^{-5} \ in^2$$

Then apply Equation (1-6) to calculate the resistance of the 12 inch wire at 25 MHz.

$$R_{25MHz} = \frac{12 \left(6.61 \cdot 10^{-7}\right)}{3.14 \cdot 10^{-5}} = 0.253 \; \Omega$$

We can see that the resistance at 25 MHz is more than 10 times greater than the resistance at DC.

Flat Ribbon Inductance

Flat ribbon style conductors are very common in RF and microwave engineering. Flat ribbon conductors are encountered in RF systems in the form of low inductance ground straps. Flat ribbon conductors can also be encountered in Microwave Integrated Circuits (MIC) as gold bonding straps. When a very low inductance is required the flat ribbon or copper strap is a good choice. The flat ribbon inductance can be calculated from the empirical Equation (1-8).

$$L = K\ell \left[\ln\left(\frac{2\ell}{W+T}\right) + 0.223\left(\frac{W+T}{\ell}\right) + 0.5 \right] \quad nH \qquad (1\text{-}8)$$

Where:

$\ell =$ The length of the wire

$K =$ dimensions in cm and $K=5.08$ for dimensions in inches

$W =$ the width of the conductor

$T =$ the thickness of the conductor

1.4 Physical Resistors

The resistance of a material determines the rate at which electrical energy is converted to heat. In Table 1-2 we have seen that the resistivity of materials is specified in Ω-meters rather than Ω/meter. This facilitates the calculation of resistance using Equation (1-6). When working with low frequency or logic circuits we are used to treating resistors as ideal resistive components.

Example 1.4: Plot the input impedance of a 50 Ohm ideal resistor from 0 to 2 GHz.

Solution: In LTspice connect a 50 Ohm AC source to an ideal 50 Ohm resistor as shown in Figure 1-4.

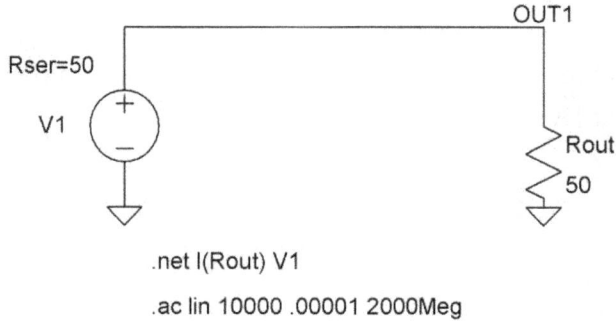

OUT1

Rser=50

V1

Rout
50

.net I(Rout) V1

.ac lin 10000 .00001 2000Meg

Figure 1-4 Schematic of an Ideal 50 Ω resistor

Simulate the schematic (see the simulation procedure in Example 2.1) and display the input impedance over a frequency range of 0 to 2 GHz, as shown in Figure 1-5. The plot shows a constant resistance at all frequencies.

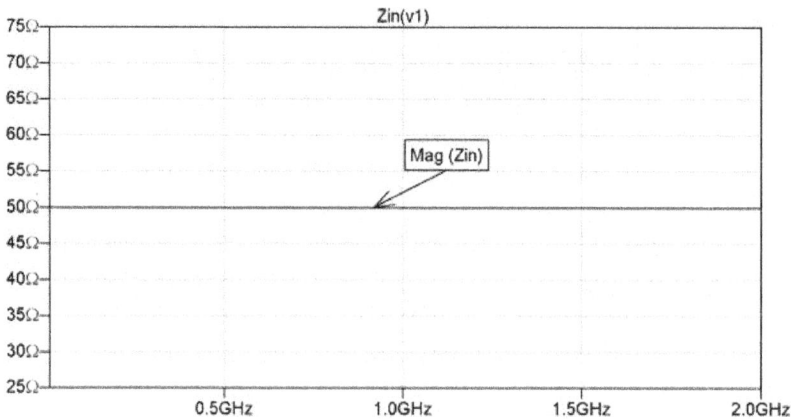

Zin(v1)

Mag (Zin)

Figure 1-5 Ideal 50Ω resistor impedance versus frequency

At RF and microwave frequencies however, resistors also possess inductive and capacitive elements. The stray inductance and capacitance associated with a resistor are often called parasitic elements.

Example 1.5: Plot the input impedance of a 50 Ohm leaded resistor from 0 to 2 GHz.

Solution: In LTspice connect a 50 Ohm AC source to the 50 Ohm leaded resistor as shown in Figure 1-6. For a 1/8 watt leaded resistor it is common for each lead to have about 10 nH of inductance. The body of the resistor may exhibit 0.5 pF capacitance between the leads. The leaded resistor model is shown in Figure 1-6.

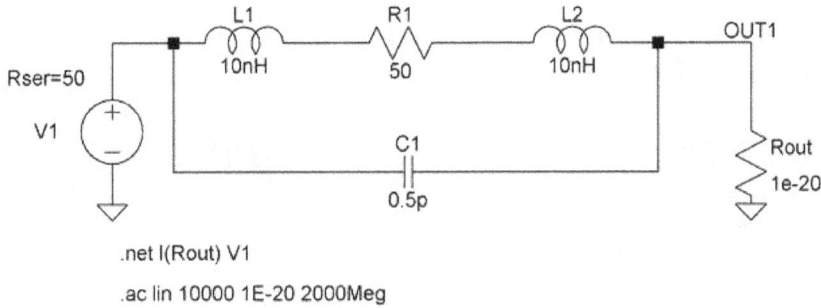

Figure 1-6 Schematic of the 50 Ohm leaded resistor with parasitic elements

Simulate the schematic in LTspice and plot the input impedance as a function of frequency from 0 to 2 GHz, as shown, in Figure 1-7.

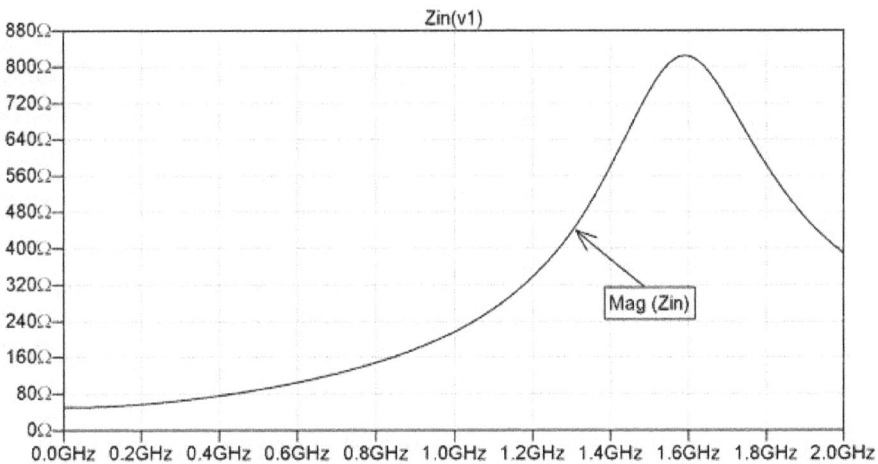

Figure 1-7 Leaded-resistor impedance versus frequency

Chip Resistors

Thick film resistors are used in most contemporary electronic equipment. The thick film resistor, often called chip resistor, comes close to eliminating much of the inductance that plagues the leaded resistor. The chip resistor works well with popular surface mount assembly techniques preferred in modern electronic manufacturing. Figure 1-8 shows a typical thick film chip resistor along with a cross section of its design.

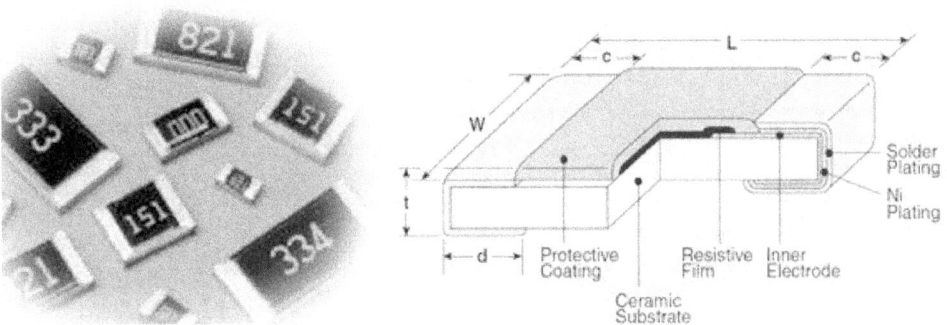

Figure 1-8 Thick film chip resistors (*courtesy of KOA Speer Electronics*)

There are many types of chip resistors designed for specific applications. Common sizes and power ratings are shown in Table 1-3.

Size	Length x Width	Power Rating
0201	20mils x 10mils	50mW
0402	40mils x 20mils	62mW
0603	60mils x 30mils	100mW
0805	80mils x 50mils	125mW
1206	120mils x 60mils	250mW
2010	200mils x 100mils	500mW
2512	250mils x 120mils	1W

Table 1-3 Standard thick film resistor size and approximate power rating

The thick film resistor is comprised of a carbon based film that is deposited onto the substrate. Contrasted with a thin film resistor that is typically etched onto a substrate or printed circuit board, the thick film resistor can usually handle higher power dissipation. The ends of the chips have metalized wraps that are used to attach the resistor to a circuit board. Some manufacturers may provide models that can be incorporated into the software. There are also companies that specialize in developing CAD models of components such as Modelithics, Inc. Modelithics has a wide variety of component model libraries that can be incorporated into the simulation software.

1.5 Physical Inductors

In the previous sections we introduced the topic of inductance. The inductance of straight cylindrical wire and flat ribbon conductors were discussed. The primary method of increasing inductance is not to simply keep increasing the length of a straight conductor but rather form a coil of wire. Forming a coil of wire increases the magnetic flux linkage and greatly increases the overall inductance. Because of the greater surrounding magnetic flux, inductors store energy in the magnetic field. Lumped element inductors are used in bias circuits, impedance matching networks, filters, and resonators. As we will see throughout this section inductors are realized in many forms including: air-core, toroidal and very small chip inductors. The concept of Q factor is introduced and will come up frequently in RF and microwave circuit design. It is a unit-less figure of merit that is used in circuits in which both reactive and resistive elements coexist. The individual air wound inductors in Figure 1-9, are actually networks made up of resistors, inductors, and capacitors.

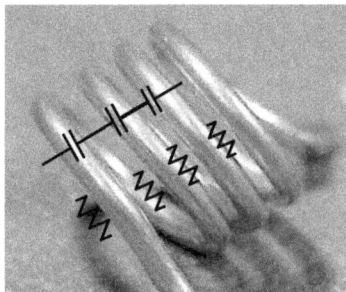

Figure 1-9 Air wound inductor showing the wire resistance and inter-winding capacitance

Basically, the higher the Q factor, the less loss or resistance exists in the energy storage property. The inductor quality factor Q is defined as:

$$Q = \frac{X}{R_S}$$

(1-9)

Where:

X is the reactance of the inductor
R_s is the resistance in the inductor

At low RF frequencies the resistance comes primarily from the resistivity of the wire and as such is quite low. At higher frequencies the skin effect and inter-winding capacitance begin to influence the resistance and reactance thus causing the Q factor to decrease. In most applications we want as high a component Q factor as possible. We can increase the Q factor of inductors by using larger diameter wire or by silver plating the wire. In a multi-turn coil, the windings can be separated to reduce the inter-winding capacitance which in turn will increase the Q factor. Winding the coil on a magnetic core can increase the Q factor.

Air Core Inductors

Forming a wire on a removable cylinder is the basic realization of the air core inductor. When designing an air-core inductor, use the largest wire size and close spaced windings to result in the lowest series resistance and high Q. The basic empirical equation is given by Equation (1-10).

$$L = \frac{(17)N^{1.3}(D+D1)^{1.7}}{(D1+S)^{0.7}}$$

(1-10)

Where:

N = Number of turns of wire
D = Core form diameter in inches
$D1$ = Wire diameter in inches
L = Coil inductance in nH
S = Spacing between turns in inches

Example 1.6: As an interesting comparison with Example 1.1 calculate the amount of inductance that we can realize in that same three inches of wire if we wind it around a core to form an inductor. Choose a core form of 0.095 inches diameter as a convenient form to wrap the wire around.

Solution: First we need to calculate the approximate number of turns that we can expect to have with the three inch length of wire. We know that the circumference of a circle is related to the diameter by the following equation.

$$Circumference = \pi (Diameter) = \pi (0.095) = 0.2985 \ inches$$

With a circumference of 0.2985 inches we can calculate the approximate number of turns that we can wrap around the 0.095 inch core with three inches of wire.

$$N = \frac{3}{0.2985} = approximately \ 10 \ turns$$

From Equation (1-10) we can see that the spacing between the turns, S, has a strong effect on the value of the inductance that we can expect from the coil. When hand winding the coil, it may be difficult to maintain an exact spacing of zero inches between the turns. Therefore it is useful to solve Equation (1-10) in terms of a variety of coil spacing so that we can see the effect on the inductance. We can solve Equation (1-10) in MATLAB for a variety of coil spacing, as follows.

1. Enter design parameters

D=.095; D1=.0126; N=10; Spacing = [0;.002;.004;.006;.008;.010];

2. Calculate the coil length and inductance in Nano Henries

Inductance_nH = (17*(N^1.3)*((D+D1)^1.7))/((D1+Spacing)^.7)

Coil_Length = (D1*N)+(Spacing.*(N-1))

Note that the coil spacing is variable and Spacing has been defined as an array variable. Placing a semicolon between the values makes the array organized in a column format. This is handy for viewing the results in tabular format. Placing a comma between the values would organize the array in a row format. As an aid in forming the coil, the overall coil length is also calculated. The coil length is simply the summation of the overall wire thickness times the number of turns and the spacing between the turns.

Add the calculated inductance, spacing, and coil length to the table as shown in Table1-4.

Index	Inductance_nH	InductanceCalculator.Spacing	InductanceCalculator.Coil_Length
1	163.784	0	0.126
2	147.735	2e-3	0.144
3	135.037	4e-3	0.162
4	124.701	6e-3	0.18
5	116.097	8e-3	0.198
6	108.806	0.01	0.216

Table 1-4 Coil inductance, spacing and coil length

The table shows that the coil inductance with no spacing between the turns is 163.78 nH and it is 0.126 inches long. Contrast this to the 93.1 nH inductance with the same three inches of wire in example1.1. We can clearly see the dramatic impact of the magnetic flux linkage in increasing the inductance by forming the wire into a coil. The Figure also shows the strong influence of the inter-winding capacitance in influencing the inductance of the coil. Just 10 mils spacing between the turns reduces the coil's inductance from 163.78 nH to 108.8 nH. It is clearly important to consider the turn spacing when analyzing the inductor's performance. In practice this is an effective means to tune the inductor's value in circuit. When designing and building the inductor it is necessary to solve Equation (1-10) for the number of turns given a desired value of inductance. The procedure in MATLAB script follows.

1. Enter design parameters

D = 0.095; D1 = 0.0126; Inductance_nH = 163.784;

Spacing = [0;.002;.004;.006;.008;.010]

2. Calculate the number of turns and coil length.

N = (((D1+Spacing)^.7)*Inductance_nH)/(17*((D+D1)^1.7)))^.7692
Coil_Length = (D1*N) + (Spacing.*(N-1))

Note that there is one subtle difference namely the calculation of the coil length. In this case both variables, Spacing and N, are array variables. When multiplying array variables use a period in front of the multiplication sign to signify that this is an operation on arrays. This makes sure that the correct array index is maintained between the variables. In the previous equations this notation was not necessary because N was a constant. The tabulated results are shown in Table 1-5.

Index	N	Number_of_Turns.Spacing	Number_of_Turns.Coil_Length
1	9.999	0	0.126
2	10.825	2e-3	0.156
3	11.599	4e-3	0.189
4	12.332	6e-3	0.223
5	13.029	8e-3	0.26
6	13.696	0.01	0.3

Table 1-5 Number of turns, coil spacing and length

Note the required parameters: number of turns, wire diameter, core diameter, and coil length. The coil spacing cannot be entered directly but is accounted for by the overall coil length for a given number of turns.

Set up a linear frequency sweep over a range of 0 to 1.3 GHz. Plot the impedance of the inductor across the frequency range as shown in Figure 1-17. Note the interesting spike, or increase in impedance that occurs around 1.170 GHz. This is the parallel, self-resonant frequency, of the inductor.

The inductor is not an ideal component or a pure inductance but rather a network that includes parasitic capacitance and resistance.

Example 1.7: Create a simple RLC network that gives an equivalent impedance response similar to Figure 1-7. One such circuit is shown in Figure 1-10.

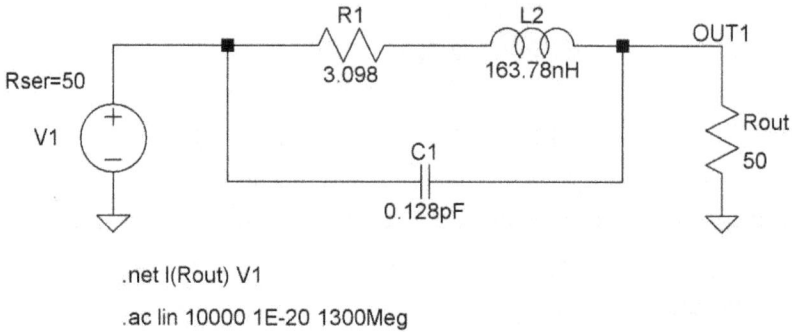

.net I(Rout) V1

.ac lin 10000 1E-20 1300Meg

Figure 1-10 Equivalent ideal element network of the air core inductor

The simulated response is given in Figure 1-11.

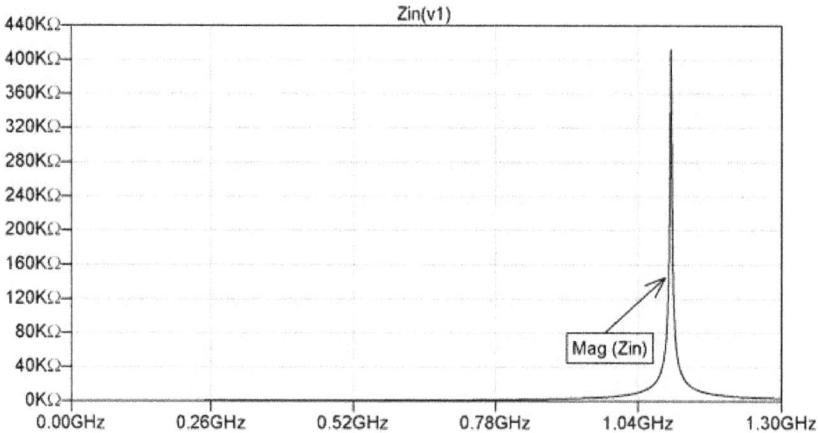

Figure 1-11 Input impedance of equivalent air core inductor

Resonant circuits are covered in detail in chapter 4 but it is important to understand that each individual component such as the air core inductor has its own resonant frequency. The resonant frequency is the frequency at which the inductive reactance and capacitive reactance are equal and cancel one another. When this condition occurs in the inductor it is a parallel resonant circuit which results in a very high real impedance. If we plot the reactance along with the impedance a very interesting response is obtained. This response is shown in Figure 1-12.

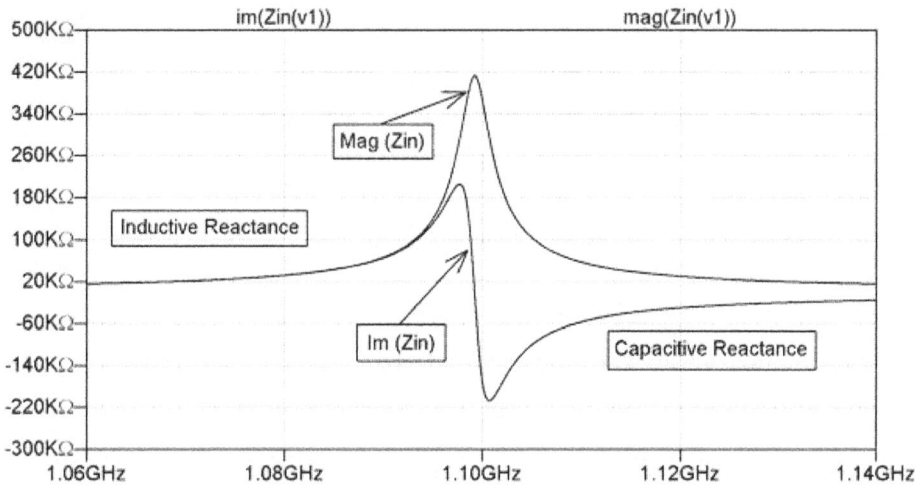

Figure 1-12 Impedance and reactance of the equivalent air core inductor

The reactance, up to the resonant frequency, is positive but beyond resonance the reactance becomes negative. From basic circuit theory we know that a negative reactance is associated with a capacitor. Therefore above the Self Resonant Frequency, SRF, the inductor actually becomes a capacitor. In practice we want to make sure that our inductor really behaves like an inductor. A good design practice is to keep this self-resonant frequency about four times higher than the frequency of operation. However using the inductor near its resonant frequency might make a good choke. A choke is a high reactance inductor often used to feed voltage to a circuit in which all RF energy is blocked from the DC side of the circuit. The inductor manufacturer will typically specify SRF of the inductor. It is important to remember that the inductor's SRF is the parallel resonant frequency; not the series resonant frequency.

1.6 Inductor Q Factor

Example 1.8: Calculate the Q factor of the air core inductor based on Equation (1-9).

Solution: The plot of inductor Q factor is shown in Figure 1-13.

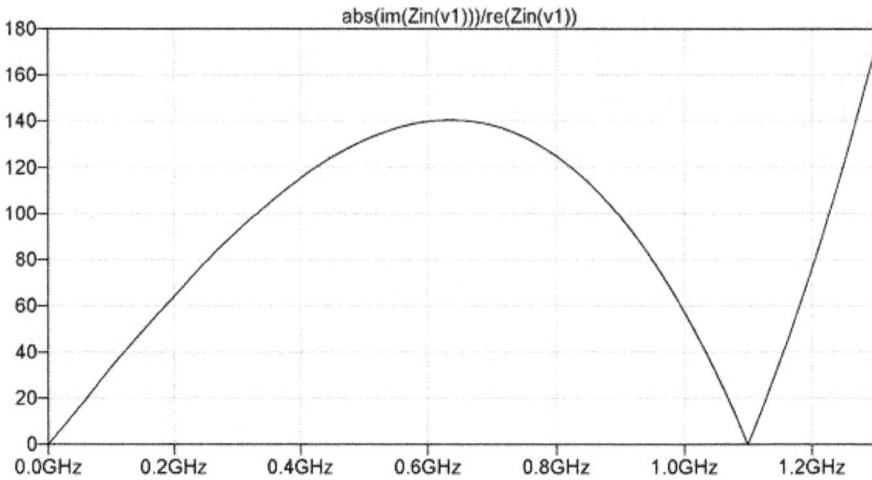

Figure 1-13 Air core inductor model Q factor versus frequency

It is interesting to note that the Q factor peaks at a frequency well below the self-resonant frequency of the inductor. The actual frequency at which the Q factor peaks will vary among inductor designs but is usually ranges from 2 to 5 times less than the SRF. Close winding spacing results in inter-winding capacitance, which lowers the self-resonant frequency of the inductor. Thus there is a tradeoff between maximum Q factor and high self resonant frequency. It is also noteworthy that the Q factor goes to zero at the self-resonant frequency. Figure 1-13 shows the Q factor monotonously increasing beyond the self-resonant frequency. This is erroneous and is due to the fact that the model used to simulate the inductor's performance is invalid beyond the self-resonance. The Air Core inductor model uses a simplified network similar to the one shown in Figure 1-10. Beyond self resonance the complexity and number of ideal elements need to increase in order to accurately model the inductor. A resistor needs to be added in series with the capacitor to begin to model the Q factor because the inductor is becoming a capacitor above self resonance. For most practical work however the native model will work fine because we should be using the inductor well below the self-resonant frequency. A technique commonly used by microwave engineers to increase the Q factor of an inductor is to silver plate the wire. This can be modeled by setting Rho = 0.95 in the inductor model.

Chip Inductors

The inductor core does not have to be air. Other materials may be used as the core of an inductor. Similar in size to the chip resistor there is a large assortment of chip inductors that are popular in surface mount designs. The chip inductor is a form of dielectric core inductor. There are a variety of modeling techniques used for chip inductors. One of the more popular modeling techniques is with the use of S parameter files. The subject of S parameters is covered in chapter 3. At this point consider the S parameter file as an external data file that contains an extremely accurate network model of the component. Most component manufacturers provide S parameter data files for their products. It is a good practice to always check the manufacturer's website for current S parameter data files. CoilCraft, Inc. is one manufacturer of chip inductors. A typical chip inductor is built with extremely small wire formed on a ceramic form as shown in Figure 1-14.

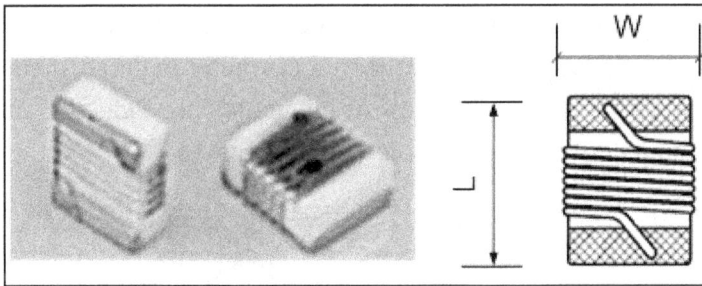

Figure 1-14 chip inductors (*courtesy of Coilcraft, Inc.*)

Chip inductors are manufactured in many sizes as shown in Table 1-6.

Size	Length x Width
0201	20mils x 18mils
0302	34mils x 15mils
0402	44mils x 20mils
0603	69mils x 30mils
0805	90mils x 50mils
1008	105mils x 80mils
1206	140mils x 56mils
1812	195mils x 100mils

Table 1-6 Standard chip Inductor size

The characteristic differences among the various sizes are more difficult to quantify than the chip resistors. A careful study of the data sheets is required for the proper selection of a chip inductor. In general the larger chip inductors will have higher inductance values. Often the smaller chip inductors will have higher Q factor. The impedance and self-resonant frequency can vary significantly across the sizes as well as the current handling capability. The plot of the Q factor derived from the S parameter file shows that the maximum Q factor is 41.262 near 192 MHz. Manufacturers will often plot the Q vs frequency on a logarithmic scale. It is very easy to change the x-axis to a logarithmic scale on the rectangular graph properties window. This allows us to have a visual comparison to the manufacturer's catalog plot.

1,7 Magnetic Core Inductors

We have seen that the inductance of a length of wire can be increased by forming the wire into a coil. We can make an even greater increase in the inductance by replacing the air core with a magnetic material such as ferrite or powdered iron. Two popular types of magnetic core inductors are the rod core and toroidal core inductors shown in Figure 1-15.

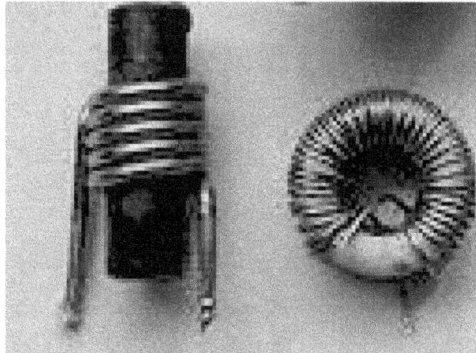

Figure 1-15 Rod and toroidal magnetic core inductors

The magnetic field around an inductor is characterized by the magnetic force H, and the magnetic flux B. They are related by the level of the applied signal and the permeability, μ, of the core material. This relationship is given by Equation (1-11).

$$B = \mu H \qquad\qquad (1\text{-}11)$$

Where:

   B = Flux density in Gauss
   H = Magnetization intensity in Oersteds
   μ = Permeability in Webers/Ampere-turn

This relationship is nonlinear in that as H increases, the amount of flux density will eventually level off or saturate. We will consider the linear region of this relationship throughout the discussion of this text. In an iron core inductor the permeability of the magnetic core is much higher than an air core and produces a high flux density. This magnetic flux density for each type of inductor is shown in Figure 1-16..

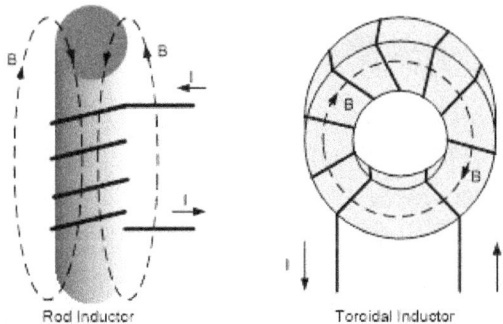

        Rod Inductor   Toroidal Inductor

Figure 1-16 Magnetic flux densities for rod and toroidal inductors

The rod inductor has magnetic flux outside of the core as well as inside the core. Rod inductors that are used in tuned circuits generally require a metal shield around the inductor to contain this magnetic flux so that it does not interfere or couple to adjacent circuits and other inductors. The toroidal inductor flux remains primarily inside the core material. This suggests that the toroidal inductor experiences less loss and should have higher Q factor. This also gives the toroidal inductor a self-shielding characteristic and does not require a metallic enclosure.

Because of its self-shielding properties and high Q factor the toroidal inductor is one of the most popular of all magnetic core inductors.

However, one advantage of the rod inductor is that it is much easier to tune. The coil can be wound on a hollow plastic cylindrical form in which the magnetic rod can be placed inside. The rod is then free to move longitudinally which can tune the inductance value. Core materials are characterized by their permeability. Core permeability can vary quite a bit with frequency and temperature and can be confusing to specify for a given application. The stability of the permeability can change with the magnetic field due to DC current or RF drive through the inductor. As the frequency increases the permeability eventually reduces to the same value as air. Therefore iron core inductors are used only up to about 200 MHz. In general powdered iron can handle higher RF power without saturation and permanent damage. Ferrite cores have much higher permeability. The higher permeability of ferrite results in higher inductance values but lower Q factors.

This characteristic can be advantageous in the design of RF chokes and broad band transformers. For inductors used in tuned circuits and filters, however, the higher Q factor of powdered iron is preferred. The powdered iron cores are manufactured in a variety of mixes to achieve different characteristics. The iron powders are made of hydrogen reduced iron and have greater permeability and lower Q factor. These cores are often used in RF chokes, electromagnetic interference (EMI) filters, and switched mode power supplies. Carbonyl iron tends to have better temperature stability and more constant permeability over a wide range of power. At the same time the Carbonyl iron maintains very good Q factor making them very popular in RF circuits. These characteristics lead to the popularity of toroidal inductors of Carbonyl iron for the manufacture of RF inductors. There is a wide variety of sizes and mixtures of Carbonyl iron that are used in the design of toroidal inductors. A few of the popular sizes that are manufactured by Micrometals Inc. are shown in Table 1-7.

Core Designator	OD, inches	ID, inches	Height, inches
T30	0.307	0.151	0.128
T37	0.375	0.205	0.128
T44	0.440	0.229	0.159
T50	0.500	0.303	0.190
T68	0.690	0.370	0.190
T80	0.795	0.495	0.250
T94	0.942	0.560	0.312
T106	1.060	0.570	0.437
T130	1.300	0.780	0.437
T157	1.570	0.950	0.570
T200	2.000	1.250	0.550
T300	3.040	1.930	0.500
T400	4.000	2.250	0.650

Table 1-7 Partial listing of popular toroidal cores with designators

The inductance per turn of a toroidal inductor is directly related to its permeability and the ratio of its cross section to flux path length as given by Equation (1-12).

$$L = \frac{4\,\pi\,N^2\,\mu\,A}{length} \quad nH \qquad (1\text{-}12)$$

Where:

L_{nH} = inductance

μ = permeability

A = cross sectional area

length = flux path length

N = number of turns

As Equation (1-12) shows the inductance is proportional to the square of the number of turns. A standard specification used by toroid manufacturers for the calculation of inductance is the inductive index, A_L. The inductive index is typically given in units of nH/turn. The inductance can then be defined by Equation (1-13).

$$L = N^2 A_L \qquad nH \tag{1-13}$$

Some manufacturers specify the A_L in terms of uH or mH. To convert among the three quantities use the following guideline.

$$\frac{1\,nH}{turn} = \frac{10\ uH}{100\ turns} = \frac{1\ mH}{1000\ turns} \tag{1-14}$$

The various powdered iron mixes are optimized for good Q factor and temperature stability over certain frequency bands. A partial summary of some popular mixtures is shown in Table 1-8. Powdered Iron cores have a standard color code and material sub-type designator. The toroid is painted with the appropriate color so that the mixture can be identified. A given A_L is dependent on both the size of the toroid and the material mix.

Material Mix Designator	Material Permeability	Magnetic Material	Color Code	Frequency Range	Temperature Stability (ppm/°C)
-17	4.0	Carbonyl	Blue/Yellow	20 – 200 MHz	50
-10	6.0	Carbonyl W	Black	10 – 100 MHz	150
-6	8.5	Carbonyl SF	Yellow	2.0 – 30 MHz	35
-7	9.0	Carbonyl TH	White	1.0 – 20 MHz	30
-2	10.0	Carbonyl E	Red	0.25 – 10 MHz	95
-1	20.0	Carbonyl C	Blue	0.15 – 2.0 MHz	280
-3	35.0	Carbonyl HP	Grey	0.02 – 1.0 MHz	370

Table 1-8 Partial listing of powdered iron core mixes and suggested frequency range

Because of the complex properties of the core material, the determination of the Q factor of a toroidal inductor can be difficult. It is not simply the magnetic material properties alone, but also the wire winding loss as well that determines the overall Q factor. These losses can then vary greatly with frequency, flux density, and the toroid and wire size. The optimal Q factor occurs when the winding losses are equal to the core losses [4]. In general for a given inductance value and core mix a larger toroid will produce larger Q factors. Conversely for a given toroid size higher Q factor is achieved at higher frequency as the permeability decreases.

The skin effect of the wire in the windings can have a significant impact on the achievable Q factor. Just as we have seen with the air core inductor, the inter-turn capacitance will also have a limiting effect on the resulting Q factor as well as the self-resonant frequency. Manufacturers often provide a set of Q curves that the designer can use as a design guide for determining the toroidal inductor Q factor. Figure 1-17 shows a typical set of optimal Q curves for Carbonyl W core material at various toroid sizes.

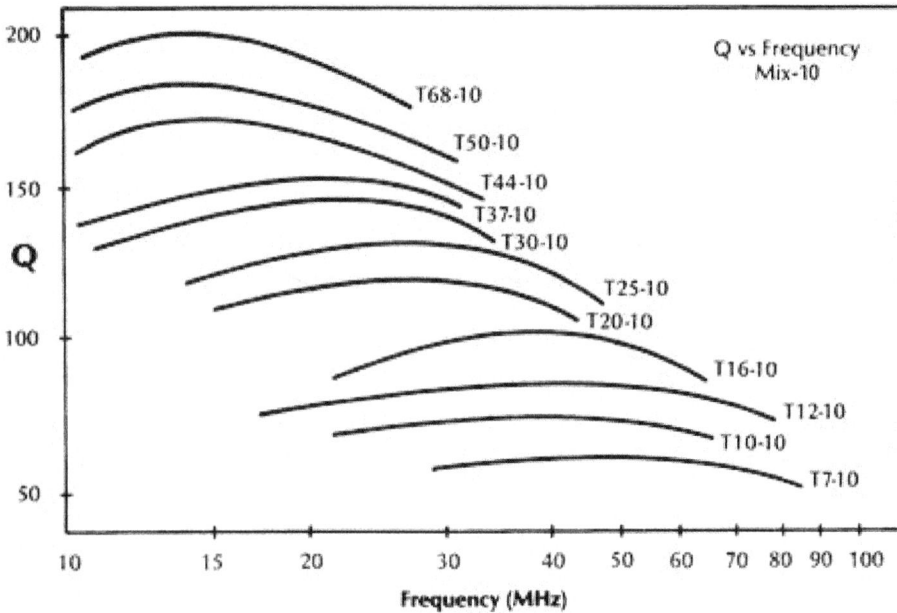

Figure 1-17 Optimal Q factor versus toroid size for core mix -10 (*courtesy Micrometals Inc.*)

Example 1.9: Design a 550 nH inductor using the Carbonyl W core of size T30. Determine the number of turns and model the inductor in .

*Solu*tion: From the manufacturer's data sheet the A_L value is 2.5 for a T30-10 toroidal core. Rearranging Equation (1-13) to solve for the number of turns we find that 14.8 turns are required.

$$N = \sqrt{\frac{L}{A_L}} = \sqrt{\frac{550\ nH}{2.5}} = 14.8$$

To reduce the winding loss we want to use the largest diameter of wire that will result in a single layer winding around the toroid. Equation (1-15) will give us the wire diameter.

$$d = \frac{\pi \ ID}{N + \pi} \tag{1-15}$$

Where:

> d = Diameter of the wire in inches
> ID = Inner diameter of the core in inches
> N = Number of turns

Therefore,

$$d = \frac{\pi \ (0.151)}{14.8 + \pi} = \frac{0.4744}{19.942} = 0.0238 \ \ inches$$

From Appendix A, AWG#23 wire is the largest diameter wire that can be used to wind a single layer around the T30 toroid. Normally AWG#24 is chosen because this is a more readily available standard wire size. The toroidal inductor model in requires a few more pieces of information. The model requires that we enter the total winding resistance, core Q factor, and the frequency for the Q factor, F_q. As an approximation, set F_q to about six times the frequency of operation. In this case set F_q to (6)(25) MHz = 150 MHz. Then tune the value of Q to get the best curve fit to the manufacturer's Q curve. We know that we have 14.8 turns on the toroid but we need to calculate the length of wire that these turns represent. The approximate wire length around one turn of the toroid is calculated from the following equation.

$$Length = \left[(2) \, Height + (OD - ID) \right] (\#turns) \tag{1-16}$$

Using the dimensions for the T30 toroid from Table 1-10 we can calculate the total length of the wire as 6.10 inches.

$$\left[(2)(0.128) + (0.307 - 0.151) \right] \cdot 14.8 = 6.10 \ \ inches$$

1.8 Physical Capacitors

The capacitor is an electrical energy storage component. The amount of energy that can be stored is dependent on the type and thickness of the dielectric material and the area of the electrodes or plates. Capacitors take on many physical forms throughout electrical circuit designs. These range from leaded bypass capacitors in low frequency applications to monolithic forms in millimeter wave applications. Table 1-9 summarizes many of the applications of the dielectric type capacitors in industry in which capacitors are found. The table also shows some of the types of materials, such as Aluminum, Tantalum, Polyester, Ceramic, Polystyrene, and Silver, in which the capacitors are manufactured.

Application	Dielectric Type	Notes
Audio Frequency Coupling	Aluminum Electrolytic Tantalum Polyester/ Polycarbonate	Very High Capacitance High Capacitance for given size Medium capacitance, low cost
Power Supply Filtering	Aluminum Electrolytic	High Capacitance, high ripple current
RF Coupling	Ceramic NPO (COG) Ceramic X7R Polystyrene	Small, low loss, low cost Small, low cost, higher loss than COG Very low loss in RF range, larger than ceramic
Tuned Circuits, Resonators	Silver Mica Ceramic NPO (COG)	

Table 1-9 Applications for various types of capacitors

As Table 1-9 shows, the ceramic capacitors dominate the higher RF and microwave frequency applications. Two of the most popular of the ceramic capacitors are the single layer and multilayer ceramic capacitors, as shown in Figure 1-18. These capacitors are available with metalized terminations so that they are compatible with a variety of surface mount assembly techniques from hand soldering to wire bonding and epoxy attachment.

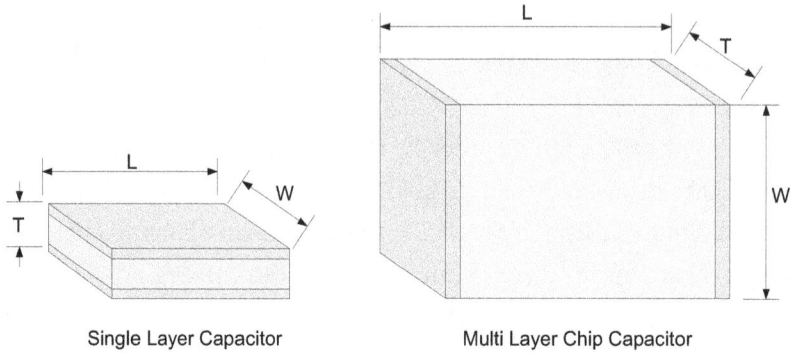

Figure 1-18 Single layer and multi-layer chip capacitor dimensions

Single Layer Capacitor

The single layer capacitor is one of the simplest and most versatile of the surface mount capacitors. It is formed with two plates that are separated by a single dielectric layer as shown in Figure 1-19. Most of the electric field (E) is contained within the dielectric however there is a fraction of the E field that exists outside of the plates. This is known as the fringing field.

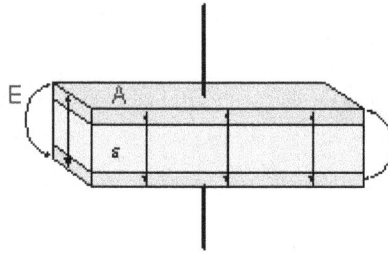

Figure 1-19 Single layer parallel plate capacitor

The capacitance formed by a dielectric material between two parallel plate conductors is given by Equation 1-17.

$$C = (N-1)\left(\frac{KA\,\varepsilon_{\mathrm{r}}}{t}\right)(FF) \quad pF \tag{1-17}$$

Where,

   A = plate area

   ε_r = relative dielectric constant

   t = separation

   K = unit conversion factor; 0.885 for cm and 0.225 for inches

FF = fringing factor; 1.2 when mounted on microstrip

N = number of parallel plates.

Example 1.10: Consider the design of a single layer capacitor from a dielectric that is 0.010 inches thick and has a dielectric constant of three. Each plate is cut to 0.040 inches square.

Solution: When the capacitor is mounted with at least one plate on a large printed circuit board track, a value of 1.2 is typically used in calculation. The single layer capacitor can be modeled as the Thin Film Capacitor.

$$C = (2-1)\left(\frac{(.225)(0.04 \cdot 0.04)(3)}{0.010}\right)(1.2) = 0.13\ pF$$

The ceramic dielectrics used in capacitors are divided into two major classifications. Class 1 dielectrics have the most stable characteristics in terms of temperature stability. Class 2 dielectrics use higher dielectric constants which result in higher capacitance values but have greater variation over temperature. The temperature coefficient is specified in either percentage of nominal value or parts per million per degree Celsius (ppm/°C). Ceramic materials with a high dielectric constant tend to dominate RF applications with a few exceptions. NPO (negative-positive-zero) is a popular ceramic that has extremely good stability of the nominal capacitance versus temperature.

Dielectric Material	Dielectric Constant
Vacuum	1.0
Air	1.004
Mylar	3
Paper	4 - 6
Mica	4 - 8
Glass	3.7 - 19
Alumina	9.9
Ceramic (low ε_r)	10
Ceramic (high ε_r)	100 – 10,000

Table 1-10 Dielectric constants of materials

Multilayer Capacitors

Multilayer capacitors, shown in Figure 1-20, are very popular in surface mount designs. They are physically larger than single layer capacitors and can be attached by hand or by automatic pick-and-place machines. As Figure 1-20 shows, the multilayer chip capacitor is a parallel array of capacitor plates in a single package. Due to this type of construction the chip capacitor can handle higher voltages than the single layer capacitor. The insulation resistance of the capacitor is its ability to oppose the flow of electricity and is a function of the dielectric material and voltage. The insulation resistance is typically specified as a minimum resistance value in MΩ at a specified working voltage. The working voltage rating, WVDC, is the maximum DC voltage at which the capacitor can operate over the lifetime of the capacitor. The AC voltage rating is approximately one half of the WVDC value. The dielectric withstand voltage (DWV) is the electrical strength of the dielectric at 2.5 times the rated voltage. This is a maximum short term over-voltage rating and is usually specified as a length of time that the dielectric can withstand the 2.5 times the WVDC value without arcing through.

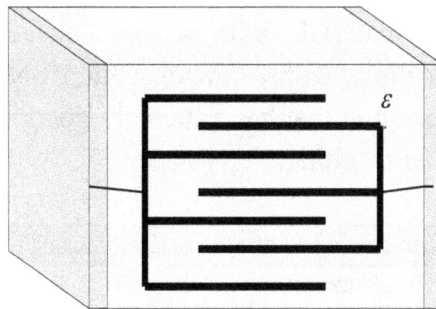

Figure 1-20 Multilayer chip capacitor construction

There is no physical model for the multilayer chip capacitor in LTspice. The designer must rely on S parameter files or Modelithics models as we have used for the chip resistor. Two of the major manufacturers of multilayer chip capacitors are American Technical Ceramics, ATC, and Dielectric Laboratories Inc., DLI. These manufacturers provide S parameter files for their capacitors that are readily available on the company websites.

They also provide helpful software applications that can aid the designer in making decisions for the proper selection of chip capacitors in specific applications.

Capacitor Q Factor

RF losses in the dielectric material of a capacitor are characterized by the dissipation factor. The dissipation factor is also referred to as the loss tangent and is the ratio of energy dissipated to the energy stored over a period of time. It is essentially the capacitor's efficiency rating. The dissipation factor and other ohmic losses lead to a parameter known as the Equivalent Series Resistance, ESR. The dissipation factor is the reciprocal of the Q factor. Just as we have seen with resistors and inductors, the physical model of a capacitor is a network of R, L, and C components.

Example 1.11: Calculate the Q factor versus frequency for the physical model of an 8.2 pF multilayer chip capacitor shown in Fig. 1-21.

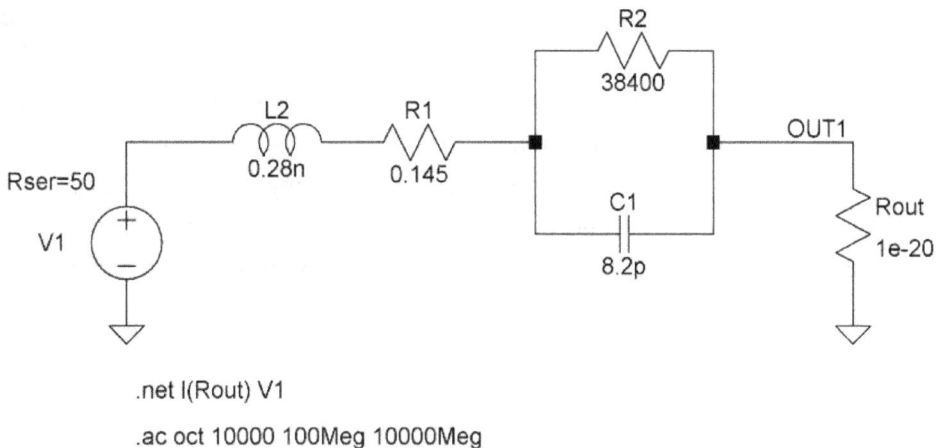

.net I(Rout) V1

.ac oct 10000 100Meg 10000Meg

Figure 1-21 Physical model of the 8.2 pF chip capacitor

The input impedance of the physical model is shown in Figure 1-22.

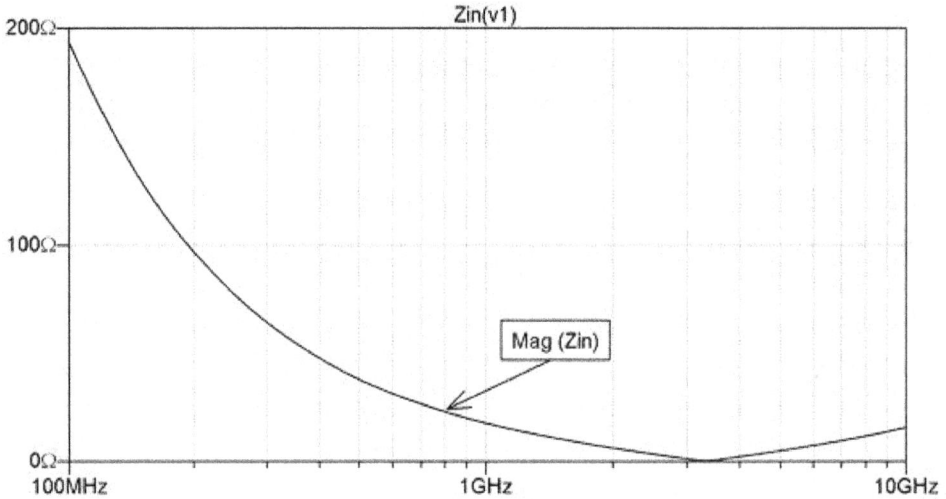

Figure 1-22 Physical 8.2 pF chip capacitor impedance

The capacitor has a series inductance and resistance component along with a resistance in parallel with the capacitance. The parallel resistor sets the losses in the dielectric material. The series resistance and inductance represent any residual lead inductance and ohmic resistances.

Solution: The values entered for the physical model and the Q factor can be obtained from the capacitor manufacturer. The plot of Figure 1-22 shows the impedance of the capacitor versus frequency. Note that the impedance decreases as would be expected until the self-resonant frequency is reached. Above the self-resonant frequency the impedance begins to increase suggesting that the capacitor is behaving as an inductor. The self-resonant frequency is due to the series inductance resonating with the capacitance. At resonance the reactance cancels leaving only the resistances *R1* and *R2*. The parallel resistance, *R2*, can be converted to an equivalent series resistance by Equation (1-18).

$$R2' = \frac{1}{1+Q^2} \ R2 \tag{1-18}$$

These two series resistances can then be added to find the equivalent series resistance, ESR, as defined by Equation (1-19).

$$ESR = R1 + R2' \tag{1-19}$$

The capacitor Q factor is then calculated by Equation (1-20). X_T is the total series reactance of the inductive and capacitive reactance.

$$Q = \frac{X_T}{ESR} \tag{1-20}$$

As Figure 1-23 shows, the 8.2 pF multilayer chip capacitor has a series resonant frequency of 3313 MHz. A marker is placed on the trace indicating the impedance at the SRF as 0.149 –. Because the reactance is cancelled at the SRF, this impedance essentially becomes the ESR of the capacitor. R2 of Figure 1-23 is extremely frequency dependent. This means that the capacitor's Q factor is also extremely frequency dependent. An improved model for analyzing the capacitor's characteristics over a wide frequency range is to use the model for capacitor with Q. Using this model the Q factor of the capacitor can be made proportional to the square root of the applied frequency. The inductance, capacitance, and Q factor can be calculated from the impedance using the following equations.

Reactance = im(Zin1)

Resistance = re(Zin1)

Inductance = abs((Reactance)/(freq*2*pi))

Qfactor = abs(Inductance/Resistance)

Capacitance = 1/(abs(Reactance)*freq*2*pi)

The modified model is shown in Figure 1-23.

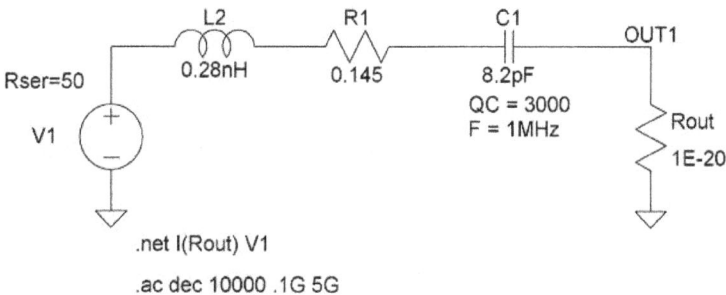

Figure 1-23 Modified physical model of 8.2 pF chip capacitor

The capacitor Q factor of the modified model versus frequency is plotted in Figure 1-24.

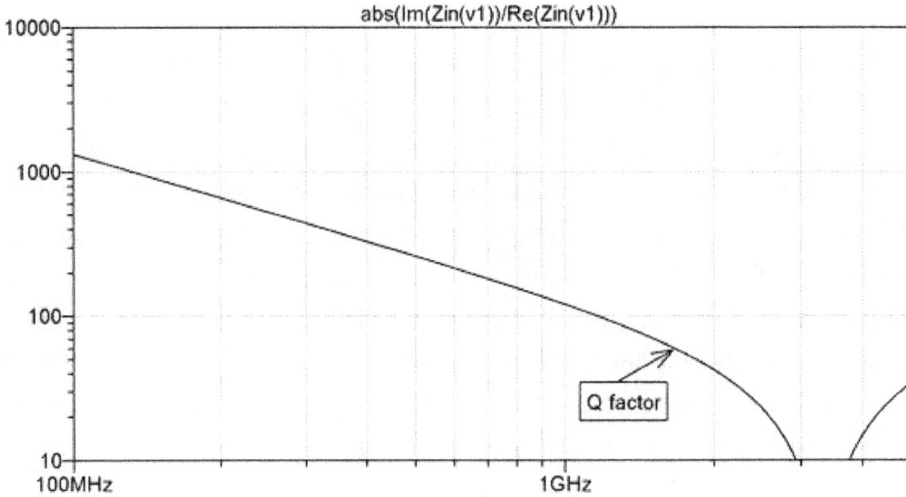

Figure 1-24 Calculated Q factor of 8.2 pF chip capacitor

Figure 1-24 shows the large dependence of the capacitor Q on frequency. The Q factor goes to zero at the self-resonant frequency. Above the self-resonant frequency the Q factor is undefined.

Calculate the effective capacitance from the total reactance of the model using Equation (1-21).

$$C = \frac{1}{2\pi\, F\, X_T} \qquad\qquad (1\text{-}21)$$

The plot of Figure 1-25 reveals some interesting characteristics about the chip capacitor. From 100 MHz to 300 MHz the capacitance value is fairly constant. As the frequency increases we see that the capacitance actually increases. The parasitic inductive reactance of the capacitor package actually makes the effective capacitance greater than its nominal value, as shown in Figure 1-25.

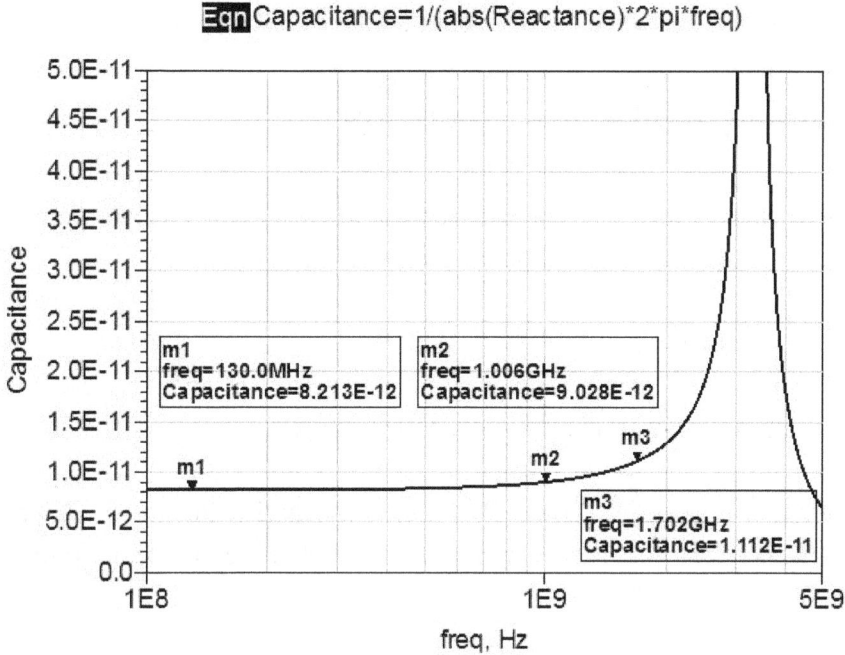

Figure 1-25 Effective capacitance of the 8.2 pF chip capacitor

This is a property of the capacitor that is not always intuitive. As the frequency approaches the self-resonant frequency (SRF) the capacitance rapidly approaches infinity. The capacitor actually becomes nearly a short circuit to RF at the self-resonant frequency. This is an important property of the capacitor that is used frequently in RF and microwave design. In RF coupling or bypass capacitor applications, capacitors are very often used at or near their self resonant frequency. The capacitor's self resonant frequency is due to the series resonant circuit. In applications requiring bypassing over a wide range of frequencies it is often necessary to use several capacitors each selected to have a uniquely spaced self resonant frequency. In filter and other tuned circuit applications where we want the chip capacitor to appear as an 8.2 pF capacitor, we clearly must stay well below the self-resonant frequency of the capacitor. A typical rule-of-thumb is to use the capacitor over a frequency range up to 35% of the self-resonant frequency. Therefore the 8.2 pF chip capacitor with a self-resonant frequency of 3321 MHz would be used as a capacitor in tuned circuits up to a frequency of about 1006 MHz. Figure 1-25 shows that at frequencies above 1006 MHz the capacitance is extremely nonlinear. However the capacitor could be used as a DC blocking, RF coupling capacitor that would efficiently pass microwave energy near its SRF of 3300 MHz.

References and Further Readings

[1]   Ali Behagi, *RF and Microwave Circuit Design*, A Design Approach Using (**ADS**), Techno Search, Ladera Ranch, CA 2017

[2]   Ali Behagi and Manou Ghanevati, *Fundamentals of RF and Microwave Circuit Design,* Practical Analysis and Design Tools, Techno Search, Ladera Ranch, California 2017

[3]   Paul Lorrain, Dale P. Corson, and Francois Lorrain, *Electromagnetic Fields and Waves*, W.H. Freeman and Company, New York, 1988

[4]   *Design Guide, Microwave Components* Inc., P.O. Box 4132, South Chelmsford, MA 01824

[5]   *Iron Powder Cores for High Q Inductors*, Micrometals, Inc.

[6]   *Capacitors for RF Applications,* Dielectric Laboratories, Inc.2777 Rt.20 East, Cazenovia, NY. 13035

[7]   R. Ludwig and P. Bretchko, *RF Circuit Design -Theory and Applications*, Prentice Hall, New Jersey, 2000

[8]   William Sinnema and Robert McPherson, *Electronic Communication,* Prentice Hall Canada, 1991

[2]   M.F. "Doug" DeMaw, *Ferromagnetic Core Design & Application Handbook*, MFJ Publishing Co., Inc. Starkville, MS. 39759, 1996

[10]   *The RF Capacitor Handbook,* American Technical Ceramics, One Norden Lane, Huntington, New York 11746

Problems

1-1.   Calculate the wavelength of an electromagnetic wave operating at a frequency of 428 MHz.

1-2.   Calculate the inductance of a 5 inch length of AWG #30 straight copper wire.

1-3.   Calculate the reactance of the wire from Problem 1-2 at 10 Hz, 10 MHz, and 10 GHz. Create a Linear analysis and display the wire impedance vs. frequency.

1-4.   Calculate the resistance of a 12 inch length of AWG #24 copper wire at DC and at 25MHz

1-5.   Find the skin depth and the resistance of a 2 meter length of copper coaxial line at 2 GHz. The inner conductor radius is 1 mm and the outer conductor is 4 mm.

1-6.   Calculate the inductance of a 5 inch length of copper flat ribbon conductor. The dimensions of the ribbon are 0.100 inches in width and 0.002 inches thick.

1-7.   Model a chip resistor (size 0603) with a resistance of 50Ω inch. Consider an application in which 50 Ω impedance must be maintained with +10%. Create a Linear Analysis and determine the maximum usable frequency of the chip resistor.

1-8.   Design an air core inductor with an inductance value of 84 nH. Use a copper wire of 0.050 inch diameter wound on a core diameter of 0.100 inch. Determine the number of turns required assuming a tight spaced winding.

1-9.   Using the inductor from Problem 1-8 and the techniques developed in Chapter 1 to examine the change in the coil inductance as the turn spacing is increased from zero to 0.10 inch in 0.002 inch increments.

1-10. Using the inductor from Problem 1-8, determine the self-resonant frequency of the inductor and comment on the maximum frequency in which the inductor may be used in a tuned circuit application.

1-11. Using the inductor from Problem 1-8, determine the maximum Q factor of the inductor and the frequency at which the maximum Q factor is obtained.

1-12. In, select a chip inductor from the CoilCraft library with an inductance value of 80nH. Determine the maximum Q factor of the inductor and comment on the maximum usable frequency of the inductor in a filter application.

1-13. Design a 1 mH toroidal inductor on a Carbonyl core size T30. Determine the maximum wire size that could be used to realize a single layer winding.

1-14. Using the inductor from Problem 1-13, model the inductor and determine the approximate self resonant frequency. Comment on the maximum usable frequency of the inductor in the front end of a radio receiver.

1-15. A 0.05 pF capacitor is required to couple a transistor to the resonator of a microwave oscillator. Design a single layer capacitor using a 0.020 inch thick dielectric with $e_r = 2.2$. Determine the dimensions of the capacitor assuming square footprint is desired.

1-16. For the single layer capacitor of Problem 1-15, determine the dimensions of the capacitor with a dielectric constant $e_r = 10.2$.

Chapter 2

Transmission Lines

2.1 Introduction

Transmission lines play an important role in designing RF and microwave networks. In chapter 1 we have seen that, at high frequencies where the wavelength of the signal is smaller than the dimension of the components, even a small piece of wire acts as an inductor and affects the performance of the network. In this chapter we present the lumped-element equivalent model of a transmission line and analytically define several important transmission line parameters such as transmission and reflection coefficients, characteristic impedance, propagation constant, attenuation constant, phase constant, voltage standing wave ratio, return loss, velocity factor and group delay. It is demonstrated how to simulate and measure these parameters using the software. Popular types of transmission lines, such as: coaxial lines, microstrip lines, strip lines, and waveguides are discussed. Several methods of characterizing reflection coefficients and the characteristic impedance of these transmission lines are examined. Field coupling between adjacent (coupled) transmission lines is introduced. The chapter concludes with the design of a microstrip directional coupler.

2.2 Plane Waves

Plane waves are the simplest form of electromagnetic waves in which the electric field intensity, E, is perpendicular to magnetic field intensity, H, and both are perpendicular to the direction of propagation (E and H are represented by vectors). Such a wave is also called transverse electromagnetic or TEM wave.

Plane Waves in a Lossless Medium

In rectangular coordinates, assuming that electric field E is a vector in the x direction and varies as it moves along the z direction, the Wave Equation for the electric field in a lossless medium can be written as:

$$\frac{\partial^2 E_x}{\partial z^2} + k^2 E_x = 0 \tag{2-1}$$

Where,

$k = \omega\sqrt{\mu\varepsilon}$ is the wave number

ω is the angular frequency in radians per second

μ is permeability of the medium in Henries per meter

ε is permittivity of the medium in Far per meter

In a lossless medium μ and ε are positive real numbers, therefore k is also positive and real. The solution to Equation (2-1) is of the following form.

$$E_x(z) = E^+ e^{-jkz} + E^- e^{+jkz} \tag{2-2}$$

Where E^+ and E^- are arbitrary constants determined by the boundary conditions. For the sinusoidal waveforms at frequency ω, Equation (2-2) can be written as:

$$\mathcal{E}_x(z,t) = E^+ \cos(\omega t - kz) + E^- \cos(\omega t + kz) \tag{2-3}$$

Where:

$E^+ \cos(\omega t - kz)$ is the wave traveling in the forward direction

$E^- \cos(\omega t + kz)$ is the wave traveling in the reverse direction

Some of the wave characteristics are as follows:

1. Phase velocity is the velocity of a fixed point on the wave that is obtained by setting the derivative of the phase, with respect to t, equal to zero:

$$\omega - k\frac{\partial z}{\partial t} = 0$$

$$v_p = \frac{\partial z}{\partial t} = \frac{\omega}{k} = \frac{1}{\sqrt{\mu\varepsilon}} \tag{2-4}$$

In free space: $\mu = \mu_0 = 4\pi \ (10^{-7})$ H/m and $\varepsilon = \varepsilon_0 = 8.854(10^{-12})$ F/m, therefore,

$$v_p = \frac{1}{\sqrt{\mu_o \varepsilon_o}} = 2.998\,(10^8) = c \qquad meters\,/\sec ond$$

Where c is the velocity of light in free space.

2. The wavelength, λ, is defined as the distance between two successive maximum or minimum points on the wave at a fixed instant of time.

Therefore, $k\lambda = 2\pi$ leads to the Equation (2-5).

$$\lambda = \frac{2\,\pi}{k} = \frac{v_p}{f} \qquad\qquad (2\text{-}5)$$

3. The plane wave impedance, η, is defined as the ratio between electric and magnetic field components travelling in the same direction. Therefore,

$$E_x^+ = E^+ \cos(\omega\,t - k\,z)$$
$$H_y^+ = H^+ \cos(\omega\,t - k\,z)$$

Then,

$$\eta = \frac{E_x^+}{H_y^+} = \frac{E^+}{H^+} \qquad\qquad (2\text{-}6)$$

Here E^+ and H^+ represent the electric and magnetic field amplitude travelling in the positive z direction in units of Volt/meter and Ampere/meter. Based on Maxwell's curl equations the plane wave impedance is given by:

$$\eta = \sqrt{\frac{\mu}{\varepsilon}} \qquad\qquad (2\text{-}7)$$

In free space the plane wave impedance is,

$$\eta_0 = \sqrt{\frac{\mu_0}{\varepsilon_0}} = 377 \ \Omega$$

Plane Waves in a Good Conductor

Metallic conductors used in microwave networks are not perfect but they are considered to be very good conductors. In a material with conductivity σ the current density due to conduction is given by:

$$J = \sigma E \qquad (2\text{-}8)$$

Where σ is the conductivity of the conductor in S/m.

In a good conductor the conductive current is much greater than the displacement current, therefore, by ignoring the displacement current, the propagation constant can be written as [1]:

$$\gamma = \alpha + j\beta = \sqrt{\frac{\omega\mu\sigma}{2}} + j\sqrt{\frac{\omega\mu\sigma}{2}} \qquad (2\text{-}9)$$

The skin depth for a good conductor is defined as the inverse of the attenuation constant:

$$\delta = \frac{1}{\alpha} = \frac{1}{\sqrt{\pi f \mu \sigma}} \qquad (2\text{-}10)$$

Where:

δ is the skin depth in meters

f is the frequency in Hertz

μ is the permeability in H/m

σ is the conductivity in S/m

In a good conductor the positive traveling wave is of the form:

$$e^{-\alpha z} \cos(\omega t - \beta z) \qquad (2\text{-}11)$$

Note that when the wave travels a distance equal to, $Z = \delta = \frac{1}{\alpha}$, the amplitude drops to $e^{-1}\cos(\omega t - \beta z) = 0.368\cos(\omega t - \beta z)$ which is 36.8% of the original signal's amplitude at $z = 0$.

2.3 Lumped Element Representation of Transmission Lines

A lumped model of a small section of parallel wire transmission line, of physical length dz, is shown in Figure 2-1. Any length transmission line can be considered a cascade of many sections of the length dz.

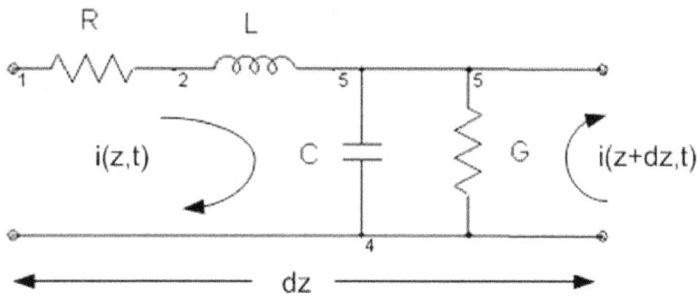

Figure 2-1 Lumped model representation of transmission line section

In Figure 2-1, R is the per unit length series resistance of both lines in Ω/m, L is the per unit length series inductance of both lines in Henries/meter, G is the per unit length shunt conductance of the line in Siemens/meter, and C is the per unit length capacitance of the line in Farads/meter. At higher frequencies, where the wavelength of the signal is smaller than the physical dimension of the network, the voltages and currents along a uniform transmission line are functions of position and time. For a TEM wave traveling in the z direction, the voltage and current are given by:

$$V(z,t) = re\left[V(z)e^{j\omega t}\right] \tag{2-12}$$

$$I(z,t) = re\left[I(z)e^{j\omega t}\right] \tag{2-13}$$

The quantities *V(z)* and *I(z)* are complex functions of *z* along the transmission line and ω is the frequency of the source in radians per second.

2.4 Transmission Line Equations and Parameters

For sinusoidal steady-state excitations, Kirchhoff's voltage and current laws along the line yield the following equations:

$$\frac{dV(z)}{dz} + (R + j\omega L)\, I(z) = 0 \qquad\qquad (2\text{-}14)$$

$$\frac{dI(z)}{dz} + (G + j\omega C)\, V(z) = 0 \qquad\qquad (2\text{-}15)$$

By taking the derivative of both sides of Equation (2-14) with respect to z, and substituting for $\frac{dI(z)}{dz}$ from Equation (2-15), we have:

$$\frac{d^2V(z)}{dz^2} - (R + j\omega L)(G + j\omega C)\, V(z) = 0 \qquad\qquad (2\text{-}16)$$

Similarly, by taking the derivative of both sides of Equation (2-15) with respect to z, and substituting for $\frac{dV(z)}{dz}$ from Equation (2-14), we have:

$$\frac{d^2I(z)}{dz^2} - (R + j\omega L)(G + j\omega C)\, I(z) = 0 \qquad\qquad (2\text{-}17)$$

By defining the complex propagation constant:

$$\gamma = \sqrt{(R + j\omega L)\,(G + j\omega C)} \qquad\qquad (2\text{-}18)$$

Equations (2-16) and (2-17) can be redefined as:

$$\frac{d^2V(z)}{dz^2} - \gamma^2 V(z) = 0 \qquad\qquad (2\text{-}19)$$

$$\frac{d^2I(z)}{dz^2} - \gamma^2 I(z) = 0 \qquad (2\text{-}20)$$

Equations (2-19) and (2-20) are known as voltage and current wave equations respectively. It is easy to show that the solution to Equation (2-19) is of the form:

$$V(z) = Ae^{-\gamma z} + Be^{\gamma z} \qquad (2\text{-}21)$$

Where:

A and B are constants

$Ae^{-\gamma z}$ is the incident voltage traveling in the +z direction

$Be^{\gamma z}$ is the reflected voltage traveling in the -z direction

Similarly, it can be shown that the solution to Equation (2-20) is of the form:

$$I(z) = \frac{\gamma}{(R + j\omega L)} \left[Ae^{-\gamma z} - Be^{\gamma z} \right] \qquad (2\text{-}22)$$

Where:

$\dfrac{\gamma}{R + j\omega L} Ae^{-\gamma z}$ is the incident current traveling in the +z direction

$\dfrac{\gamma}{R + j\omega L} Be^{\gamma z}$ is the reflected current traveling in the −z direction

Definition of Attenuation and Phase Constant

The complex propagation constant defined in Equation (2-18), is written as:

$$\gamma = \alpha + j\beta \qquad (2\text{-}23)$$

Where the real part α is defined as the attenuation constant in Nepers per meter (1 Neper = 8.686 dB) and the imaginary part β is defined as the phase constant in radians per meter.

Definition of Transmission Line Characteristic Impedance

The characteristic impedance of a transmission line, Z_0, is defined as the ratio of the incident voltage to incident current. Therefore by dividing the incident voltage by the incident current, in Equations (2-21) and (2-22), and replacing γ from Equation (2-18), the transmission line characteristic impedance is written as:

$$Z_o = \frac{(R + j\omega L)}{\gamma} = \sqrt{\frac{R + j\omega L}{G + j\omega C}} \qquad (2\text{-}24)$$

Definition of Reflection Coefficient

It is important to understand the concepts of incident (forward) and reflected (backward) wave propagation in transmission lines. In high power RF systems there can be potentially dangerous high voltage peaks that can occur at points along the transmission line when the incident and reflected waves are in phase and add together. The voltage reflection coefficient, $\Gamma(z)$, of a transmission line along the z axis is defined as the ratio of reflected to incident voltage as shown in Equation (2-25).

$$\Gamma(z) = \frac{Be^{\gamma z}}{Ae^{-\gamma z}} = \frac{B}{A}e^{2\gamma z} = \Gamma_o e^{2\gamma z} \qquad (2\text{-}25)$$

Where, Γ_o is the load reflection coefficient at $z = 0$, namely:

$$\Gamma_o = \Gamma(0) = \frac{B}{A} \qquad (2\text{-}26)$$

Notice when there is no reflection the load reflection coefficient $\Gamma_o = 0$.

Definition of Voltage Standing Wave Ratio, VSWR

VSWR is the ratio of the maximum to minimum value of the standing wave. The VSWR is a very common quantity for describing the percentage of power reflected by a given load impedance. The forward and reflected waves travel in opposite directions to form a standing wave pattern. The maximum value of the standing wave is given by Equation (2-27) while the minimum value is given by Equation (2-28).

$$|V(z)|_{max} = |A|\left(1+|\Gamma_o|\right)$$

(2-27)

$$|V(z)|_{min} = |A|\left(1-|\Gamma_o|\right)$$

(2-28)

The ratio of the maximum to minimum voltage is related to the reflection coefficient as shown in Equation (2-29).

$$VSWR = \frac{|V(z)|_{max}}{|V(z)|_{min}} = \frac{1+|\Gamma_o|}{1-|\Gamma_o|}$$

(2-29)

Solving the equation for the magnitude of the reflection coefficient, Γo results in Equation (2-30).

$$|\Gamma_o| = \frac{VSWR-1}{VSWR+1}$$

(2-30)

Notice that when the reflection coefficient is zero, VSWR = 1. This VSWR is commonly presented as 1:1 ratio.

Definition of Return Loss

When the transmission line is mismatched to the load, a portion of the incident power is reflected back to the source. This can be considered a loss of power absorbed by the load. Therefore the return loss, RL, in dB, is defined as:

$$RL\ (dB) = -20\log \left| \Gamma \right|$$ (2-31)

For a matched line:

$|\Gamma| = 0$ and RL = ∞ dB whereas for $|\Gamma| = 1$, RL = 0 dB

Lossless Transmission Line Parameters

From Figure 2-1 we can see that a transmission line is considered lossless when R = G = 0. For a lossless transmission line the propagation constant reduces to:

$$\gamma = j\beta = j\omega \sqrt{LC}$$ (2-32)

Therefore,

$$\beta = \omega \sqrt{LC}$$ (2-33)

And

$$\alpha = 0$$ (2-34)

The characteristic impedance of a lossless transmission line i.e. R = G = 0 is obtained from Equation (2-24) as:

$$Z_o = \sqrt{\frac{L}{C}}$$ (2-35)

Similarly, the voltage reflection coefficient in Equation (2-25) reduces to:

$$\Gamma(z) = \Gamma_o\, e^{2j\beta z}$$ (2-36)

Where: $\Gamma 0 = B/A$, the load reflection coefficient.

Notice from Equations (2-32) through Equation (2-34) for lossless transmission lines the attenuation constant is zero and the phase constant is linearly proportional to frequency. The propagation constant is purely imaginary, and the characteristic impedance is a positive real number.

Lossless Transmission Line Terminations

When a lossless transmission line of length d and characteristic impedance Z_0 is terminated in an arbitrary load Z_L, the input impedance of the line is given by.

$$Z_{in} = Z_o \frac{Z_L + j Z_o \tan \beta d}{Z_o + j Z_L \tan \beta d} \qquad (2\text{-}37)$$

Where β is the phase constant of the line

In the following sections we discuss the input impedance of terminated transmission lines.

1. Transmission Line Terminated in Z_0

When a lossless transmission line of characteristic impedance Z_0 is terminated in a load equal to Z_0, Equation (2-37) shows that the input impedance becomes equal to Z_0. Such a line behaves like an infinitely long transmission line with no reflection. In this case all of the incident power is absorbed by the load.

2. Transmission Line Terminated in a Short Circuit

When a lossless transmission line of characteristic impedance Z_0 is terminated in a short circuit, $Z_L = 0$, Equation (2-37) shows that the input impedance becomes a purely imaginary number equal to:

$$Z_{in} = j Z_o \tan \beta d \qquad (2\text{-}38)$$

Depending on the length of the line the input impedance takes any possible reactive value from minus to plus infinity.

3. Transmission Line Terminated in an Open Circuit

When a lossless transmission line of characteristic impedance Z_0 is terminated in an open circuit, $Z_L = \infty$, Equation (2-37) shows that the input impedance becomes equal to:

$$Z_{in} = -jZ_o \cot \beta d \qquad (2\text{-}39)$$

In this case Z_{in} is also purely imaginary. Depending on the length of the line the input impedance takes any possible reactive value from minus to plus infinity.

4. Half Wavelength Transmission Lines

For a lossless transmission line of characteristic impedance Z_0 with the length $d = \lambda/2$, the input impedance from Equation (2-37) becomes equal to:

$$Z_{in} = Z_L \qquad (2\text{-}40)$$

This means that the input impedance of a transmission line of one half wavelength is equal to the load impedance regardless of the line characteristic impedance.

5. Quarter Wavelength Transmission Lines

For a lossless transmission line with the length $d = l/4$, the input impedance from Equation (2-37) becomes equal to:

$$Z_{in} = \frac{Z_o^{\,2}}{Z_L} \qquad (2\text{-}41)$$

In this case the transmission line transforms the load impedance to a different impedance as defined by Equation (2-41). Such a line is called a quarter-wave transformer. Equation (2-41) can be rewritten to solve for the characteristic impedance of a quarter-wave matching section:

$$Z_o = \sqrt{Z_{in}\, Z_L} \qquad (2\text{-}42)$$

Simulation of Reflection Coefficient and VSWR

Simulation softwares usually have built-in functions to directly display the common measurements of VSWR, return loss, and reflection coefficient. The following example explores the simulation of input reflection coefficient, S[1,1], and VSWR.

Example 2.1: For the series RLC elements in Figure 2-2 measure the reflection coefficients and VSWR from 100 to 1000 MHz in 100 MHz steps.

Solution: Solution: To generate a table of tabulated numbers, follow the procedure outlined below:

1. Construct the schematic diagram in LTspice as shown in Figure 2-2.

2. From the main window select: Simulate > Edit Simulation Cmd to open the Edit Simulation Command window as shown in Figure 2-3.

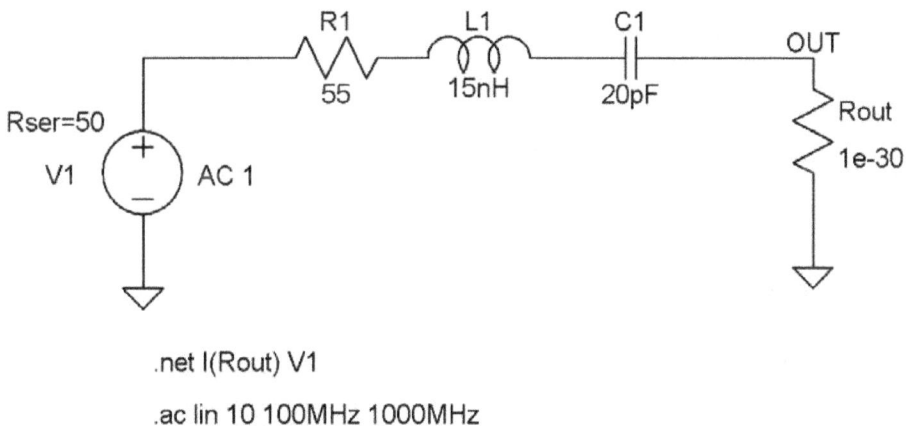

.net I(Rout) V1

.ac lin 10 100MHz 1000MHz

Figure 2-2 Schematic of the series RLC resonator

3. Select AC Analysis and complete the Edit Simulation Cmd as shown in Figure 2-3. Then click OK and run simulation.

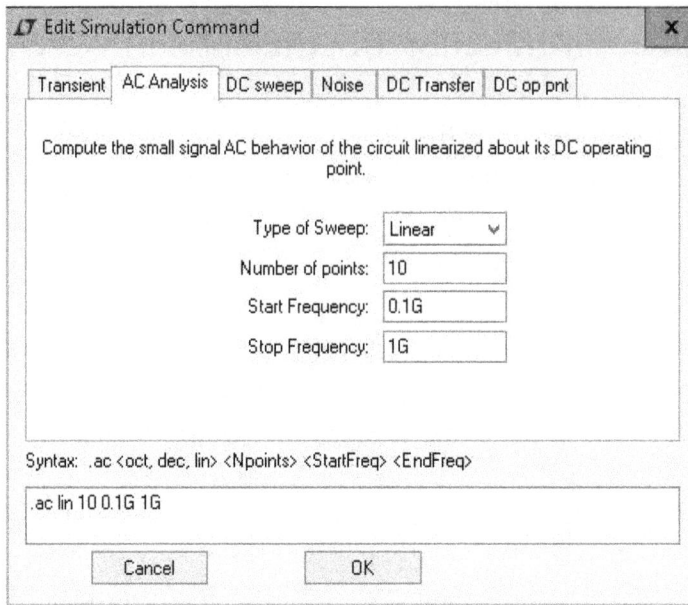

Figure 2-3 AC analysis setup

4.   Right click on the plot window and select Add Trace. Type in the Equation (2-29) under the "Expression(s) to add," as shown in Figure 2-4.

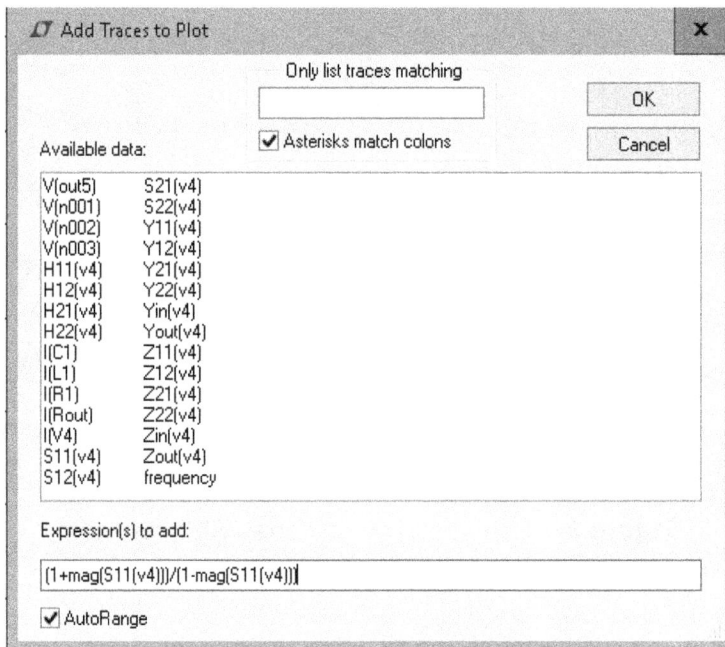

Figure 2-4 Window Add Traces to Plot

Press Ok to generate Figure 2-5.

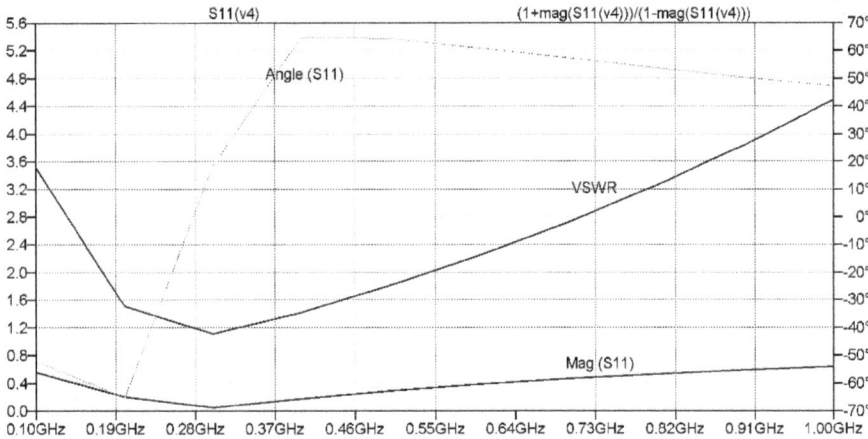

Figure 2-5 S11 Magnitude and Angle plus VSWR

5. Select the S11 traces and Format [Real, Imaginary] or Polar: (dB,deg), as shown in Figure 2-6.

6. Press Ok to save the text file. Use the Browse tab to save the text file in the desired directory.

7. Go to the above directory and open the file to access data.

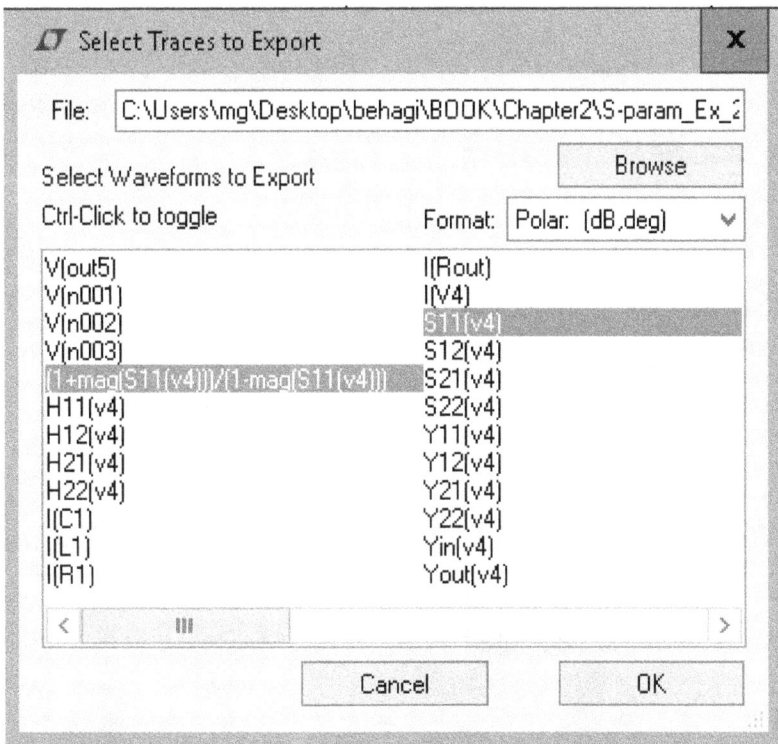

Figure 2-6 Select traces to add in polar format

8. Organize the tabulated data as shown in Table 2-1 and Table 2-2.

Freq. (Hz)	VSWR(dB)		Mag,S11(dB)		Angle,S11(Deg)
1.00E+08	10.92		-5.0837		-52.175
2.00E+08	3.541		-13.933		-65.29
3.00E+08	0.877		-25.943		18.317
4.00E+08	3.048		-15.206		64.687
5.00E+08	5.159		-10.797		64.343
6.00E+08	7.024		-8.321		61.006
7.00E+08	8.712		-6.682		57.291
8.00E+08	10.26		-5.506		53.694
9.00E+08	11.71		-4.62		50.344
1.00E+09	13.05		-3.931		47.27

Table 2-1: Tabulated VSWR and S11 (Magnitude in dB)

Freq.(Hz)	VSWR		Real,S11		Imag.,S11
1.00E+08	3.51401		0.34154659		-0.43992229
2.00E+08	1.50334		0.084053375		-0.18265703
3.00E+08	1.10626		0.047892137		0.015854803
4.00E+08	1.42028		0.074246399		0.156977608
5.00E+08	1.81111		0.124931429		0.260087866
6.00E+08	2.24497		0.185967242		0.335577977
7.00E+08	2.72658		0.250364556		0.389844097
8.00E+08	3.26009		0.314124589		0.427531726
9.00E+08	3.84853		0.37493197		0.452313137
1.00E+09	4.49418		0.431541452		0.467160375

Table 2-2: Tabulated VSWR and S11 (Real and Imaginary

Return Loss, VSWR, and Gamma (S11)

Return Loss, VSWR, and Reflection Coefficient are all different ways of characterizing the wave reflection. These definitions are often used interchangeably in practice. Therefore, it is important to be able to convert from one form of reflection to another. Return Loss is often used to characterize components such as filters, amplifiers, and networks.

VSWR is normally used in systems such as radio and TV transmitters while reflection coefficient is normally used in device characterization such as transistors, capacitors, inductors, etc. Equations (2-29), (2-30), and (2-31) give the conversions among these three parameters. Note, however, that the reflection coefficient is a vector quantity whose magnitude is all that is required to determine the VSWR or return loss. Therefore when calculating Γ from the VSWR or return loss we can only find the magnitude and not the angle of the reflection coefficient. Another useful parameter is the mismatch loss. The mismatch loss is a measure of power that is reflected by the load.

$$Mismatch \;\; Loss \;\; (dB) = -10\log\left(1 - |\Gamma|^2\right) \qquad (2\text{-}43)$$

freq	VSWR1	dB(S(1,1))	S(1,1)
100.0 MHz	3.514	-5.084	0.557 / -52.176
200.0 MHz	1.503	-13.933	0.201 / -65.292
300.0 MHz	1.106	-25.945	0.050 / 18.321
400.0 MHz	1.420	-15.207	0.174 / 64.690
500.0 MHz	1.811	-10.796	0.289 / 64.345
600.0 MHz	2.245	-8.321	0.384 / 61.007
700.0 MHz	2.727	-6.682	0.463 / 57.292
800.0 MHz	3.260	-5.506	0.531 / 53.694
900.0 MHz	3.849	-4.620	0.588 / 50.344
1.000 GHz	4.494	-3.931	0.636 / 47.270

Table 2-3 Tabular output of VSWR, and reflection coefficient

Example 2.2: Generate a Table showing the return loss, reflection coefficient, and reflected power as function of VSWR.

Solution: Create a schematic with a resistor as shown in Figure 2-7. Make the resistance value a tunable variable. Set the Linear Analysis at a fixed frequency of 100 MHz. Add a Parameter Sweep to sweep the value of the resistor {R} from 50 Ohm to 500 Ohms as listed below.

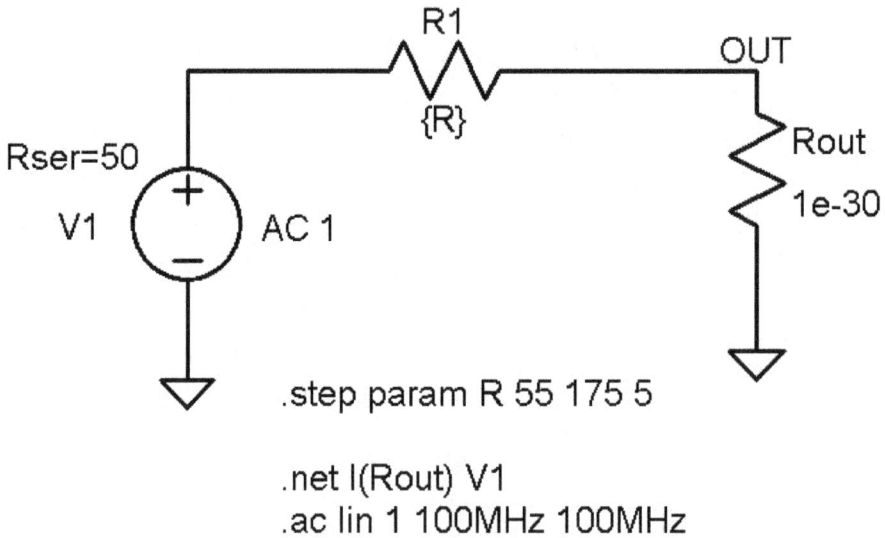

Figure 2-7 Schematic and parameter sweep for VSWR table

Edit the Parameter Sweep and select only the following resistor values for generating the required Table.

R= 50.5, 51, 55, 60, 65, 70, 75, 80, 85, 90, 95, 100, 105, 110, 115, 120, 125, 150, 175, 200, 225, 250, 300, 350, 400, 450, and 500 Ohms.

Use the the following equations for calculation of the mismatch and power loss.

ReflCoef = (VSWR - 1)/(VSWR + 1)
Mismatch = 1 − ((ReflCoef)^2)
Powerloss = (1 − Mismatch)*100

As you notice we present the mismatch loss as a percentage of the available power that is reflected by the load. Add the equation block variable, power loss, to the output table. The results of the Parameter Sweep can be saved in a text f ile which can be read into Exel for a more attractive formatted table as shown in Table 2-4.

VSWR	|Γ|	Return Loss (dB)	% Reflected Power	VSWR	|Γ|	Return Loss (dB)	% Reflected Power
1.01	0.005	46.06	0.002	2.40	0.412	7.71	16.955
1.02	0.010	40.09	0.010	2.50	0.429	7.36	18.367
1.10	0.048	26.44	0.227	3.00	0.500	6.02	25.000
1.20	0.091	20.83	0.826	3.50	0.556	5.11	30.864
1.30	0.130	17.69	1.701	4.00	0.600	4.44	36.000
1.40	0.167	15.56	2.778	4.50	0.636	3.93	40.496
1.50	0.200	13.98	4.000	5.00	0.667	3.52	44.444
1.60	0.231	12.74	5.325	6.00	0.714	2.92	51.020
1.70	0.259	11.73	6.722	7.00	0.750	2.50	56.250
1.80	0.286	10.88	8.163	8.00	0.778	2.18	60.494
1.90	0.310	10.16	9.631	9.00	0.800	1.94	64.000
2.00	0.333	9.54	11.111	10.00	0.818	1.74	66.942
2.10	0.355	9.00	12.591	20.00	0.905	0.87	81.859
2.20	0.375	8.52	14.063	200.00	0.990	0.09	98.020
2.30	0.394	8.09	15.519	2000.00	0.999	0.01	99.800

Table 2-4 Relationship among return loss, VSWR, and \|Γ\|

As you can see from the Table 2-4 if we can keep the VSWR less than 1.25:1 we will have less than 1% power loss due to reflective impedance mismatch.

2.5 RF and Microwave Transmission Media

There are many physical transmission lines that are encountered in RF and microwave circuits and systems. In addition it should not be overlooked that free space is also a transmission medium. Therefore, Equation (2-35) could be used to describe the characteristic impedance, Z_o, of free space. In order to define L and C we need to consider the inductive and capacitive properties of free space. Permeability is the ability of a transmission media to support a magnetic field. Considered as a density in free space the permeability is related to inductance and defined by Equation (2-44).

$$\mu_o = 4\pi \cdot 10^{-7} \quad Henries/meter \tag{2-44}$$

Permittivity is the ability of a transmission media to support an electric field. Considered as a density in free space the permittivity is closely related to capacitance in Equation (2-45).

$$\varepsilon_o = \frac{1}{36\pi} \cdot 10^{-9} \quad Farads/meter \tag{2-45}$$

Therefore we can rewrite equation (2-35) as:

$$Z_o = \sqrt{\frac{\mu_o}{\varepsilon_o}} = \sqrt{\frac{4\pi \cdot 10^{-7}}{\frac{1}{36\pi} \cdot 10^{-9}}} = 377\ \Omega \tag{2-46}$$

Equation (2-46) shows that the characteristic impedance is related to the inductance and capacitance of the transmission media. The velocity of propagation is also related to inductance and capacitance. It can be shown that the time required for a sine wave to propagate through a unit length of lossless transmission line is related by Equation (2-47).

$$t = \sqrt{LC} \qquad (2\text{-}47)$$

The velocity of propagation is related to the wavelength in one period T.

$$v = \frac{\lambda}{T} = \frac{1}{t} = \frac{1}{\sqrt{LC}} \qquad (2\text{-}48)$$

Where: T is the time period of the sine wave.

Using the free space permeability and permittivity the velocity of propagation through free space is:

$$v_o = \sqrt{\mu_o \varepsilon_o} = 2.998 \cdot 10^8 \quad meters/\text{second} \quad (2\text{-}49)$$

From Equations (2-46) and (2-49) we can see that the characteristic impedance and velocity of propagation can be defined for any transmission media based on the inductive and capacitive properties of that media.

Physical Transmission Lines

Free space can be used as a transmission line for the propagation of radio signals across long distances but may not be as useful for the point to point connection and isolation of specific RF signals. For this purpose engineers use a variety of different transmission line media. This text is focused on a few of the most commonly used transmission media in RF and microwave circuit and system design including, microstrip, stripline, coaxial, and waveguide transmission lines. As a general differentiator, transmission media may be divided among pure TEM, quasi TEM, and non TEM propagation modes. TEM refers to the transverse electromagnetic mode of wave propagation. Signals traveling through space propagate in the TEM mode. This simply means that the magnetic field, electric field, and the direction of propagation are all orthogonal to one another. This relationship is depicted in Figure 2-8.

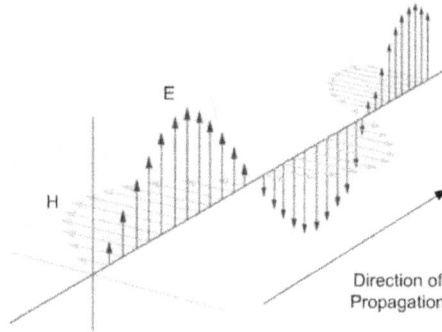

Figure 2-8 TEM wave propagation

The velocity with which a wave travels will almost always be slower in a physical transmission line than it is in free space. Engineers frequently need to trim a transmission line for a specific wavelength so it is essential that they know the velocity of propagation in the transmission line. Knowing the distributed inductive and capacitive properties of the media, Equation (2-48) could be used to determine the propagation velocity. However it is common for many transmission line manufacturers to specify a velocity factor. The velocity factor is a ratio of the actual transmission line velocity v, to the velocity of free space.

$$v_f = \frac{v}{v_o} \tag{2-50}$$

Similar to the velocity factor, manufacturers of transmission line media typically express the permittivity of the dielectric material as a relative dielectric constant. The relative dielectric constant, ε_r, is the ratio of the actual material dielectric constant to the dielectric constant of free space.

$$\varepsilon_r = \frac{\varepsilon}{\varepsilon_o} \tag{2-51}$$

Depending on the type of transmission media the manufacturer may choose to specify either the velocity factor, v_f , or relative dielectric constant ε_r. The two quantities are related by Equation (2-52).

$$v_f = \frac{1}{\sqrt{\varepsilon_r}} \qquad (2\text{-}52)$$

There are many types of physical transmission lines ranging from twisted wire pairs to fiber optic cables. There are many good reference sources to examine the characteristics and application of transmission lines. In this chapter we will focus on just a few of these transmission lines that the RF and microwave engineer will frequently encounter. They include:

Coaxial Transmission Lines

Microstrip Transmission Lines

Stripline Transmission Lines

Waveguides

Note that coaxial, microstrip, and stripline transmission lines require two conductors to transfer power from a given source to its load while waveguides require only a single hollow conductor. A brief introduction to each of the above transmission media is given here

2.6 Coaxial Transmission Line

Coaxial transmission line, also known as coaxial cable, is one of the most common types of transmission line used in electrical equipment interconnection. A typical coaxial line consists of an inner conductor inside another cylindrical conductor and a dielectric material in between, as shown in Figure 2-9. The outer conductor, or the shield, is usually grounded to minimize RF interference and radiation loss. The coaxial transmission line supports pure TEM propagation as shown in Figure 2-9. The dielectric material is typically a form of Teflon. Teflon has very good physical strength as well as low loss and high temperature operation. For very high power and low loss applications the dielectric material may be air. In this case there must be some dielectric spacer installed at certain intervals to support the center conductor and maintain concentricity. It is also common to introduce nitrogen into the air dielectric to help prevent condensation from forming inside the cable. Condensation would greatly increase the loss and lower the voltage breakdown. Coaxial cables that are used in very high power applications generally require a larger diameter.

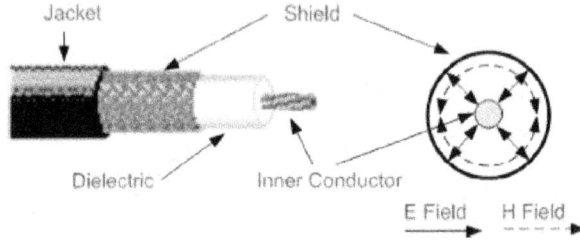

Figure 2-9 Coaxial cable construction and field orientation

The reflective losses are mostly due to the impedance mismatch between the source and load impedance connected to the coaxial cable. There also exists a dissipative loss in a coaxial cable that varies significantly with frequency. When a good impedance match exists between the source and the load it is only the cable loss that is significant. The losses in coaxial cable are primarily due to the conductor loss and the dielectric material loss. Because of its smaller cross sectional area the center conductor dominates the conductor loss contribution. The center conductor's diameter and resistivity determine the conductor loss. Skin effects which cause the RF currents to propagate near the surface of the conductor require that the center conductor have very good conductivity. Sometimes a copper conductor may have a silver plating to improve the surface conductivity. The characteristics of the dielectric material and its thickness determine the losses in the dielectric. Table 2-5 gives a brief listing of some commonly used flexible coaxial cables. Manufacturers typically specify the cable loss in dB per 100 foot lengths at various frequencies. This table gives the manufacturer's (Belden) part number along with an RG designation. The RG designation is an attempt by the U.S. government under MIL-C-17 specification to have a standard designation for the manufacture of coaxial cables. The R means that the cable is intended for RF frequency usage. The G means that the cable is manufactured to general specifications. The number then identifies the unique specifications that the cable is designed to meet. Minor differences in the specification have a letter appended to the numerical designator. The characteristic impedance and velocity factor are also specified. Two physical characteristics included in Table 2-5 are the outer dimension (O.D.) and the shield percentage. The term O.D. is referred to as either the outer dimension or overall diameter including the outer jacket. The characteristic impedance of a coaxial line is determined by the diameter of the inner conductor (d) and the outer conductor (b) along with the relative dielectric constant, as given by:

$$Z_o = \frac{60}{\sqrt{\varepsilon_r}} \cdot \ln\left(\frac{b}{d}\right) \qquad (2\text{-}53)$$

Loss in dB/100ft. @ frequency in MHz										
Coaxial Cable	100 MHz	200 MHz	400 MHz	700 MHz	900 MHz	Z_0	V_f	Capac. per ft.	O.D. inch	Shield
Belden 9913	1.30	1.80	2.70	3.60	4.20	50	.84	24.6 pF	0.405	100%
Belden 9914	1.60	2.40	3.50	5.00	5.70	50	.82	24.8 pF	0.403	100%
Belden 8214 RG-8/U	1.80	2.70	4.20	5.80	6.70	50	.78	26.0 pF	0.405	97%
Belden 8238 RG-11/U	2.00	2.90	4.20	5.80	6.70	75	.66	20.5 pF	0.405	97%
Belden 8267 RG-213/U	1.90	2.70	4.10	6.50	7.60	50	.66	30.8 pF	0.405	97%
Belden 8242 RG-9/U	2.10	3.00	4.80	6.50	7.60	51	.66	30.0 pF	0.420	98%
Belden 9258 RG-8/X	3.70	5.40	8.00	11.10	12.80	50	.80	25.3 pF	0.242	95%
Belden 84142 RG-142	3.90	5.60	8.20	11.00	12.50	50	.695	29.2 pF	0.195	98%
Belden 9273 RG-223/U	4.10	6.00	8.80	12.00	13.80	50	.66	30.8 pF	0.212	95%
Belden 8240 RG-58/U	4.50	6.80	10.00	14.00	16.00	51.5	.66	29.9 pF	0.195	95%
Belden 8259 RG-58A/U	4.90	7.30	11.50	17.00	20.00	50	.66	30.8 pF	0.193	95%
Belden 9259 RG-59/U	3.00	4.50	6.60	8.90	10.1	75	.78	17.3 pF	0.242	95%
Belden 8241 RG-59/U	3.40	4.90	7.00	9.70	11.1	75	.66	20.5 pF	0.242	95%
Belden 8216 RG-174/U	8.40	12.50	19.00	27.00	31.00	50	.66	30.8 pF	0.101	90%
Belden 9228 RG-62A/U	2.70	3.80	5.30	7.30	8.20	93	.84	13.5 pF	0.242	95%

Table 2-5 Sample table of coaxial cable specifications

2.7 Microstrip Transmission Lines

Microstrip is a planar transmission line media in which the transmission line is etched onto the top side of a printed circuit board. The bottom side of the printed circuit board is completely metalized and grounded. The printed circuit board has a low loss dielectric material that is suitable for microwave transmission. Figure 2-10 shows a sampling of microstrip lines that represent specific circuits. The top side of the dielectric substrate is shown with the copper conductor, microstrip transmission lines.

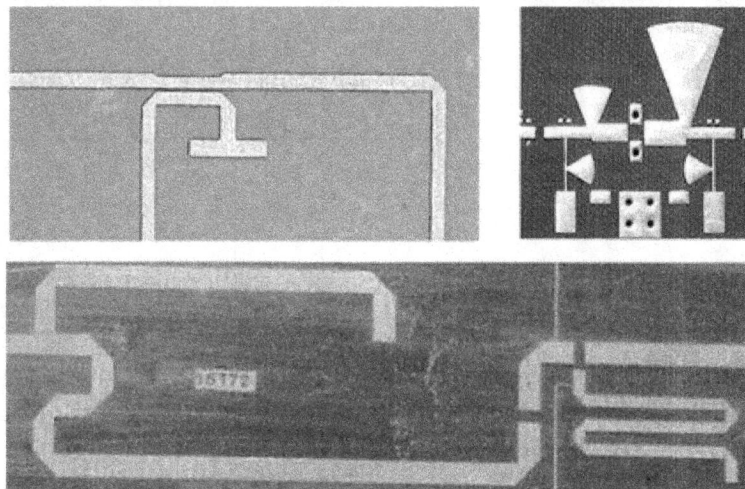

Figure 2-10 Examples of microstrip transmission lines on various dielectrics

Unlike the coaxial cable the center conductor is not shielded. This means that a portion of the electric and magnetic field is in the air space above the microstrip line as shown in Figure 2-11. Thus the propagation in microstrip is not purely TEM but rather quasi TEM. This also leads to the fact that the dielectric constant beneath the microstrip line is slightly less than the relative dielectric constant of the material. This is known as the effective dielectric constant, ε_{eff}, and is a function of the width of the microstrip line, W and the height of the substrate, h as shown in Figure 2-11. The thickness of the microstrip conductor, t, has a minor effect on ε_{eff} and is omitted from the computation. The effective dielectric constant is calculated in the empirical Equations (2-54) and (2-55).

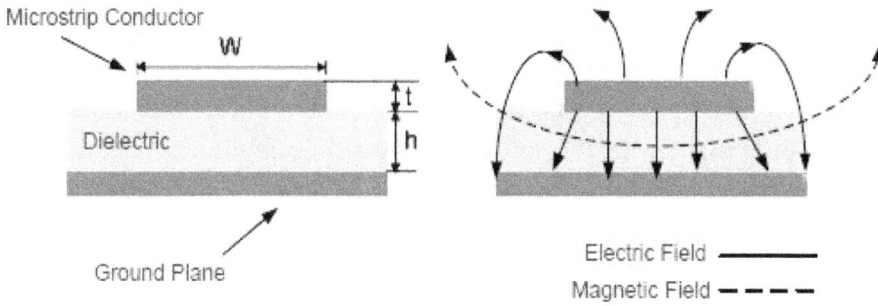

Figure 2-11 Microstrip transmission line and electromagnetic field lines

For $\left(\dfrac{W}{h}\right) < 1$

$$\varepsilon_{\textit{eff}} = \frac{\varepsilon_r + 1}{2} + \frac{\varepsilon_r - 1}{2} \cdot \left[\left(1 + 12\left(\frac{h}{W}\right)\right)^{-0.5} + 0.04\left(1 - \left(\frac{W}{h}\right)\right)^2 \right] \qquad (2\text{-}54)$$

For $\left(\dfrac{W}{h}\right) \geq 1$

$$\varepsilon_{\textit{eff}} = \frac{\varepsilon_r + 1}{2} + \frac{\varepsilon_r - 1}{2} \cdot \left(1 + 12\left(\frac{h}{W}\right)\right)^{-0.5} \qquad (2\text{-}55)$$

Similarly the characteristic impedance also has two solutions based on the ratio of the line width to substrate height [6].

For $\left(\dfrac{W}{h}\right) < 1$

$$Z_O = \frac{60}{\sqrt{\varepsilon_{\textit{eff}}}} \ \ln\left(8 \cdot \frac{h}{W} + 0.25 \cdot \frac{W}{h}\right) \qquad (2\text{-}56)$$

For $\left(\dfrac{W}{h}\right) \geq 1$

$$Z_o = \frac{120\pi}{\sqrt{\varepsilon_{\it eff}} \cdot \left[\frac{W}{h} + 1.393 + 0.6667 \ \ln \left(\frac{W}{h} + 1.444 \right) \right]} \tag{2-57}$$

Equations (2-56) and (2-57) do not consider the conductor thickness of the line. The effective width, W_e, of a microstrip line is the equivalent width of the line with the conductor thickness taken into account. The effective line width is defined by Equations (2-58) and (2-59).

For $\left(\frac{W}{h} \right) \geq \frac{1}{2\pi}$

$$W_e = W + \frac{t}{\pi} \left(1 + \ln \frac{2h}{t} \right) \tag{2-58}$$

For $\left(\frac{W}{h} \right) < \frac{1}{2\pi}$

$$W_e = W + \frac{t}{\pi} \left(1 + \ln \left(\frac{4\pi \cdot W}{t} \right) \right) \tag{2-59}$$

Microstrip Transmission Line Properties

For a desired impedance, the physical transmission line width can be determined based on the dielectric constant and substrate thickness. Note that there are two additional parameters that can affect the line width calculation for a given line impedance. These are the conductor thickness, t, and the cover or box height. When using a thicker conductor for higher current handling it is recommended to model the actual conductor thickness, t, as it will have a slight impact on the line width. If the cover height becomes too close to the microstrip line it will add enough capacitance to lower the impedance of the line. In practice a minimum cover height must be maintained so that the cover does not add significant capacitance to the microstrip transmission line. Before adding any microstrip transmission lines to a workspace schematic, a dielectric substrate must be defined. A substrate may be added to the schematic.

The substrate's dielectric constant and height are the most important parameters to enter into the parameters window. The default settings can be accepted for the remaining parameters for most circuits. The dielectric loss tangent and conductor resistivity will dominate the microstrip losses of resonators, filters or other critical low loss circuits. Substrates can be divided into hard substrate and soft substrate types.

Hard substrates are very stiff and brittle and must be cut with diamond tipped saws or lasers. Soft substrates are very pliable and can be cut with a razor knife or scissors. All of the circuit designs and components discussed in this book are based on soft substrates. Hard substrates are normally used in semiconductor design i.e., transistors, and MMICs. Alumina is a hard substrate that is often used in microwave integrated circuit (MIC) designs. Because the top half of the microstrip is exposed it is very easy to attach discrete components. This has made microstrip the most popular form of transmission line for designing and building microwave components.

There are many variants of the basic microstrip transmission line including, stripline, suspended line, inverted line, and coplanar waveguide.

2.8 Stripline Transmission Lines

Stripline is a transmission line using planar dielectric material similar to microstrip. The major difference is that the top half of the line also consists of a dielectric of the same material as the bottom conductor. Therefore the transmission line is shielded in much the same way as the coaxial transmission line. As such the stripline transmission line supports a pure TEM propagation mode. Because of the presence of the top dielectric it is much more difficult to integrate discrete components such as transistors, and chip inductors and capacitors. As a pure TEM transmission line, stripline does offer superior performance in distributed filters, directional couplers, and power combiners.

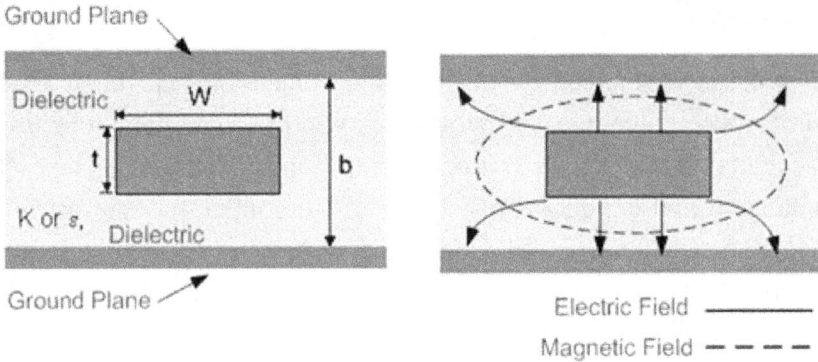

Figure 2-12`: Cross section view of stripline t and field lines

Because the propagation mode is pure TEM, there is no effective dielectric constant, just the relative dielectric constant of the substrate material. Soft substrates are almost entirely used for stripline transmission circuits. One popular variation is when an air dielectric is used. This is referred to as suspended substrate stripline (SSS) and is characterized by very low loss. Suspended substrate stripline is often used in filter and multiplexer circuit designs. The characteristic impedance of a stripline transmission line is given by the empirical Equation (2-60).

$$Z_o = \frac{94.2}{\sqrt{\varepsilon_r}} \ln \left[\frac{1 + \frac{W}{b}}{\frac{W}{b} + \frac{t}{b}} \right] \qquad (2\text{-}60)$$

In practice a 0.060 stripline media is realized by clamping or fusing two 0.030 dielectric halves together. The circuit pattern is etched on one of the 0.030 dielectrics similar to a microstrip circuit pattern. The top half dielectric would then have all metal removed on the side which meets the bottom half circuit patterns. A special bonding film can be used to effectively glue the halves together or the dielectrics could be clamped together using sufficient pressure.

2.9 Waveguide Transmission Lines

A waveguide is a special form of transmission line that is used primarily at microwave frequencies above 2 GHz. The most important difference between waveguide and other forms of transmission line is the low loss and capability to transmit very high power. Waveguides do not have a separate conductor, ground plane, or shield as the previous transmission line structures. Rather waveguide is a hollow tube in which the wave propagates through. Even though the propagation is in the air space, enclosing an RF wave in a metallic boundary causes the wave to propagate quite differently than it would in free space. Waveguide tubes can be circular or rectangular but the rectangular tube is much more popular due to its ease of manufacture. Figure 2-13 shows a typical rectangular waveguide section with flanges on each end. The flanges allow a connection to be made to a mating piece of waveguide or waveguide component. The broad dimension is labeled as the 'a' dimension while the narrow side is labeled the 'b' dimension. The TEM propagation mode does not exist in the waveguide. The propagation must be either transverse electric (TE) or transverse magnetic (TM).

The propagation modes are quite complex with many subcategories of TE and TM propagation in existence.

Figure 2-13: Rectangular waveguide section showing inside dimensions

In TE mode the electric field is transverse to the direction of propagation. This means that there is no component of the electric field in the direction of propagation. In the TM mode there is no magnetic field component in the direction of propagation. Depending on how the RF energy is launched into the waveguide there are many variations of both TE and TM propagation modes. These sub-modes are designated with the subscripts m and n ($TE_{m,n}$). The sub-mode notation describes the field patterns that exist in the waveguide. The dominant sub-mode is the $TE_{1,0}$ mode. This is the lowest frequency that the waveguide will support. Figure 2-14 shows the field pattern for the $TE_{1,0}$ mode in a rectangular waveguide. The 1,0 index signifies that there is one field variation along the broad dimension and no field variations along the narrow dimension. The remaining discussion of rectangular waveguide properties is based on the $TE_{1,0}$ mode.

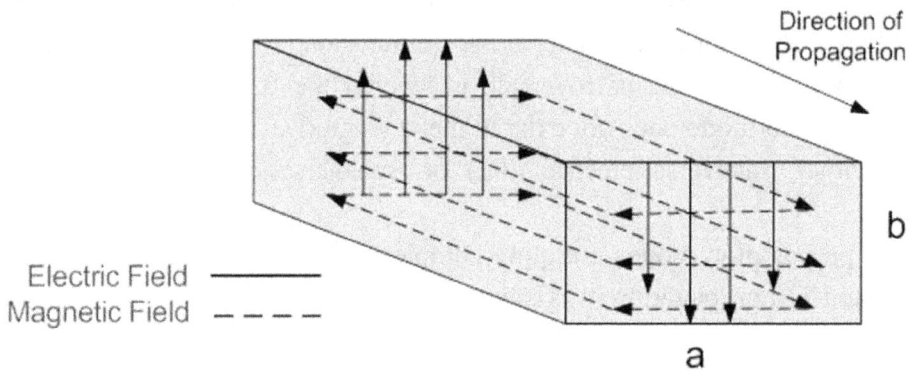

Figure 2-14 TE_{10} dominant mode field pattern in rectangular waveguide

The wave pattern in a waveguide leads to an interesting condition where at or below a certain frequency the wave bounces from side-to-side or top-to-bottom in the waveguide tube and no longer travels in the direction of propagation. This frequency is known as the cutoff frequency. Below the cutoff frequency the waveguide transmits very little energy. As such the waveguide has a natural high pass filter characteristic. The cutoff frequency of the $TE_{1,0}$ in rectangular waveguide is given by Equation (2-61),

$$f_{c_{1,0}} = \frac{c}{2a} = \frac{3 \cdot 10^8}{2a}$$

(2-61)

Where: c in Equation (2-61) is the speed of light and a is the wider dimension of the waveguide.

The wavelength in a rectangular waveguide is then defined as in Equation (2-62):

$$\lambda_g = \frac{\lambda}{\sqrt{\left(1 - \left(\frac{\lambda}{(2a)^2}\right)\right)}}$$

(2-62)

The characteristic impedance of waveguide is based on the propagation mode. The characteristic impedance must be defined separately for the transverse electric and magnetic fields as given by Equations (2-63) and (2-64).

$$Z_{o\,(TE_{m,n})} = \frac{\eta}{\sqrt{1 - \left(\frac{f_c}{f}\right)^2}}$$

(2-63)

$$Z_{o\,(TM_{m,n})} = \eta \sqrt{1 - \left(\frac{f_c}{f}\right)^2}$$

(2-64)

The term, $,\eta$ is the intrinsic impedance of the medium. In air $\eta = 377\ \Omega$. Z_0 is rarely used in practice. Waveguide is typically characterized by its cross section dimensions as shown in Table 2-6. Table 2-6 shows a listing of some of the more popular waveguides by WR designator.

Frequency Band, GHz	U.S. (EIA) Designator	British WG Designator	Cut Off Freq. in GHz $TE_{1,0}$	a dimension inches	b dimension inches
1.12 - 1.70	WR 650	WG 6	0.908	6.500	3.250
1.45 - 2.20	WR 510	WG 7	1.158	5.100	2.550
1.70 - 2.60	WR 430	WG 8	1.375	4.300	2.150
2.20 - 3.30	WR 340	WG 9A	1.737	3.400	1.700
2.60 - 3.95	WR 284	WG 10	2.080	2.840	1.340
3.30 - 4.90	WR 229	WG 11A	2.579	2.290	1.145
3.95 - 5.85	WR 187	WG 12	3.155	1.872	0.872
4.90 - 7.05	WR 159	WG 13	3.714	1.590	0.795
5.85 - 8.20	WR 137	WG 14	4.285	1.372	0.622
7.05 - 10.0	WR 112	WG 15	5.260	1.122	0.497
8.2 - 12.4	WR 90	WG 16	6.560	0.900	0.400
9.84 - 15.0	WR 75	WG 17	7.873	0.750	0.375
11.9 - 18.0	WR 62	WG 18	9.490	0.622	0.311
14.5 - 22.0	WR 51	WG 19	11.578	0.510	0.255
17.6 - 26.7	WR 42	WG 20	14.080	0.420	0.170
21.7 - 33.0	WR 34	WG 21	17.368	0.340	0.170
26.4 - 40.0	WR 28	WG 22	21.100	0.280	0.140
32.9 - 50.1	WR 22	WG 23	26.350	0.224	0.112
39.2 - 59.6	WR 19	WG 24	31.410	0.188	0.094
49.8 - 75.8	WR 15	WG 25	39.900	0.148	0.074
60.5 - 91.9	WR 12	WG 26	48.400	0.122	0.061

Table 2-6 Standard rectangular waveguide characteristics

To interface waveguide with coaxial, microstrip, or stripline transmission lines a special transformer, known as an adapter, must be used. Waveguide to coax adapters come in many forms. Some adapters couple energy to the *E* field while others couple to the *H* field. The adapter shown in Fig. 2-15 is an *E* field waveguide to coax transition. The center conductor of the coaxial connector extends into the waveguide to excite the wave propagation in the waveguide. The center conductor is approximately $\lambda_g/4$ from the back wall.

Figure 2-15 Waveguide to coax transition adapter

Because of the complex EM fields that can propagate in waveguide, modern computer aided design techniques are best handled by three dimensional EM solvers. The software is limited in its ability to model waveguide propagation and waveguide components.

2.10 Group Delay in Transmission Lines

A frequently encountered concept related to the transmission line velocity factor is group delay. Group delay is a measure of the time that it takes a signal to traverse a transmission line, or its transit time. It is a strong function of the length of the line, and usually a weak function of frequency. It is expressed in units of time, picoseconds for short distances or nanoseconds for longer distances. Remember that in free space all electromagnetic signals travel at the speed of light, c, which is approximately 300,000 kilometers per second. Therefore, in free space, electromagnetic radiation travels one foot in one nanosecond, unless there is something to slow it down such as a dielectric. Mathematically the group delay is the derivative of phase versus frequency. In communication systems, the ripple in the group delay creates a form of distortion.

2.11 Transmission Line Components

There are many useful components that can be realized using transmission lines. These include power splitters, directional couplers, voltage and current insertion networks, as well as various filter networks. These networks can be realized with any of the physical transmission line structures. However, because of its popularity, we will explore many of these components in microstrip transmission line. These components are referred to as distributed components. It can be shown that the series inductance and shunt capacitance can be realized with distributed microstrip transmission lines. We will begin this section with an examination of the open and short-circuited microstrip transmission lines.

Short-Circuited Transmission Line

Equation (2-38) demonstrated that the input impedance of a lossless short-circuited transmission line is a pure imaginary function; therefore, the input reactance is given by the following equation.

$$X_{in} = Z_O \tan\theta \qquad\qquad (2\text{-}65)$$

Where $\theta = Beta(d)$ is the electrical length of the transmission line in degrees.

From Equation (2-65) we can see that this reactance can change from inductive to capacitive depending on the length of the transmission line.

Example 2.3: Plot the reactance of a loss less short-circuited transmission line as a function of the group delay or electrical length of the line.

Solution: To plot the reactance of the short-circuited transmission line, create a schematic in LTspice with a grounded transmission line. Make the group delay of the transmission line variable {t}. Setup a new S parameter simulation with a single frequency at 1500 MHz, as shown in Figure 2-16.

Td= {t} Z0=50

T1   OUT

Rser=50

V1   AC 1

Rout
1e-30

.step param t 0 0.666667n 1p

.net I(Rout) V1
.ac lin 1 1500Meg 1500Meg

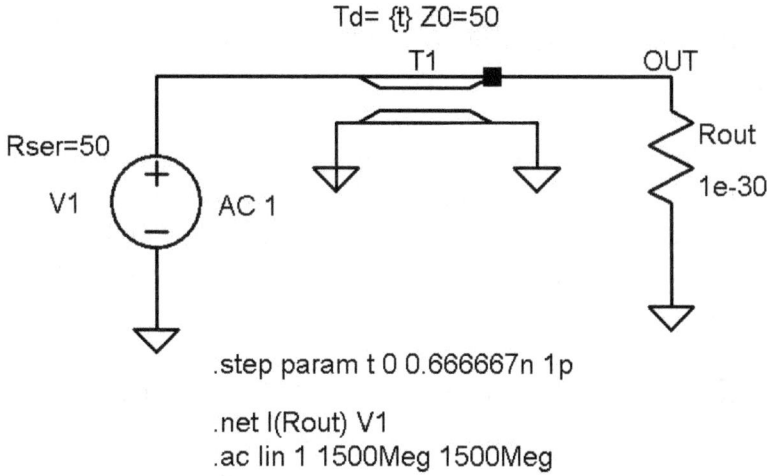

Figure 2-16 Short-circuited line reactance versus electrical length

Use a Parameter sweep to vary the group delay of the transmission line from 0 nanosecond to 0.666667 nanosecond at 1500 MHz with 1 picoseconds step (representing electrical length from 0 degree to 360 degrees). Setup a graph to plot the reactance of the shorted transmission line versus the group delay, or the electrical length in degrees, as shown in Fig. 2-17.

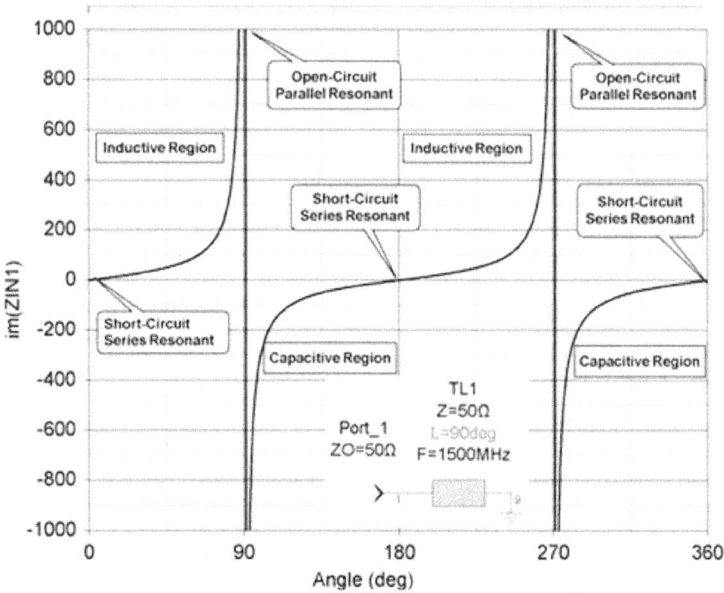

Figure 2-17 Short-circuited line reactance versus electrical length

Normally the independent variable in most linear simulations is frequency.

From the reactance plot of Figure 2-17 we can see that at one quarter wavelength, t =166 picoseconds, the short-circuited line looks like an open circuited line. From 0 to 90 degrees the line looks like an inductor. From $90°$ to $180°$ the circuit looks like a capacitor. The pattern then repeats from $180°$ to $360°$.

Modeling Short-Circuited Microstrip Lines

The short-circuited microstrip line can be modeled as a microstrip transmission line connected to a grounded via hole. A via hole is made by drilling a hole in the dielectric and metalizing the inside of the hole to form a conductive path to the ground side of the dielectric.

To create a 90 degree line in the microstrip substrate we must know the effective dielectric constant so that the wavelength in the dielectric can be calculated. The relationships between line length and electrical degrees are given by the following equations.

$$\theta = \frac{2\pi\ell}{\lambda_g} \tag{2-66}$$

$$\ell = \frac{\theta\lambda_g}{360\sqrt{\varepsilon_{eff}}} \tag{2-67}$$

Example 2.4: Calculate the input impedance of a quarter-wave short-circuited transmission line at 1.5 GHz.

Solution: Place a short-circuited transmission line element on the schematic. Figure 2-18 shows the correct method of modeling a microstrip short-circuited transmission line. Simulate the schematic and display the input impedance in Table 2-7.

Td= {t} Z0=50

T1   OUT

Rser=50

V1   (+)   AC 1   Rout

     (−)   1e-30

.step param t 0 0.666667n 1p

.net I(Rout) V1
.ac lin 1 1500Meg 1500Meg

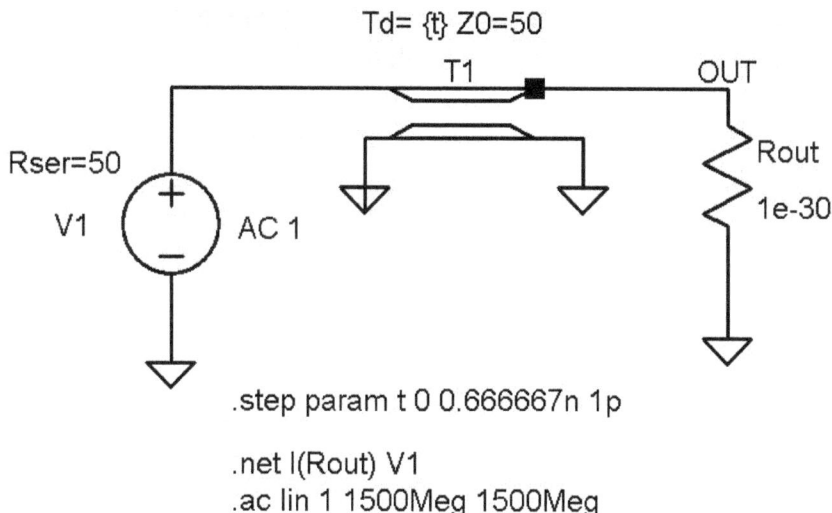

Figure 2-18 Quarter wave short circuited line schematic

freq	Zin1	Zin1
1.500 GHz	768.602 - j3.067E3	3.161E3 / -75.930

Table 2-7 Quarter wave short-circuited line impedance in two formats

As the Table 2-7 shows, the impedance of a quarter-wave section of short circuited line is quite high, close to an open circuit. This type of line section could be used as a parallel resonant circuit.

Open-Circuited Transmission Line

Equation (2-39) showed that the input impedance of a lossless open circuited transmission line is a pure imaginary function; therefore, the input reactance is given by Equation (2-68).

$$X_{in} = Z_O \cot\theta \tag{2-68}$$

Where: θ is the electrical length of the transmission line in degrees.

Example 2-5: Plot the reactance of a loss less open-circuited transmission line as a function of the electrical length of the line.

Solution: Using the same procedures of Example 2-3, use a parameter sweep to observe the behavior of this reactance as the group delay {t} is varied from 0 to 0.666667 nanosecond (corresponding to 0 to 360 degrees electrical length). Note that the 10^{30} Ω load in Figure 2-19 is to emulate an open circuit termination.

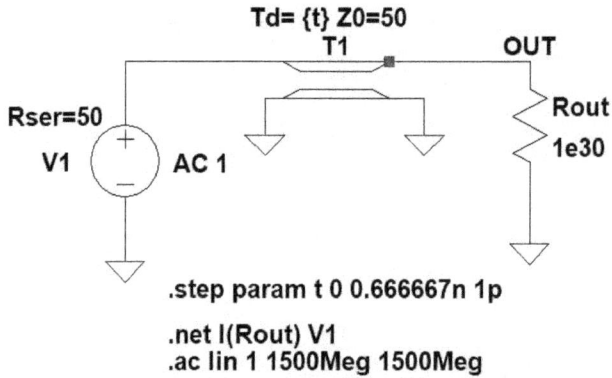

Td= {t} Z0=50
T1   OUT

Rser=50
V1   AC 1   Rout
                      1e30

.step param t 0 0.666667n 1p

.net I(Rout) V1
.ac lin 1 1500Meg 1500Meg

Figure 2-19 Quarter wave Open circuited transmission line schematic

Comparing the open circuit reactance to the short-circuited line reactance we see that from 0 to 90° the line is capacitive and from 90 degrees to 180 degrees the line is inductive. This situation repeats from 180 to 360 degrees.

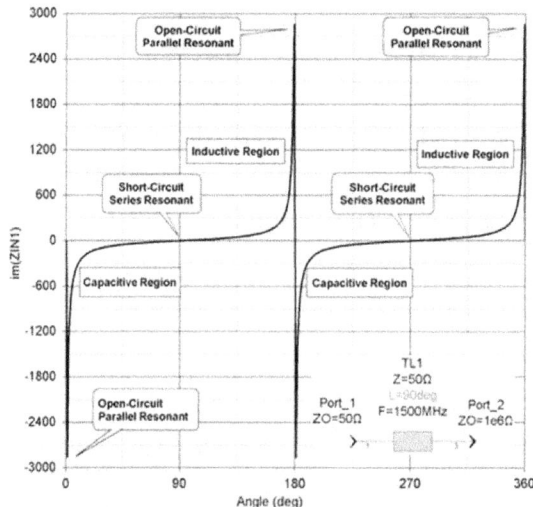

Figure 2-20 Open circuit transmission line reactance vs electrical length

Modeling Open-Circuited Microstrip Lines

Care must be used when modeling the open circuit microstrip line due to the radiation effects from the end of the transmission line. The E fields that exist in the air space of the microstrip line add capacitance to the microstrip transmission line. On an open circuit microstrip line this fringing capacitance is referred to as an end effect. The end effect makes the line electrically longer than the physical length. This requires that the physical line length be shortened to achieve the desired reactance. The 'microstrip end' model accurately accounts for this end effect capacitance for the specified substrate. The end effect is visualized in Figure 2-21.

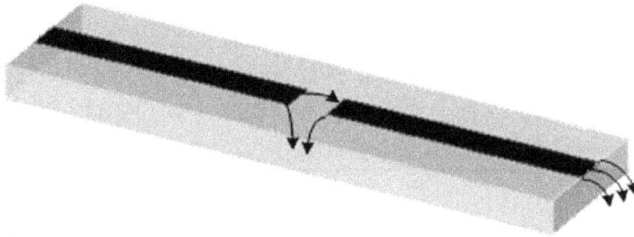

Figure 2-21 Fringing E fields in microstrip open circuit and gaps

Figure 2-21 shows the correct method of modeling a microstrip open circuit transmission line using an end effect. As this Figure shows, the impedance of a quarter-wave section of open circuit line is quite low, close to a short circuit. This type of line section could be used as a series resonant circuit.

Example 2.6: Calculate the input impedance of a quarter-wave open – circuited transmission line at 1.5 GHz.

Solution: Place an open-circuited transmission line on the schematic. A group delay of 166.67 picoseconds shows that the electrical length is 90 degrees. Figure 2-22 also shows that we have used a 1 Mega Ohm resistor to act as an open circuit for the transmission line. Simulate the schematic and display the input impedance in Table 2-8.

Td= 166.673p Z0=50

T1   OUT

Rser=50

V1   AC 1

Rout
1e6

.net I(Rout) V1
.ac lin 1 1500Meg 1500Meg

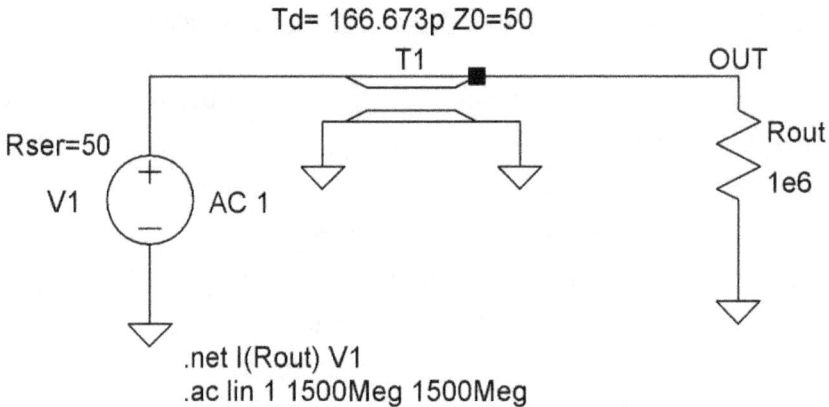

Figure 2-22 Quarter wave open circuited line schematic

The input impedance of the quarter-wave open circuited transmission line is given in Table 2-8.

freq	Zin1	Zin1
1.500 GHz	0.045 + j0.835	0.837 / 86.935

Table 2-8 Quarter wave open circuited line impedance

As the Table 2-8 shows, the input impedance of a quarter-wave section of open-circuited transmission line is quite small, close to a short circuit. This type of line section could be used as a series resonant circuit.

Distributed Inductive and Capacitive Elements

Thus far we have dealt with only 50 Ω transmission lines. It is possible to synthesize series inductance by using short lengths of transmission lines that have considerably higher impedance than 50 Ω. It is also possible to synthesize shunt capacitors by using short lengths of transmission lines that have considerably lower impedance than 50 Ω. Typical impedances would range from approximately 20 Ω for capacitive elements and 80 Ω for inductive elements. The actual impedance used is a compromise between the substrate height and dielectric constant and the ability to physically realize the distributed element. For example an 80 Ω line on a thin substrate or a high dielectric constant substrate may be too narrow to etch on a printed circuit board. In such a case it may be necessary to use a 70 Ω impedance to realize the inductive element. These elements can be used successfully in narrow bandwidth applications. These distributed elements are very useful in the design of filters, bias feed and impedance matching networks.

Distributed Microstrip Inductance and Capacitance

For short lengths of high impedance transmission line use the following equation to calculate the length of microstrip line to synthesize a specific value of inductance.

$$Inductive\ \ Line\ \ Length = \frac{f\ \lambda_g\ L}{Z_L} \tag{2-69}$$

$$Capacitive\ \ Line\ \ Length = f\ \lambda_g\ Z_C\ C \tag{2-70}$$

Where:

f = frequency at which inductance is calculated
L = nominal inductance value
C = nominal capacitance value
Z_L = impedance of inductive transmission line
λ_g = wavelength using the effective dielectric constant
Z_C = impedance of the capacitive transmission line

Example 2.7: Convert the lumped elements bias feed capacitors and inductors in Figure 2-23 to distributed elements.

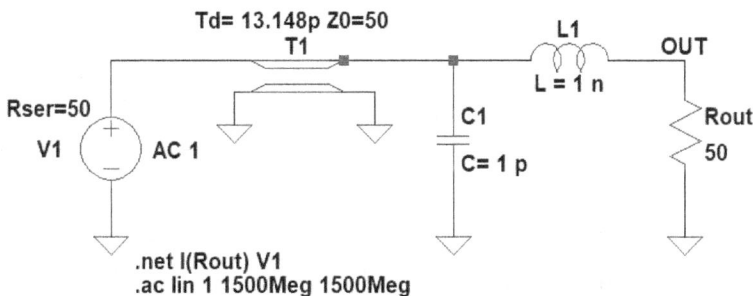

Figure 2-23: Lumped capacitor and inductor bias feed

Solution: Use microstrip substrate having εr =3, t = 1.4 mils, H = 30 mils, and Equations (2-69) and (2-70) to convert the lumped elements to 20 Ohm and 80 Ohm microstrip lines. Use APCAD software for electrical length calculations and then convert the electrical lengths to delay in pico seconds.

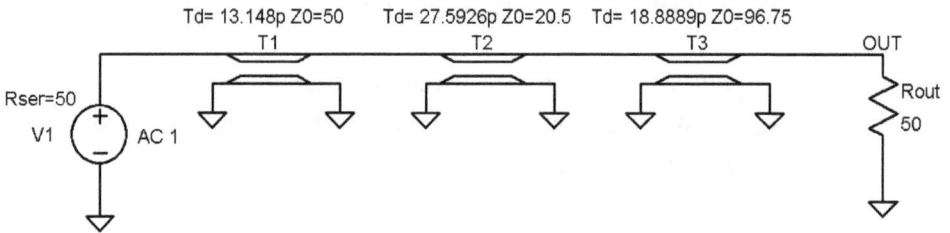

Figure 2-24 Schematic for distributed capacitive and inductive lines

The printed circuit layout in Figure 2-25 shows the line width relationship among the 50 Ω, 20 Ω, and 80 Ω transmission lines.

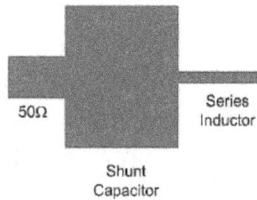

Figure 2-25 Capacitive and inductive lines in PCB layout

Step Discontinuities

Note the change in geometry of Figure 2-25 as the impedance transitions from 50 Ω to 20 Ω and then to 80 Ω. These changes are known as discontinuities. Discontinuities result in fringing capacitance and parasitic inductance that modify the frequency response of the circuit. At RF frequencies (up to about 2 GHz) the effects of discontinuities are minimal and sometimes neglected. As the operation frequency increases, the effects of discontinuities can significantly alter the performance of a microstrip circuit.

Microstrip Bias Feed Networks

Another useful purpose for high impedance and low impedance microstrip transmission lines is the design of bias feed networks. Often it is necessary to insert voltage and current to a device that is attached to a microstrip line. Such a device could be a transistor, MMIC amplifier, or diode. The basic bias feed or "bias decoupling network" consists of an inductor (used as an "RF Choke") and shunt capacitor (bypass capacitor). At lower RF frequencies (< 200 MHz) these networks are almost entirely realized with lumped element components. Even at these low frequencies it is very important to account for the parasitics in the components.

Example 2-8: Fig. 2-26 shows a typical series inductor, shunt capacitor, lumped element bias feed and its effect on a 50 Ω transmission line. Plot the response of the bias insertion network.

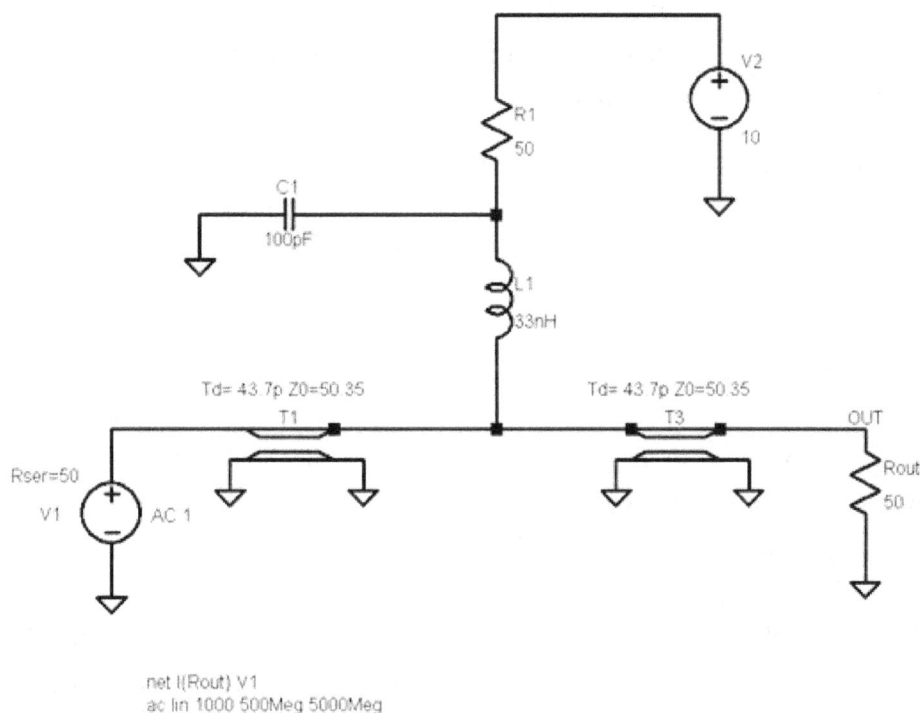

Figure 2-26 Inductor and bypass capacitor bias insertion network

Solution: Response of the bias feed network is shown in Figure 2-27.

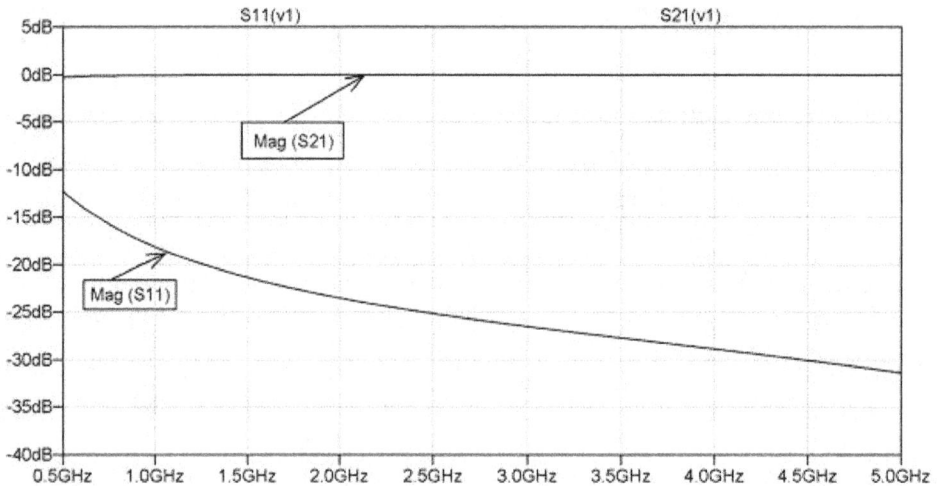

Figure 2-27: Response of the typical bias insertion network

Distributed Bias Feed Design

A high impedance microstrip line of $\lambda_g/4$ can be used to replace the lumped element inductor. Similarly a $\lambda_g/4$ of low impedance line can be used to model the shunt capacitor.

Example 2.9: Design a distributed bias feed network by calculating the physical line length of the $\lambda_g/4$ sections of 80 Ω and 20 Ω microstrip lines at a frequency of 2 GHz. Draw the schematic of the distributed bias feed network.

Solution: Use Equations (2-69) and (2-70) to convert the lumped elements in Figure 2-26 to distributed elements. Use the 80 Ω high impedance quarter wave section and a shunt capacitance as shown in Figure 2-28. In practice we use a tee junction to accurately model the electrical length of the junction and includes all parasitic effects of the discontinuity. An end-effect element is also used on the open circuit line. The response of the bias feed is characterized by the null in the return loss and very low insertion loss near the design frequency of 2 GHz. As figure 2-29 shows the return loss null occurs at 1.85 GHz suggesting that the high impedance line length should be decreased to center the design on 2 GHz.

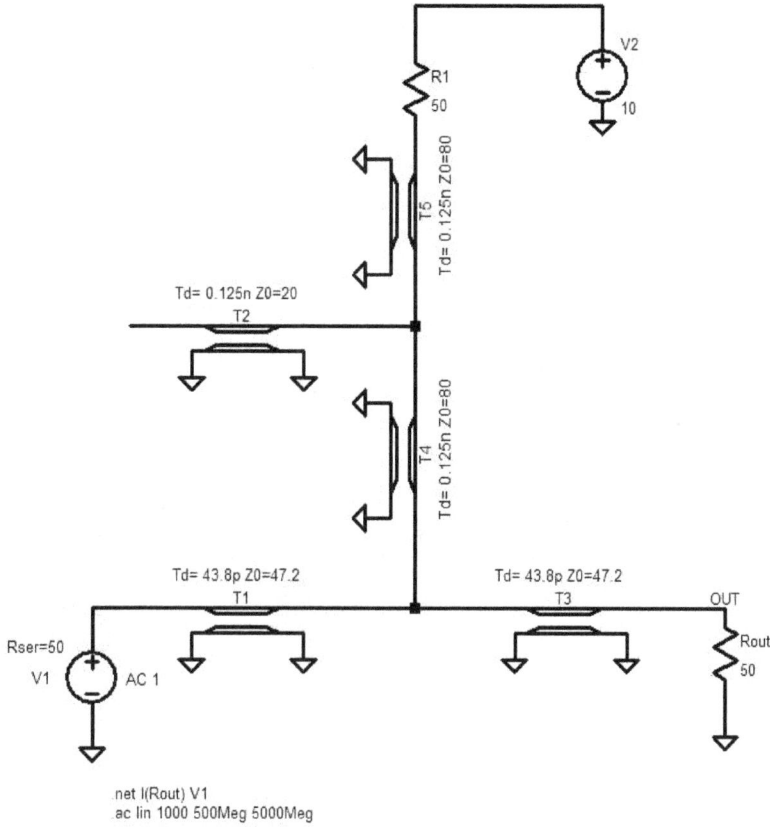

Figure 2-28 Bias feed modeled with distributed transmission lines

Frequency response of distributed feed network is shown in Figure 2-29.

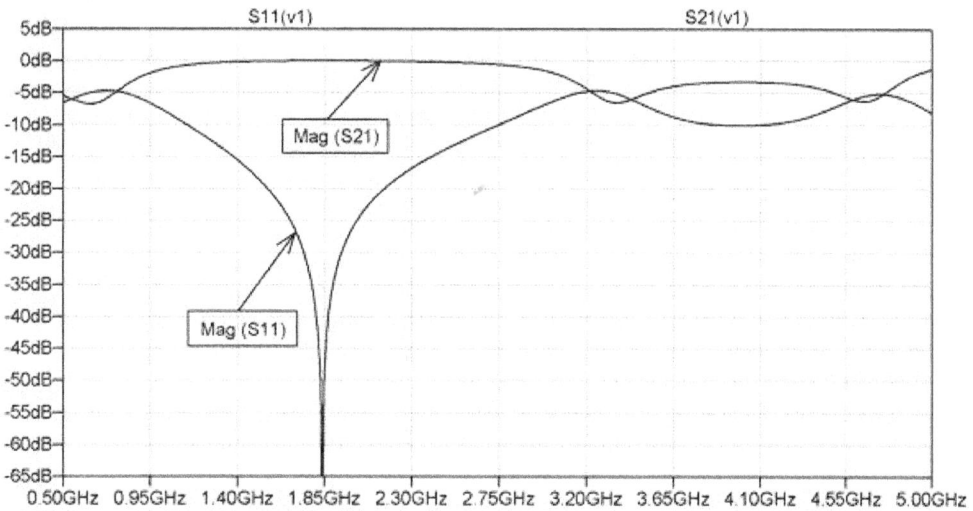

Figure 2-29 Distributed bias feed response

2.12 Coupled Transmission Lines

There are three primary methods in which coupled lines are used in microwave circuit design. These are end-coupled, edge-coupled, and broadside coupled line structures. End coupled lines are often used to realize microstrip resonators and filters. Edge coupled lines are used in both coupler and filter designs. Broadside coupled lines are popular with various coupler designs.

End Coupled   Edge Coupled   Broadside Coupled

Figure 2-30 Types of coupled line structures

Current flow in the edge-coupled and broadside coupled sections can be difficult to quantify. Because of the coupling between the lines there exist two modes of impedance required to characterize the circuit. These are known as the even mode, Z_{oe}, and odd mode, Z_{oo}, impedance. Figure 2-31 shows the field distribution on the edge coupled lines for both conduction modes. In the even mode, a common displacement current flows from the conductors as shown by the E fields of the same polarity. In the odd mode, a component of the electric field is at opposite direction with respect to the two conductors. Therefore the magnitude of the even and odd mode impedance is strongly dependent on the separation between the lines which also determine the electrical coupling between the lines. The coupling between the lines, in dB, is defined by Equation (2-71).

$$C = 20 \log \left| \frac{Z_{oe} - Z_{oo}}{Z_{oe} + Z_{oo}} \right| \tag{2-71}$$

The even and odd mode impedances are then defined by the following equations [2].

$$Z_{oo} = Z_o \sqrt{\frac{1 - 10^{\left(\frac{-C}{20}\right)}}{1 + 10^{\left(\frac{-C}{20}\right)}}} \qquad (2\text{-}72)$$

$$Z_{oe} = Z_o \sqrt{\frac{1 + 10^{\left(\frac{-C}{20}\right)}}{1 - 10^{\left(\frac{-C}{20}\right)}}} \qquad (2\text{-}73)$$

$$Z = \sqrt{Z_{oo} \, Z_{oe}} \qquad (2\text{-}74)$$

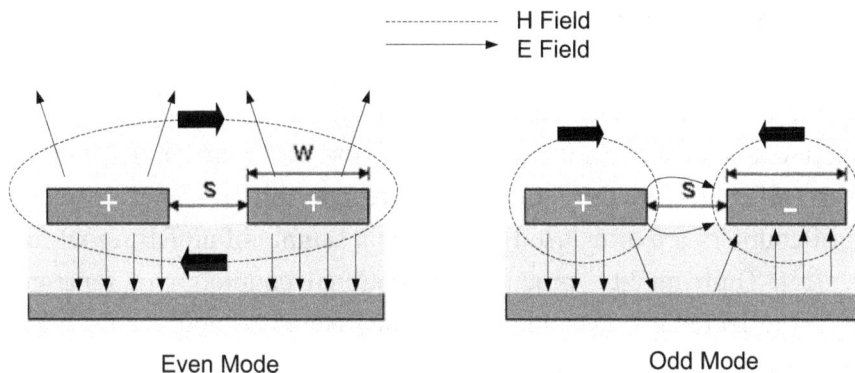

Figure 2-31 Edge coupled microstrip line field distribution

The TLINE utility has the capability to perform the calculations for coupled line impedance. The even and odd mode impedance can be calculated for a given characteristic impedance, Z_o , and coupling ratio in dB.

Even Mode	Odd Mode			
			50	Zo=sqrt(Zoe·Zoo)
69.3714	36.0379	Impedance	-9.99997	Coupling, ¼ Wave
0.0244492	0.0614251	Total Loss/in	0.139274	MinValue Cover Ht, in
0.0239248	0.061006	-Cond. Loss/in	59055.1	Highest Acc Freq
0.000524408	0.00041903	-Diel. Loss/in	4.9334	Wavelength, in
62.9196	68.7451	Velocity (%c)		
2.52597	2.116	E effective		
236.702	86.2312	Q, unloaded		

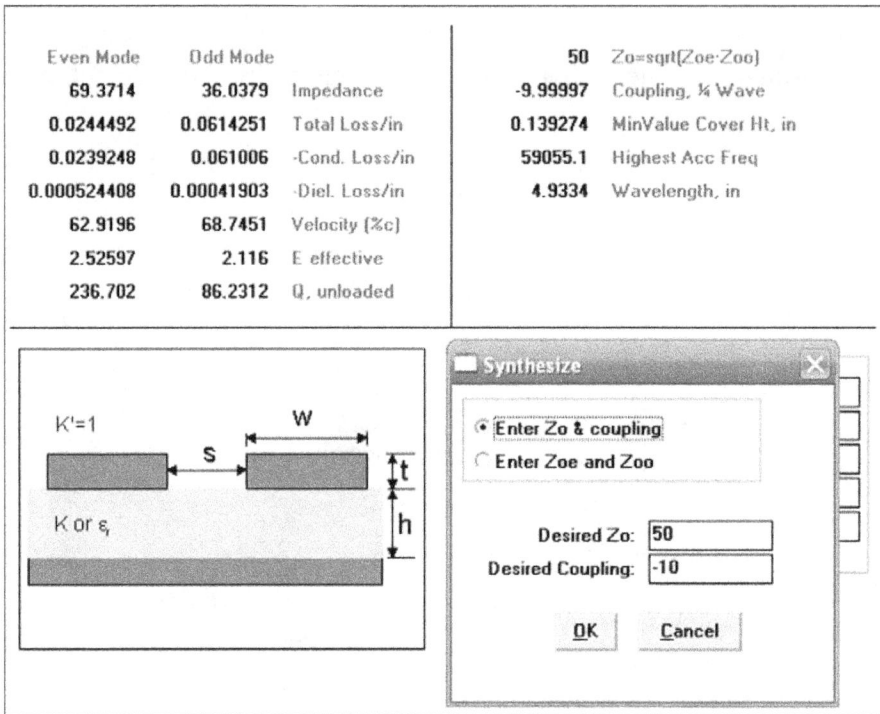

Figure 2-32 TLINE calculation of microstrip edge coupled line impedance

Directional Coupler

One important use of coupled lines is the design of directional couplers. Directional couplers are useful components for sampling an RF signal without significantly loading or perturbing the input signal. Simple directional couplers are often used to provide a sample of an RF signal for measurement. High quality, precision, directional couplers can separate incident and reflected signals and are the fundamental component used for the measurement of VSWR and return loss. Two directional couplers can be placed back-to-back to form a dual directional coupler as shown in Figure 2-33. This type of coupler forms a four port network that can provide a sample of the forward power and reflected power between a source and load. Some basic properties of directional couplers include:

Insertion Loss:

Insertion Loss is simply the ratio of the output power at P2 to the input power at P1. Expressed in dB:

$$Insertion\ Loss\ (dB)\ =\ P2_{\ dBm} - P1_{\ dBm} \qquad (2\text{-}75)$$

Coupling:

The coupling factor is the ratio of the output power at P3 to the input power at P1. In microstrip and stripline circuits the coupled port is adjacent to the input line port. In waveguide couplers the coupled port is furthest from the input port.

$$Input\ Port\ Coupling\ \ (dB) = P1_{\ dBm} - P3_{\ dBm} \qquad (2\text{-}76)$$

Isolation:

Isolation is the ratio of the output power at P4 to the input power at P1.

$$Isolation\ (dB) = P1_{\ dBm} - P4_{\ dBm} \qquad (2\text{-}77)$$

Directivity:

Directivity is the difference between the isolation and the coupling when P2 is perfectly terminated in 50 Ω. Another way to think of a coupler's directivity is its ability to separate the forward and reflected waves.

$$Directivity\ (dB) = P3_{dBm} - P4_{\ dBm} \qquad (2\text{-}78)$$

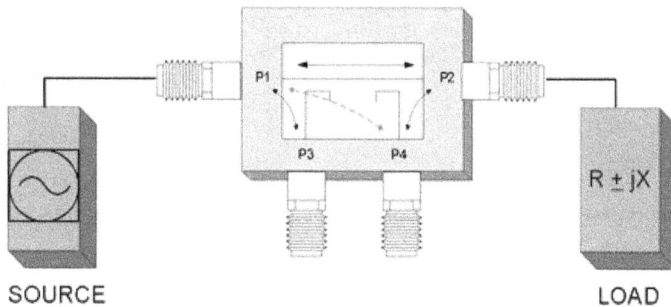

Figure 2-33 A dual directional coupler to measure VSWR, and return loss

A coupler will always have a finite amount of isolation. Ideally, all of the power input to P1 should be directed to P2 and P3. However some finite amount of power will show up at P4. This power will then add with any reflected power coming from P2 and being directed to P4. It is this finite isolation that limits the directivity of the directional coupler. Thus the power at P3 is the (Incident power at P1 − coupling factor + the (Reflected power from P2− isolation. For simple RF power sampling applications, the directivity is not that critical. But if we are using the directional coupler to measure VSWR, the directivity is very important. In VSWR measurement applications it is important to know the directivity of the directional coupler that is used to perform the measurement. A significant measurement error can exist when the coupler directivity is less than 40 dB. Figure 2-34 is a standard plot of measurement error vs. coupler directivity. The plot shows that when measuring a load that has an actual return loss of 20 dB with a directional coupler of 40 dB directivity, an error of +0.8 to -0.9 dB exists. This means that our instrument may read (-19.2 dB to -20.9 dB. We can readily see that if we used a directional coupler with directivity of 20 dB the resultant error could be approximately +2.3 dB to −3.0 dB. The asymmetry is due to whether the reflected signal is in phase with the forward signal or 180 degrees out of phase.

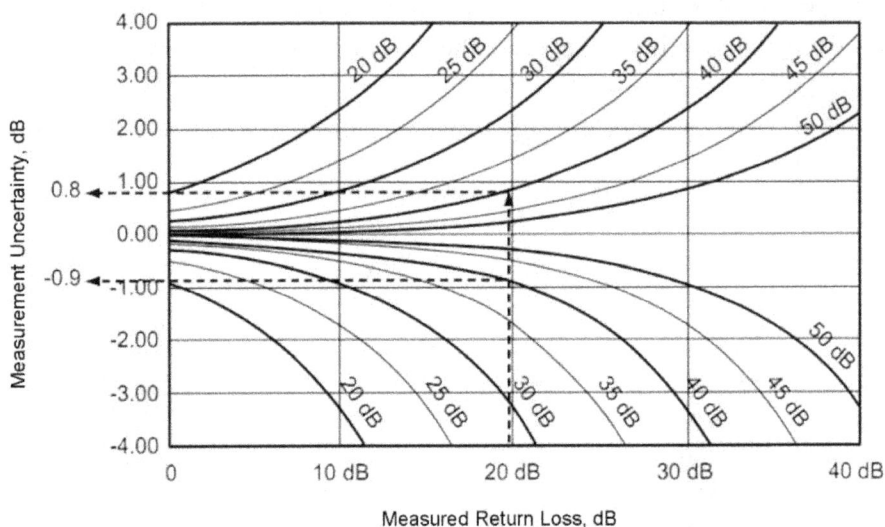

Figure 2-34 Error directivity (*Courtesy of Anritsu Corporation*)

References and Further Reading

[1]   David M. Pozar, *Microwave Engineering*, Third Edition, John Wiley and Sons, Inc. 2005

[2]   Ali Behagi and Manou Ghanevati, *Fundamentals of RF and Microwave Circuit Design,* Practical Analysis and Design Tools, Techno Search, Ladera Ranch, California 2017

[3]   Foundations for Microstrip Circuit Design, T.C. Edwards, John Wiley & Sons, New York, 1981

[4]   Ali A. Behagi, *RF and Microwave Circuit Design*, A Design Approach Using (**ADS**), Techno Search, Ladera Ranch, CA 2017

[5]   William Sinnema and Robert McPherson, Electronic Communications, Prentice-Hall Canada, Inc., Scarborough, Ontario, 1991

[6]   UHF/Microwave Experimenters Manual, American Radio Relay League, Newington, CT.1990

[7]   Reference: I. J. Bahl and D. K. Trivedi, "A Designer's Guide to Microstrip Line", Microwaves, May 1977, pp. 174-182.

[8]   Microwave Handbook Volume 1, Radio Society of Great Britain, The Bath Press, Bath, U.K., 1989.

[9]   Tatsuo Itoh, Planar Transmission Line Structures, IEEE Press, New York, NY, 1987

Problems

2-1.   Determine the VSWR of a satellite antenna with a return loss of -11.4 dB.

2-2.   The input reflection coefficient of a transistor is measured to be 0.22 at an angle of $32°$. Determine the input VSWR of the device.

2-3. Determine the impedance of a quarter-wave transformer to match a 25 Ω load to a 50 Ω source.

2-4. Design the quarter-wave transformer from Problem 2-3 using a microstrip transmission line. The frequency of operation is 2.05 GHz. The dielectric constant is 3.0 with a thickness of 0.030 in.

2-5. A radio transmitter is operating into a transmission line that measures a 3:1 VSWR. Determine the percentage of power that would be expected to reflect back into the transmitter.

2-6. A series RLC load, R = 75 Ω, L = 10 nH, C = 25 pF is connected to a 50 Ω transmission line. Setup a Linear Analysis in LTspice to sweep the frequency from 200 MHz to 2000 MHz in 200 MHz steps. Display the input reflection Coefficient, S11, and VSWR in a Table.

2-7. Create a simple schematic using the RG8 coaxial cable. Set the length to 50 ft. Calculate the insertion loss in a Table. Terminate the coaxial line with a 100 Ω resistor and display the input return loss and reflection coefficient in the same Table.

2-8. Calculate the cutoff frequency of the TE1,0 mode in a rectangular waveguide with a height of 0.200 inches and a width of 0.470 inches. Also calculate the waveguide wavelength,

2-9. Determine the physical length of a $\lambda_g/4$ open circuit microstrip transmission line with an impedance of 20 Ω. The frequency of operation is 10 GHz. Use a microstrip dielectric constant of 2.2 and a thickness of 0.010 inches. Determine whether an end-effect model element should be used.

2-10. Design a 35dB directional coupler to be used in a 100Wtransmitter operating at 4 GHz. Design the coupler in stripline. Use a dielectric constant of 3.0.

Chapter 3

Network Parameters and the Smith Chart

3.1 Introduction

At low frequencies below the VHF range, the terminal voltages V_1 and V_2 and the terminal currents I_1 and I_2 of a two-terminal network, shown in Fig. 3-1, can be related to each other by a different set of matrix parameters. The most common representations are the impedance matrix (Z parameters), the admittance matrix (Y parameters), the hybrid matrix (h parameters), and the transmission matrix (ABCD parameters).

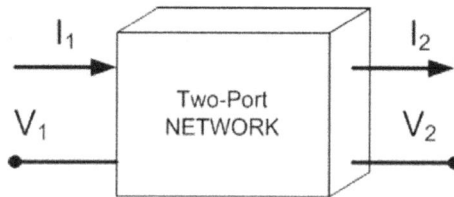

Figure 3-1 Low frequency two-port network

Z Parameters

The network representation with Z parameters, relating the input and output voltages to input and output currents, is given by the following equations.

$$V_1 = Z_{11}I_1 + Z_{12}I_2 \qquad\qquad (3\text{-}1)$$

$$V_2 = Z_{21}I_1 + Z_{22}I_2 \qquad\qquad (3\text{-}2)$$

The Z parameters, also known as impedance parameters, are determined by making the following open circuit measurements:

$$Z_{11} = \left.\frac{V_1}{I_1}\right|_{I_2=0} \qquad \textit{(Requires open circuit measurement)}$$

$$Z_{12} = \frac{V_1}{I_2}\bigg|_{I_1=0} \qquad \textit{(Requires open circuit measurement)}$$

$$Z_{21} = \frac{V_2}{I_1}\bigg|_{I_2=0} \qquad \textit{(Requires open circuit measurement}$$

$$Z_{22} = \frac{V_2}{I_2}\bigg|_{I_1=0} \qquad \textit{(Requires open circuit measurement)}$$

The Z parameters are very useful when two-terminal networks are connected in series. In this case the overall Z parameters are simply the algebraic sum of the individual Z parameters.

Y Parameters

Similarly, the input and output currents can be related to input and output voltages by Y parameters as shown in Equations (3-3) and (3-4).

$$I_1 = Y_{11}V_1 + Y_{12}V_2 \qquad\qquad (3\text{-}3)$$

$$I_2 = Y_{21}V_1 + Y_{22}V_2 \qquad\qquad (3\text{-}4)$$

The Y parameters, also known as admittance parameters, are determined by making the following short circuit measurements.

$$Y_{11} = \frac{I_1}{V_1}\bigg|_{V_2=0} \qquad \textit{(Requires short circuit measurement)}$$

$$Y_{12} = \frac{I_1}{V_2}\bigg|_{V_1=0} \qquad \textit{(Requires short circuit measurement)}$$

$$Y_{21} = \frac{I_2}{V_1}\bigg|_{V_2=0} \quad \textit{(Requires short circuit measurement)}$$

$$Y_{22} = \frac{I_2}{V_2}\bigg|_{V_1=0} \quad \textit{(Requires short circuit measurement)}$$

The Y parameters are useful when two-terminal networks are connected in parallel. In this case the overall Y parameters are simply the algebraic sum of the individual Y parameters.

h Parameters

The network representation by Hybrid or h parameters, relating the input voltage and output current to input current and output voltage, is given by the following equations.

$$V_1 = h_{11}I_1 + h_{12}V_2 \tag{3-5}$$

$$I_2 = h_{21}I_1 + h_{22}V_2 \tag{3-6}$$

The h parameters can be determined from the following measurements.

$$h_{11} = \frac{V_1}{I_1}\bigg|_{V_2=0} \quad \textit{(Requires short circuit measurement)}$$

$$h_{12} = \frac{V_2}{V_1}\bigg|_{I_1=0} \quad \textit{(Requires open circuit measurement)}$$

$$h_{21} = \frac{I_2}{I_1}\bigg|_{V_2=0} \quad \textit{(Requires short circuit measurement)}$$

$$h_{22} = \frac{I_2}{V_2}\bigg|_{I_1=0} \qquad \textit{(Requires open circuit measurement)}$$

The h parameters are often used to characterize the low frequency characteristics of transistor circuits. The parameter, h_{21}, defines the forward current gain while the parameter, h_{12}, defines the reverse voltage gain of the network. The parameter, h_{11}, is the input impedance and parameter, h_{22}, is the output admittance of the network.

ABCD Parameters

Another representation relating the input voltage and current to the output voltage and current, is by ABCD parameters.

$$V_1 = AV_2 - BI_2 \qquad\qquad (3\text{-}7)$$

$$I_1 = CV_2 - DI_2 \qquad\qquad (3\text{-}8)$$

The ABCD parameters are found by the following measurements.

$$A = \frac{V_1}{V_2}\bigg|_{I_2=0} \qquad \textit{(Requires open circuit measurement)}$$

$$B = -\frac{V_1}{I_2}\bigg|_{V_2=0} \qquad \textit{(Requires short circuit measurement)}$$

$$C = \frac{I_1}{V_2}\bigg|_{I_2=0} \qquad \textit{(Requires open circuit measurement)}$$

$$D = -\frac{I_1}{I_2}\bigg|_{V_2=0} \qquad \textit{(Requires short circuit measurement)}$$

The ABCD parameters in Equations (3-7) and (3-8) are often presented in matrix form and referred to as the transmission matrix [2].

$$\begin{vmatrix} V_1 \\ I_1 \end{vmatrix} = \begin{vmatrix} A & B \\ C & D \end{vmatrix} \cdot \begin{vmatrix} V_2 \\ -I_2 \end{vmatrix}$$

The ABCD parameters are very useful to characterize networks when individual circuits are cascaded in a chain fashion. They have also been popular in the design of telephone networks. In this case the overall ABCD matrix is found by the product of the individual ABCD matrices. For example, if two networks are connected in cascade the overall ABCD matrix is the product of individual ABCD matrices.

$$\begin{vmatrix} A & B \\ C & D \end{vmatrix} = \begin{vmatrix} A_1 & B_1 \\ C_1 & D_1 \end{vmatrix} \cdot \begin{vmatrix} A_2 & B_2 \\ C_2 & D_2 \end{vmatrix} \tag{3-9}$$

At microwave frequencies, due to the difficulty in short circuit and open circuit measurements, the two-terminal network representation by Z, Y, h, or ABCD parameters is not practical. Therefore, at microwave frequencies, a new representation known as S parameters has been developed.

3.2 Development of Network S Parameters

At RF and microwave frequencies, where the wavelength of the voltage and current waveforms are comparable or smaller than the physical dimensions of the network, it is difficult to obtain perfect open and short circuit terminations. It is also difficult to measure voltage and current in high frequency circuits, therefore, the usefulness of the parameters in Section 3.1 diminish. In high frequency networks it is much easier to measure power than voltage or current. For RF and microwave networks, a form of the transmission matrix has been defined based on power measurements into the system's characteristic impedance. These parameters are known as S parameters, named after their scattering matrix form. Consider a two-port network, shown in Figure 3-2, where Z_{01} is the real characteristic impedance and V_1^+ and V_1^-, respectively, are the incident and reflected voltage waveforms at the input port. Similarly, Z_{02} is the real characteristic impedance and V_2^+ and V_2^-, respectively, are the incident and reflected voltage waveforms at the output port.

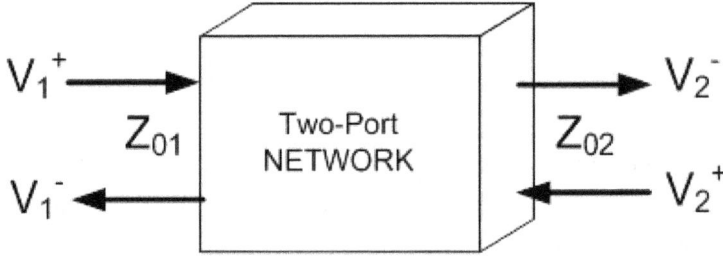

Figure 3-2 Two-port network with incident and reflected voltage waveforms

In order to obtain measurable power relations in terms of wave amplitudes, we need to define a new set of waveforms by normalizing the voltage amplitudes with respect to the square root of the respective characteristic impedances, namely:

$$a_1 = \frac{V_1^+}{\sqrt{Z_{01}}} \tag{3-10}$$

$$b_1 = \frac{V_1^-}{\sqrt{Z_{01}}} \tag{3-11}$$

$$a_2 = \frac{V_2^+}{\sqrt{Z_{02}}} \tag{3-12}$$

$$b_2 = \frac{V_2^-}{\sqrt{Z_{02}}} \tag{3-13}$$

The two-port network with normalized waveforms is shown in Figure 3-3.

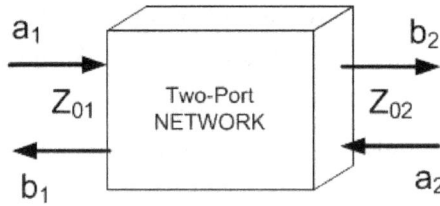

Figure 3-3 Two-port network with incident and reflected waveforms

Notice that:

$$|a_1|^2 = \frac{|V_1^+|^2}{Z_{01}} = \text{Incident power at the network input}$$

$$|b_1|^2 = \frac{|V_1^-|^2}{Z_{01}} = \text{Reflected power at the network input}$$

$$|a_2|^2 = \frac{|V_2^+|^2}{Z_{02}} = \text{Incident power at the network output}$$

$$|b_2|^2 = \frac{|V_2^-|^2}{Z_{02}} = \text{Reflected power at the network output}$$

The S parameters relate b_1 and b_2 to a_1 and a_2 by the following Equations [1].

$$b_1 = S_{11}a_1 + S_{12}a_2 \tag{3-14}$$

$$b_2 = S_{21}a_1 + S_{22}a_2 \tag{3-15}$$

In matrix form the scattering matrix is written as:

$$\begin{bmatrix} b_1 \\ b_2 \end{bmatrix} = \begin{bmatrix} S_{11} & S_{12} \\ S_{21} & S_{22} \end{bmatrix} \cdot \begin{bmatrix} a_1 \\ a_2 \end{bmatrix}$$

At RF and microwave frequencies the normalized voltage waveforms a_1, a_2, b_1, and b_2 represent vectors having both magnitude and phase. Terminating the output of the two-port network with a real load impedance that is equal to the real system characteristic impedance, $Z_{01} = Z_{02}$, forces $a_2 = 0$. Solving for the individual S parameters then gives the following relationships.

$$S_{11} = \frac{V_1^-}{V_1^+} = \frac{b_1}{a_1}\bigg|_{a_2=0} \tag{3-16}$$

$$S_{21} = \frac{V_2^-}{V_1^+} = \frac{b_2}{a_1}\bigg|_{a_2=0} \tag{3-17}$$

These measurements are better visualized using the network of Figure 3-4.

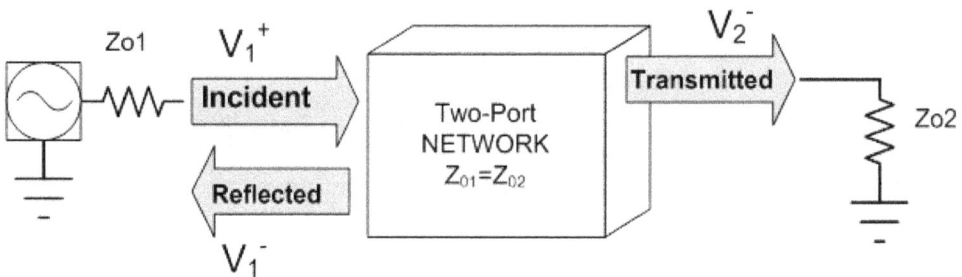

Figure 3-4 Measurement of S11 and S21 in a two-port network

Terminating the input of the two-port network with a real load impedance that is equal to the real system characteristic impedance, $Z_{01} = Z_{02}$, forces $a_1 = 0$. The S parameters S_{22} and S_{12} can be solved by Equations (3-18) and (3-19) as demonstrated in Figure 3-5.

$$S_{22} = \frac{V_2^-}{V_2^+} = \frac{b_2}{a_2}\bigg|_{a_1=0} \tag{3-18}$$

$$S_{12} = \frac{V_1^-}{V_2^+} = \frac{b_1}{a_2}\bigg|_{a_1=0} \qquad (3\text{-}19)$$

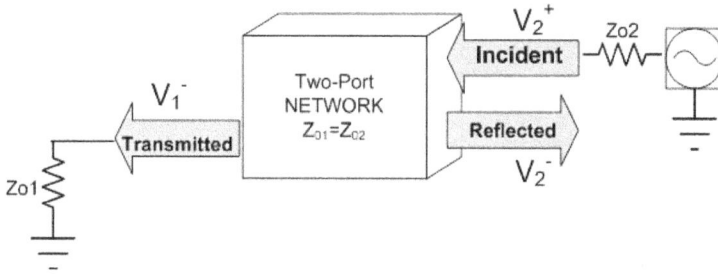

Figure 3-5 Measurement of S22 and S12 in a two-port network

S11 is often referred to as the *input reflection coefficient* and S22 as the *output reflection coefficient* of the network. S21 is the *forward transmission* and is often expressed as a gain or loss depending on whether S21, in dB, is positive or negative. S12 is the *reverse transmission* or isolation of the network. Because the S parameters are complex entities they must be measured with a Vector Network Analyzer capable of measuring both amplitude and phase. A Scalar Network Analyzer is used to measure the magnitude of two-port networks. It is very easy to cascade S parameters to calculate the overall S parameters of a multi-stage network.

3.3 Using S Parameter Files

The data file element can be edited so that the designer can browse to the location of the saved S parameter data file, as shown in Table 3-1. Table 3-1 shows the S parameter file of a C Band amplifier. The S Parameter file is simply an ASCII text file that can be edited with any text editor. For a two-port network the filename should use the .s2p extension. Similarly a one port network would use the .s1p file extension. In S parameter files only one definition line is required and must begin with a (#) symbol. Any number of comment lines can be included and must begin with the (!) symbol. The only remaining requirement is the order in which the S parameters are entered into the text file. This order must be in the form of S_{11}, S_{21}, S_{12}, and S_{22}, as shown in Table 3-1.

```
!   S-Parameter File of 200W C-Band Amplifier
# GHz  S  MA  R  50
!FREQ      |S11|               |S21|                  |S12|                 |S22|
 5.800    0.070   -52.367     65.436    59.242       0.001    49.912       0.151   -69.520
 5.804    0.067   -58.549     64.997    49.393       0.002   -91.508       0.159   -69.723
 5.807    0.078   -57.215     64.574    39.402       0.003    15.689       0.150   -70.922
 5.811    0.071   -62.285     64.270    29.318       0.004   -46.869       0.161   -68.117
 5.814    0.062   -63.434     64.010    18.336       0.003   100.039       0.155   -72.297
 5.818    0.068   -68.539     63.737     9.254       0.002    55.959       0.154   -73.473
 5.821    0.069   -68.652     63.011    -1.009       0.001   -49.662       0.153   -72.922
 5.825    0.067   -75.742     63.067   -11.077       0.000    72.992       0.155   -73.332
 5.828    0.067   -79.371     63.110   -20.816       0.001   -46.926       0.149   -73.844
 5.832    0.066   -80.000     62.883   -30.369       0.001   163.273       0.153   -76.746
 5.835    0.064   -80.039     62.700   -40.990       0.002    -0.501       0.146   -75.992
 5.839    0.064   -83.637     62.461   -50.729       0.001    14.107       0.146   -77.883
 5.842    0.065   -93.328     62.278   -61.002       0.002   168.805       0.153   -75.441
 5.846    0.063   -89.715     61.721   -70.473       0.004   -33.932       0.151   -80.746
 5.849    0.066   -91.605     62.152   -80.652       0.002    84.551       0.151   -79.625
 5.853    0.060   -96.484     61.416   -90.824       0.001    78.816       0.143   -83.137
```

Table 3-1 C band amplifier S parameter data file

As Table 3-1 shows the definition line contains four descriptive parameters for the data file. The available options are summarized below [5]. Only a single space is required between each entry on a line of the S parameter file. For better visual presentation a tab space can be used between entries.

GHz: Units for the swept frequency data column. Frequency can be GHz, MHz, KHz, or Hz.

S: Defines network parameter type. Parameters can be S, Y, Z, or h parameters format.

MA: Magnitude-Angle format for the parameter data. Available formats are DB for dB-angle, MA for magnitude-angle, or RI for real-imaginary format.

R 50: Reference resistance. This is the characteristic impedance in which the parameters have been measured.

A linear analysis can be setup to analyze the amplifier's S parameter data file over a frequency range of 5800 MHz to 5820 MHz in 1 MHz steps. Setup a tabular output and display each of the four S parameters. On the Table Properties window, set the units to absolute (Abs) and magnitude-angle format as shown in Table 3-1. The resulting table is shown in Table 3-2. Note that even though the S Parameters in the data file are recorded in increments greater than 1 MHz, has interpolated the values between each data point and can output the S parameters in 1 MHz increments.

	F (MHz)	mag(S[1,1]) (dB)	ang(S[1,1])	mag(S[2,1]) (dB)	ang(S[2,1])	mag(S[1,2]) (dB)	ang(S[1,2])	mag(S[2,2]) (dB)	ang(S[2,2])
1	5800	0.07	-52.367	65.436	59.242	1e-3	49.912	0.151	-69.52
2	5800.2	0.07	-52.676	65.414	58.75	1.05e-3	42.841	0.151	-69.53
3	5800.4	0.07	-52.985	65.392	58.257	1.1e-3	35.77	0.152	-69.54
4	5800.6	0.07	-53.294	65.37	57.765	1.15e-3	28.699	0.152	-69.55
5	5800.8	0.069	-53.603	65.348	57.272	1.2e-3	21.628	0.153	-69.561
6	5801	0.069	-53.913	65.326	56.78	1.25e-3	14.557	0.153	-69.571
7	5801.2	0.069	-54.222	65.304	56.287	1.3e-3	7.486	0.153	-69.581
8	5801.4	0.069	-54.531	65.282	55.795	1.35e-3	0.415	0.154	-69.591
9	5801.6	0.069	-54.84	65.26	55.302	1.4e-3	-6.656	0.154	-69.601
10	5801.8	0.069	-55.149	65.238	54.81	1.45e-3	-13.727	0.155	-69.611
11	5802	0.068	-55.458	65.217	54.318	1.5e-3	-20.798	0.155	-69.621
12	5802.2	0.068	-55.767	65.195	53.825	1.55e-3	-27.869	0.155	-69.632
13	5802.4	0.068	-56.076	65.173	53.333	1.6e-3	-34.94	0.156	-69.642
14	5802.6	0.068	-56.385	65.151	52.84	1.65e-3	-42.011	0.156	-69.652
15	5802.8	0.068	-56.694	65.129	52.348	1.7e-3	-49.082	0.157	-69.662
16	5803	0.068	-57.004	65.107	51.855	1.75e-3	-56.153	0.157	-69.672
17	5803.2	0.068	-57.313	65.085	51.363	1.8e-3	-63.224	0.157	-69.682
18	5803.4	0.067	-57.622	65.063	50.87	1.85e-3	-70.295	0.158	-69.693
19	5803.6	0.067	-57.931	65.041	50.378	1.9e-3	-77.366	0.158	-69.703
20	5803.8	0.067	-58.24	65.019	49.885	1.95e-3	-84.437	0.159	-69.713
21	5804	0.067	-58.549	64.997	49.393	2e-3	-91.508	0.159	-69.723
22	5804.2	0.068	-58.46	64.969	48.727	2.067e-3	-84.362	0.158	-69.803
23	5804.4	0.068	-58.371	64.941	48.061	2.133e-3	-77.215	0.158	-69.883
24	5804.6	0.069	-58.282	64.912	47.395	2.2e-3	-70.069	0.157	-69.963
25	5804.8	0.07	-58.193	64.884	46.729	2.267e-3	-62.922	0.157	-70.043

Table 3-2 Tabular display of the S Parameter data file

Scalar Representation of the S Parameters

When working with a two-port network, such as an amplifier, the engineer is usually more interested in the magnitude in (dB's) of the S parameters rather than the absolute units. This is referred to as the Scalar representation of the S parameters and is readily measured on scalar network analyzers. Because the S parameters are based on voltage waveforms they must be multiplied by 20log to convert to dB format.

$$|S_{11}(dB)| = -20\log|S_{11}|$$ *Input return loss*

$$S_{21}(dB) = 20\log|S_{21}|$$ *Insertion gain (+) or insertion loss (-)*

$$S_{12}(dB) = 20\log|S_{12}|$$ *Reverse gain (+) or isolation (-)*

$$|S22(dB)| = -20\log|S_{22}|$$ *Output return loss*

With the C Band amplifier, setup a rectangular graph and display the S Parameters in Scalar format. Assign each S parameter to the graph using the dB-magnitude format. The display is shown in Figure 3-6.

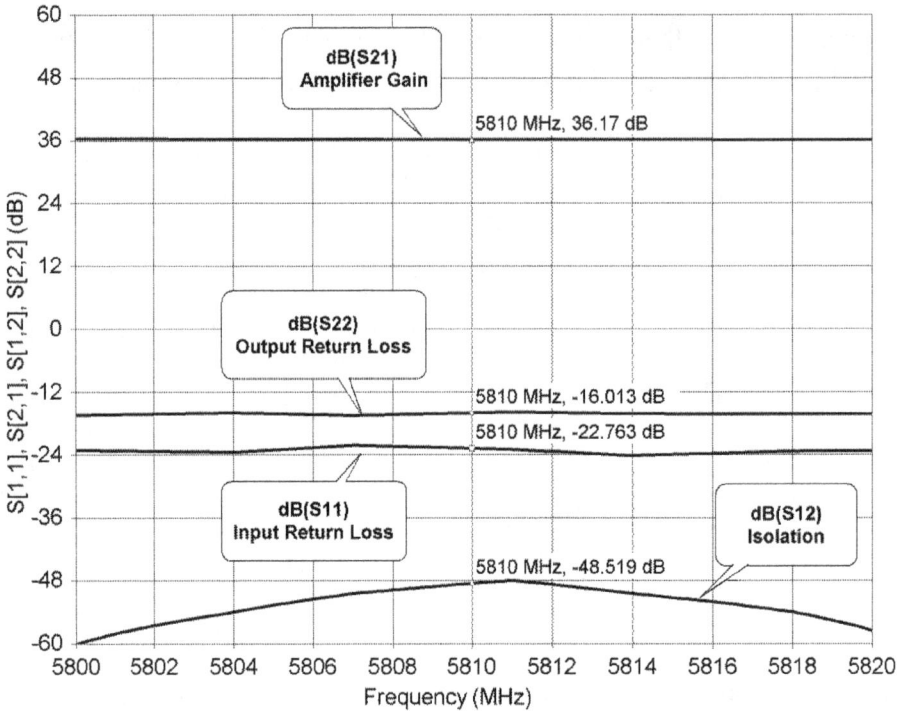

Figure 3-6 Scalar display of frequency swept amplifier S Parameters

3.4 Development of the Smith Chart

The most frequently used graphical tool used to visualize the vector properties of S parameters is the Smith Chart. The Smith Chart conformally maps the familiar rectangular impedance coordinates onto a polar plane. Essentially the reactive, normally the vertical axis, has been bent around in such a way that ± infinity is included within the boundary of the graph. Therefore any positive complex impedance can be plotted on the standard Smith Chart shown in Figure 3-7. Negative impedances or gain is outside of the standard Smith Chart. A compressed Smith Chart must be used to plot negative impedances. Figure 3-7 shows the standard Smith Chart graph with impedance coordinates. A rectangular axis has been overlaid to show the relationship to the rectilinear grid system. The Smith Chart can be normalized to any characteristic impedance. The chart of Figure 3-7 is normalized to 50 Ω. The normalized impedance is a pure resistance that is a single point at the center of the chart. The purely real impedances exist along the horizontal axis from 0 Ω to infinity. Note the locations of the short circuit (0 Ω) and open circuit (infinity) on the real axis. The concentric circles that intersect the real axis are known as the constant resistance circles. The constant reactance circles appear as arcs on the standard Smith Chart. As shown in Figure 3-7 the reactance circles on the top half of the chart represent inductive reactance while the circles on the bottom half of the chart represent capacitive reactance. Any impedance defined by rectangular coordinates (R + jX) can be plotted as a point where the R value on the constant resistance circle intersects the X value on the reactance circle.

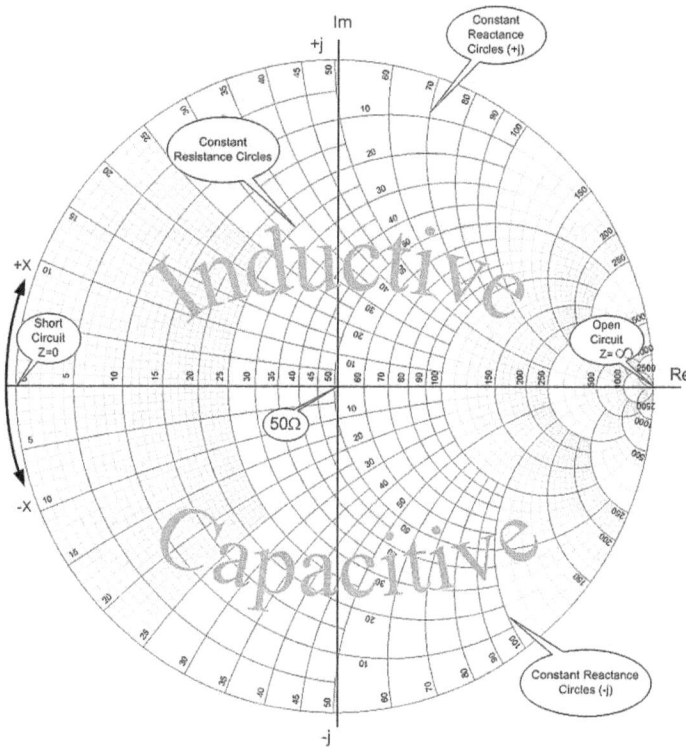

Figure 3-7 Standard Smith Chart with impedance coordinates

Example 3.1: Plot the Impedance Z = 25 + j25 on the Smith Chart.

Solution: Open Smith V3.10 and click on the "Keyboard" Tab. Data Point window opens up. Select impedance (Ω) > Insert 25 for "re" and 25 for "im". Insert 1 under "frequency" and select "GHz" for frequency, as shown in Figure 3-9(a). Press OK. Impedance 25 + j25 Ω is shown as DP1 on Figure 3-9(b).

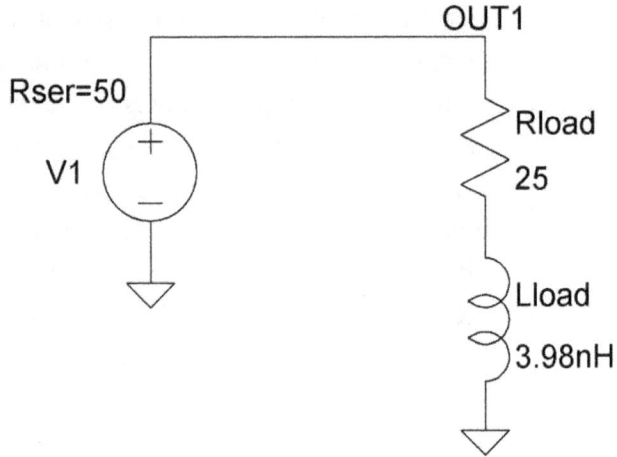

Figure 3-8 Schematic for plotting the impedance on the Smith chart

Simulate the schematic and display the load impedance on the Smith chart, as shown in Figure 3-9.

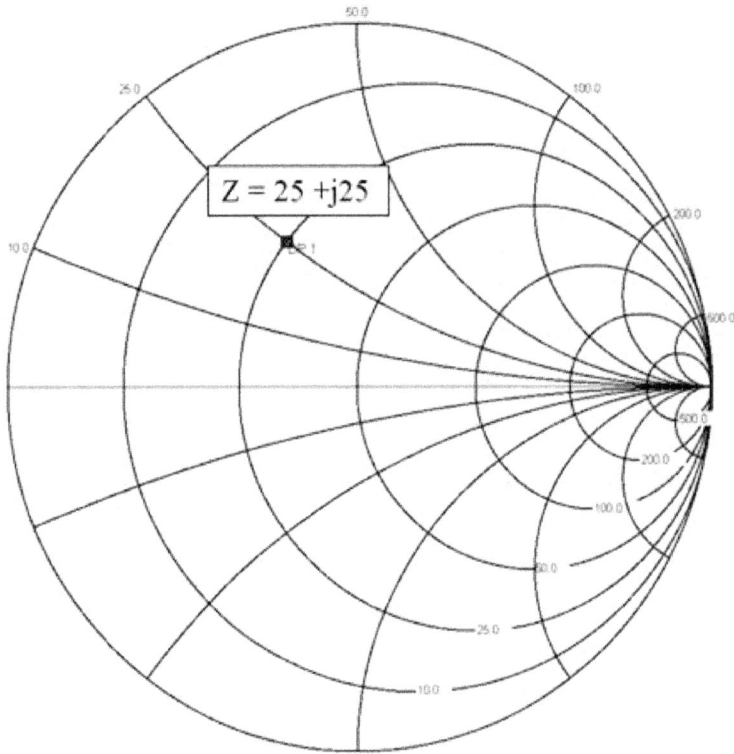

Figure 3-9 Plotting impedance on the Smith Chart at DP1

The impedance on a Smith Chart is often presented in its normalized form. This means that the actual impedance is divided by the value of the characteristic impedance. The Smith Chart allows the selection of either normalized or actual impedance. Equation (3-20) shows how the reflection coefficient, Γ, is related to the actual impedance on the Smith Chart.

$$\Gamma = \frac{\left(Z_{actual} - Z_o\right)}{\left(Z_{actual} + Z_o\right)} \tag{3-20}$$

If we define $z = \frac{\left(Z_{actual}\right)}{\left(Z_o\right)}$, then Γ can be found from Equation (3-21):

$$\Gamma = \frac{\left(z - 1\right)}{\left(z + 1\right)} \tag{3-21}$$

The normalized impedance is read as $z = 0.5 + j0.5\ \Omega$. The reflection coefficient on the Smith Chart is the vector from the center of the chart to the normalized impedance. The transmission coefficient is the vector from the origin ($Z = 0$) to the normalized impedance. The reflection S parameters, S_{11} and S_{22}, are measured as a reflection coefficient on the Smith Chart. The transmission S parameters, S_{21} and S_{12}, are measured as transmission parameters on the Smith Chart.

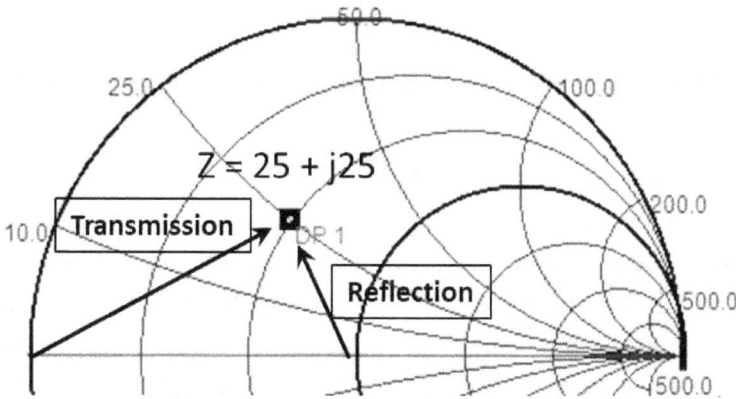

Figure 3-10 Impedance plotted with normalized impedance coordinates

Knowing that the reflection coefficients and S parameters are vector quantities, there must be a method to measure the angular portion of the vector. Figure 3-11 shows the angular measurement convention on the Smith Chart. Note that the reflection coefficient of a 50 Ω resistance (center of the chart) is equal to zero. A total reflection, like that due to a perfect short or open circuit, has a reflection coefficient equal to one. Therefore all positive impedances result in a reflection coefficient between 0 and 1. The reflection coefficient of the 25 + j25 Ω impedance can be determined by displaying the magnitude and angle of the S parameter, S11. As Figure 3-11 shows the reflection coefficient of this impedance is about 25 Ohm at an angle of +116.565 degrees.

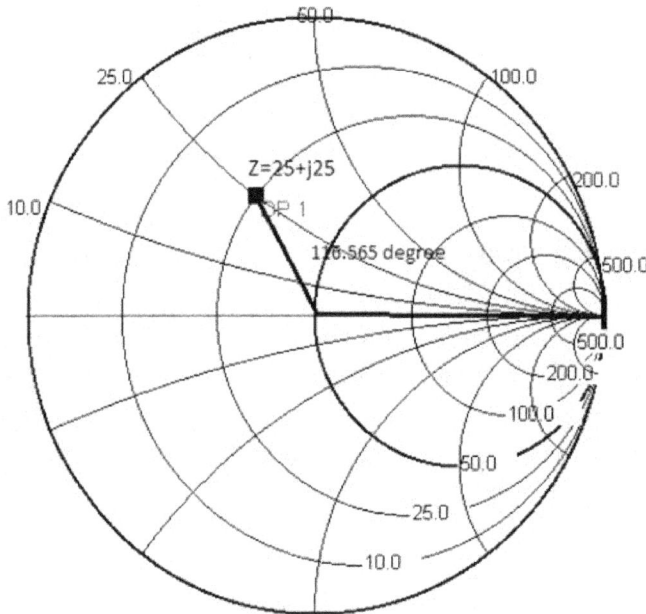

Figure 3-11 Angular measurement of reflection coefficients

Admittance on the Smith Chart

Admittance circles can also be displayed on the Smith Chart. The admittance circles can be enabled on the Smith Chart by their selection on the graph's Properties page. The admittance circles consist of constant conductance and susceptance circles which are inverted from the impedance circles. Subsequent chapters dealing with the subject of impedance matching will make frequent use of the admittance parameters. Having both impedance and admittance parameters displayed on the Smith Chart makes it very easy to design impedance matching networks that include both series and parallel (shunt) elements. It also becomes very easy to convert series impedance to its parallel admittance equivalent. The admittance of a network is the inverse of the impedance as given by Equation (3-22).

$$Y = \frac{1}{Z} = G \pm jB \qquad (3\text{-}22)$$

Where,

G = conductance in mhos

B = susceptance in mhos

The equivalent admittance can be read directly from the admittance circles on the Smith Chart. For the normalized 0.5 + j0.5 Ω series impedance the admittance is read directly from the chart as 1.0 - j1.0 mho. It is important to remember that there is an inversion in the sign of the imaginary component when converting from impedance to admittance or vice versa.

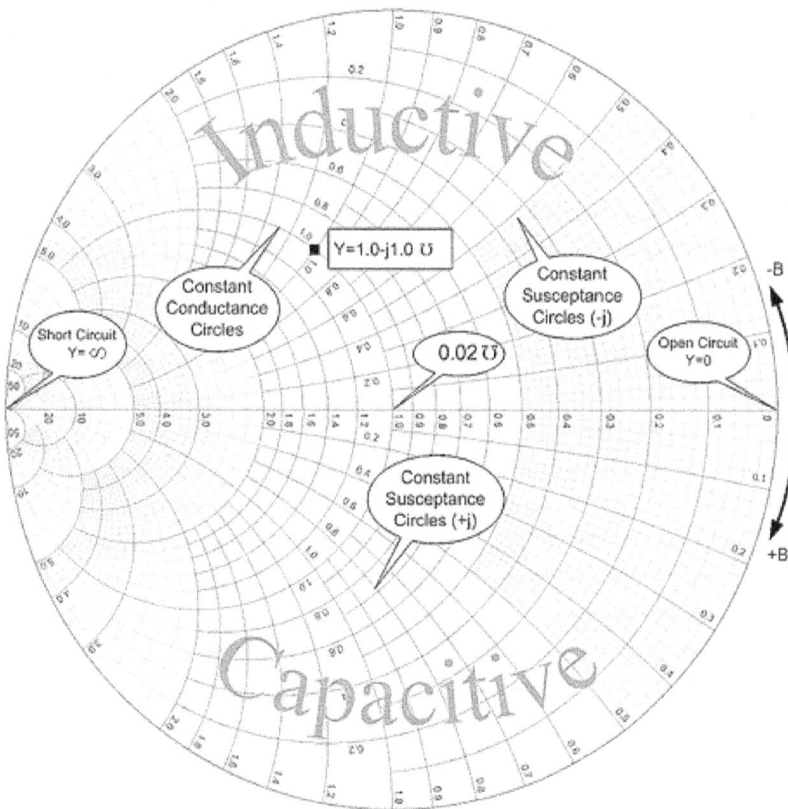

Figure 3-12 Admittance circles on the Smith Chart

3.5 Lumped Element Movements on the Smith Chart

Lumped element movements on the Smith Chart form the basis for impedance matching. The Smith Chart is a wonderfully intuitive tool, for visualization of moving from one impedance to another, without involving circuit synthesis mathematics. Understanding the basic movements around the Smith Chart will build a foundation for the circuit designs covered in this text. It is helpful to display both the impedance and admittance coordinates simultaneously on the Smith Chart.

Adding a Series Reactance to an Impedance

Adding a series reactance to an impedance point on the Smith Chart causes the resulting impedance to move along the constant resistance circle in which the impedance intersects. A series inductance will move the impedance in a clockwise direction while a series capacitance will move the resulting impedance in a counter-clockwise direction on the constant resistance circle. The reactance that is added to the impedance by the series element can be read from the Smith Chart by finding the difference between the lines of reactance that intersect the start point and end point on the constant resistance circle. This reactance can then be converted to a capacitance or inductance value using Equations (3-23) and (3-24).

$$C(series) = \frac{1}{\omega Xn} \tag{3-23}$$

$$L(series) = \frac{Xn}{\omega} \tag{3-24}$$

Where,

X = reactance measured along the arc length
ω = the design frequency, $2\pi f$
n = impedance normalizing value (50 Ω)

Example 3.2: Measure the amount of capacitance required to move the impedance Z = 25 + j25 Ω from point A to point B on the Smith Chart, as shown in Figure 3-13.

Solution: The amount of reactance required in the inductor can be measured from the reactance lines that intersect the start point (A) and end point (B).

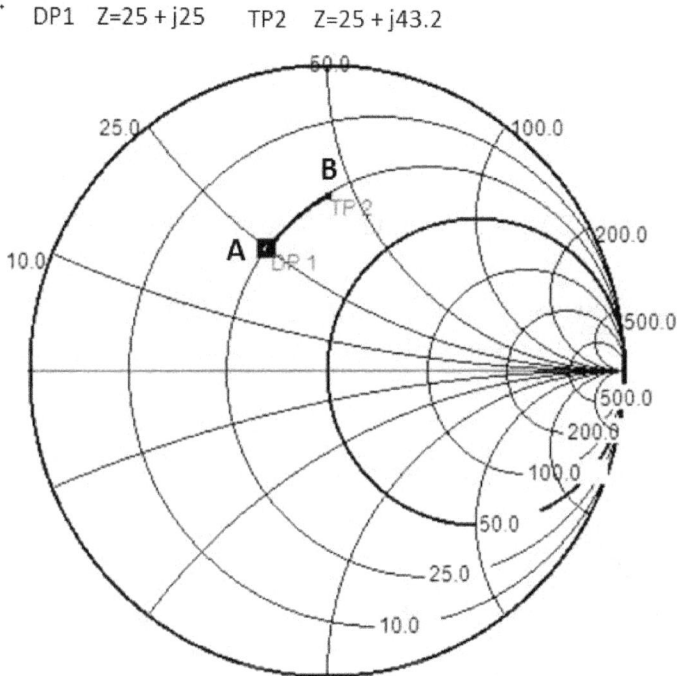

Figure 3-13 Moving point A to point B

Using a design frequency of 1GHz and the Equation (3-24), the series inductance is calculated to be 2.86 nH.

$$L(series) = (43-25)/(2(3.14)(10^9)) = 2.86 \ nH$$

To verify that a 2.86 nH inductor moves the impedance to point B on the Smith Chart. Make the inductor tunable and set the initial value to 0.1 nH. Tune the value of the inductance to move the impedance from point A to point B.

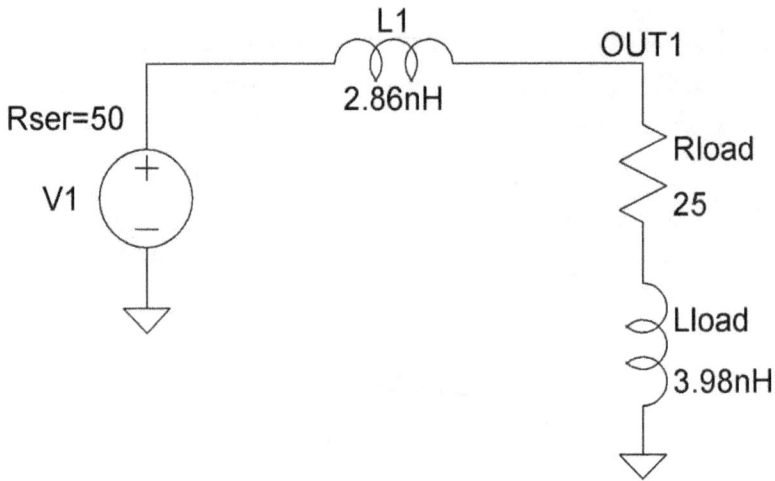

Figure 3-14 Series inductance added to 25 + j25 Ω impedance

Simulate the schematic and display the input impedance on the Smith chart, shown as TP2 in Figure 3-15.

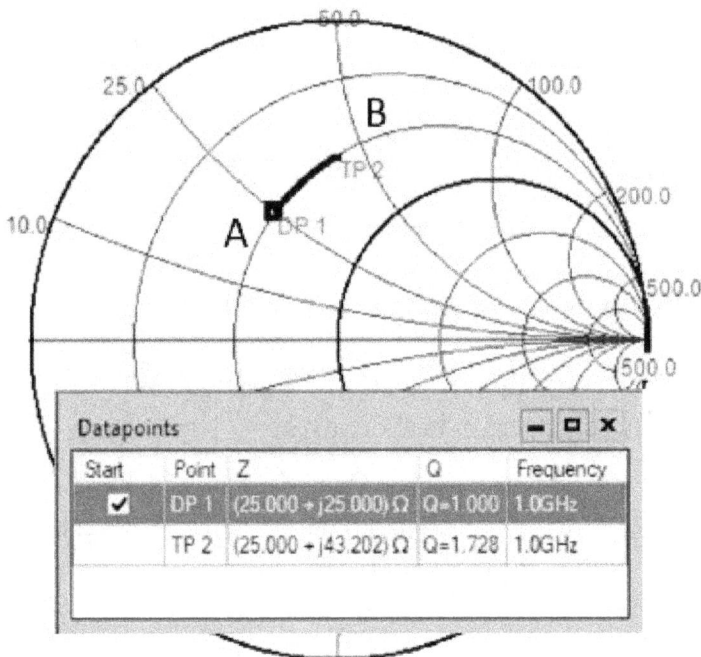

Figure 3-15 Moving point A to point B

Adding a Shunt Reactance to an Impedance

Adding a shunt element to an impedance point on the Smith Chart causes the resulting impedance to move along the constant conductance circle in which the impedance intersects. A shunt inductance will move the impedance in a counter-clockwise direction while a shunt capacitance will move the resulting impedance in a clockwise direction on the constant conductance circle. The susceptance that is added to the impedance by the shunt element can be read from the Smith Chart by finding the difference between the lines of susceptance that intersect the start point and end point on the constant conductance circle. This susceptance can be converted to a capacitance or inductance by using Equations (3-25) and (3-26).

$$L(shunt) = \frac{n}{\omega B} \qquad (3\text{-}25)$$

$$C(shunt) = \frac{B}{\omega n} \qquad (3\text{-}26)$$

Where,

B = susceptance measured along the arc length

ω = the design frequency $2\pi f$

n = impedance normalizing value (50 Ω)

Example 3.3: Continuing with the previous example measure the amount of shunt capacitance required to move from point B to point C on the real impedance axis.

Solution: Add a shunt capacitance to move the impedance from point B to point C as shown in Figure 3-17. When adding a shunt element switch from the impedance grid to the admittance coordinates. The admittance follows the constant conductance circle in which the point lies by the difference between the intersecting susceptance lines. The normalized susceptance as measured on the perimeter of the chart is 0.865 mhos. Use Equation (3-26) to calculate the capacitance.

$$C\,(shunt) = \frac{B}{\omega n} = \frac{0.865}{2 \cdot \pi \cdot 1000 \cdot 10^6 \cdot (50)} = 2.75 \ pF$$

Alternatively you can add a shunt capacitor to the circuit and make the capacitance value tunable. Start with a very low value of approximately 0.1 pf and increase the value of capacitance until the admittance is moved from point B to point C.

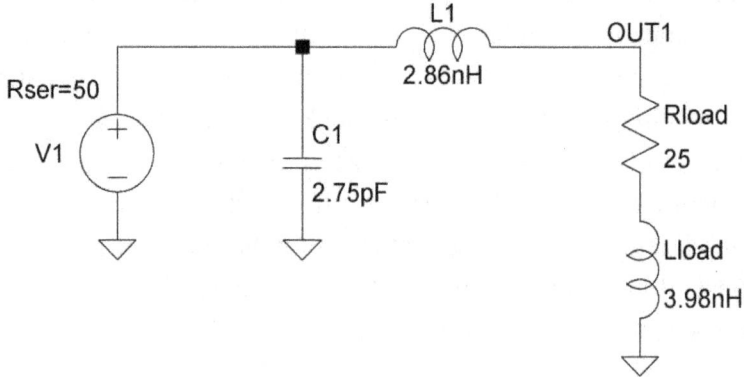

Figure 3-16 Shunt capacitance is added to the network

Display the input impedance on the Smith chart, as shown in Figure 3-17.

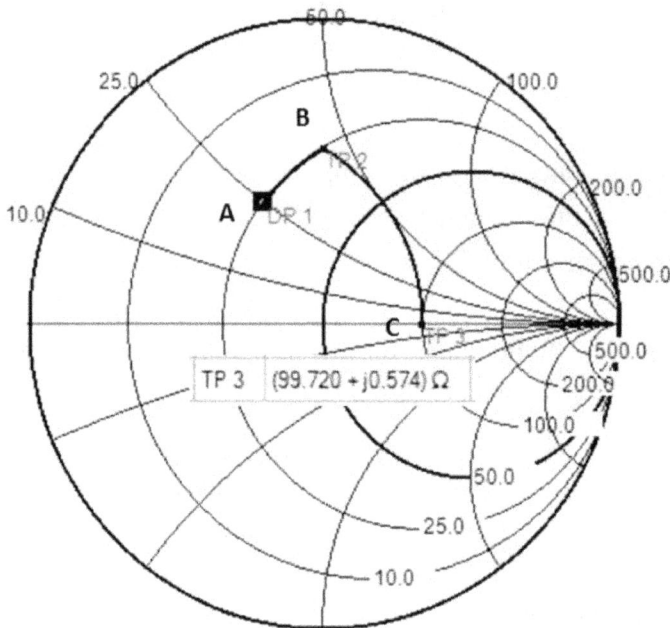

Figure 3-17 Moving from point B to point C on the real axis

The resulting admittance at point C is 0.574 mhos. This is the resulting normalized admittance looking into port one of the network.

The actual admittance is found by multiplying the normalized admittance by the characteristic admittance (1/50 Ω. Therefore the admittance at point C is 0.01 mhos or an impedance of 100Ω. These techniques form the basis for performing impedance transformations using the Smith Chart. In this example a complex impedance of 25 + j25 Ω has been transformed to a pure resistance of 100 Ω. The lumped element movement directions are summarized in Figure 3-18.

Example 3-4: Show the direction of movement on the Smith Chart.

Solution: By iterative addition of series and shunt elements an impedance point can be moved from one location to another on the Smith Chart, as shown in Figure 3-18. This process forms the basis of graphical impedance matching design

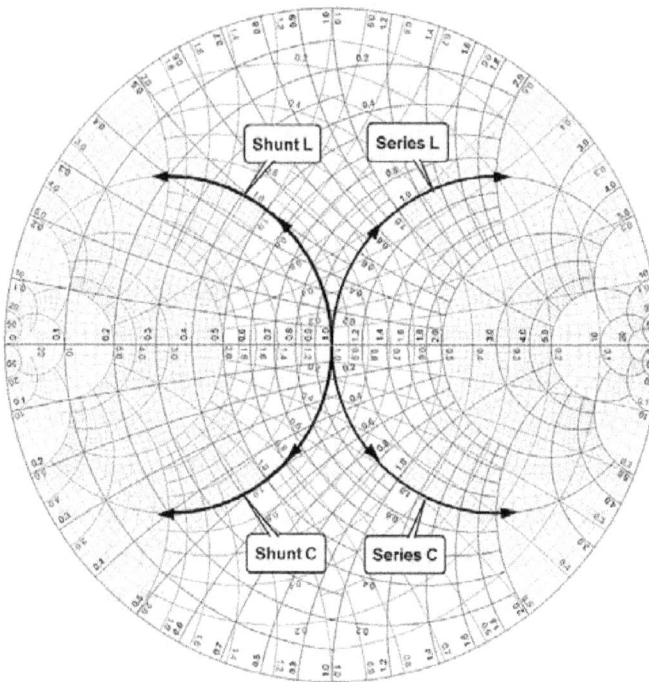

Figure 3-18 Lumped element movements on the Smith Chart

3.6 VSWR Circles on the Smith Chart

Equation (2-29) demonstrated that the VSWR of a network is related to the

magnitude of the reflection coefficient, independent of the angle. From plotting the reflection coefficient on the Smith Chart we know that the origin of the impedance vector is located in the center of the chart. This suggests that as the reflection coefficient vector rotates 360 degrees around the Smith Chart with a constant magnitude, the VSWR will remain constant. This locus of points around the center of the Smith Chart is known as the constant VSWR circle. In LTspice there is no direct way to plot constant VSWR circles, which are frequently used for LNA design.

3.7 Adding a Transmission Line in Series with an Impedance

In Example 3-2 we have seen that adding a reactance in series with an impedance point causes the impedance to follow the constant resistance circles. Adding a transmission line of the same impedance as the Smith Chart's in series with an impedance point causes the resulting impedance to follow a constant VSWR circle in which the impedance lies. The impedance moves in a clockwise direction on the constant VSWR circle.

Example 3.5: Calculate the electrical length of a series transmission line moving the impedance from point A to point B on the Smith Chart, as shown in Figure 3-20.

Solution: Add a series transmission line to the impedance and make the electrical length tunable. Increase the line length to move the impedance to point B. Since the angle of point A is about 116 degrees the electrical length of the transmission line is about 116/2 =58 degrees. The impedance on the real axis (zero reactance) on the right hand side of the Smith Chart would represent a point of maximum-voltage and minimum-current along the transmission line.

T1   OUT1

Rser=50

Td=0.162n Z0=50

V1   Rload
                                                25

                                                Lload
                                                3.98nH

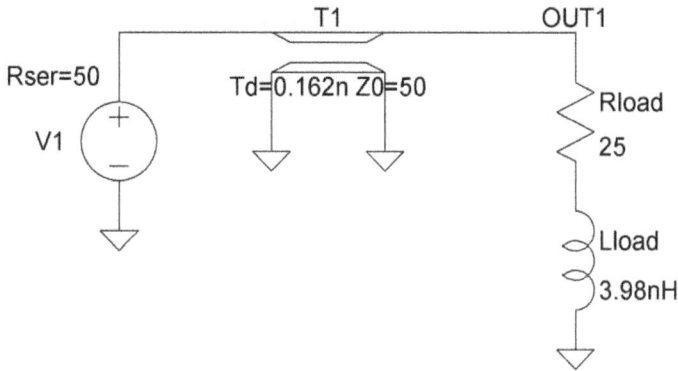

Figure 3-19 Series transmission line added to impedance

Transmission line lengths are sometimes referred to in terms of fractional wavelengths. Because one wavelength is equal to 360°, a 58° electrical length represents (58/360) = 0.161λ. Continue to add electrical length to the transmission line to reach point C on real axis on the Smith Chart. As Figure 3-20 shows, the real impedance (zero reactance) on the Smith Chart represents a minimum voltage or maximum current point

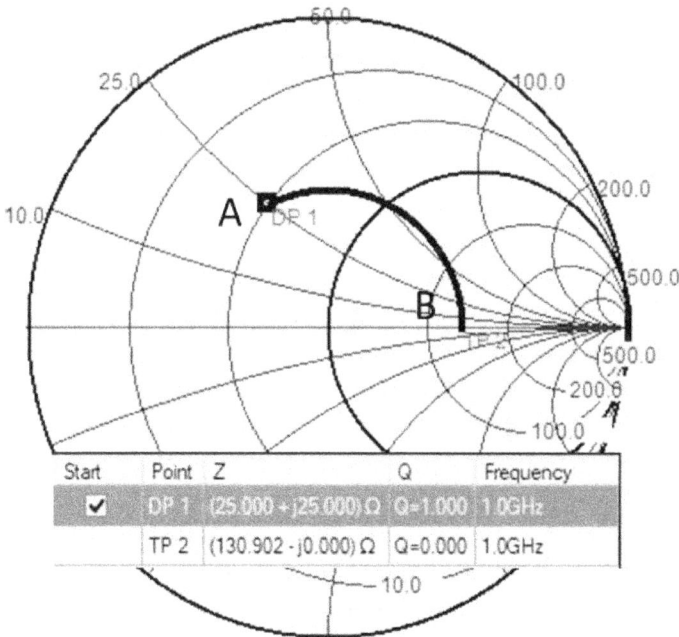

Start	Point	Z	Q	Frequency
☑	DP 1	(25.000 + j25.000) Ω	Q=1.000	1.0GHz
	TP 2	(130.902 - j0.000) Ω	Q=0.000	1.0GHz

Figure 3-20 Series **transmission line moves impedance on constant VSWR circle**

The schematic in Figure 3-21 shows that the series transmission line added to the impedance to bring it to the minimum-voltage point C.

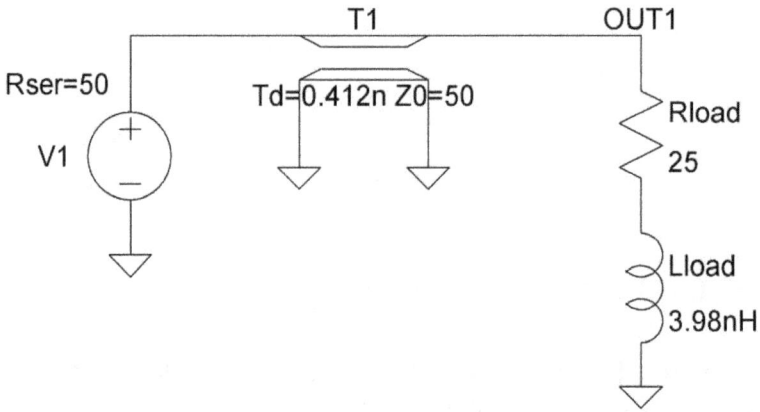

Figure 3-21 Series transmission line added to impedance

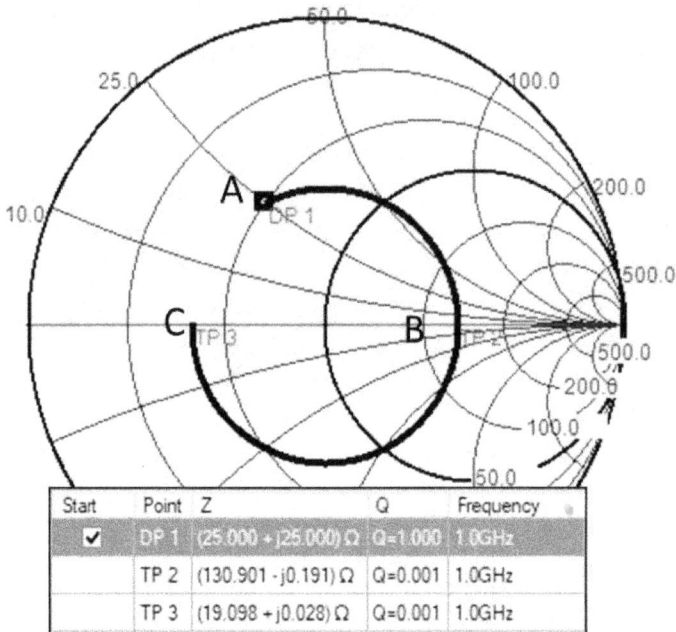

Start	Point	Z	Q	Frequency
✓	DP 1	(25.000 + j25.000) Ω	Q=1.000	1.0GHz
	TP 2	(130.901 - j0.191) Ω	Q=0.001	1.0GHz
	TP 3	(19.098 + j0.028) Ω	Q=0.001	1.0GHz

Figure 3-22 Series transmission line moves impedance to a minimum voltage point

Further increasing the length of the transmission line we find that we arrive back at point A at 180 degrees of electrical length. Therefore the electrical distance around the Smith Chart is 180° or λ/2 wavelength. The points of maximum-voltage and minimum-voltage will repeat every λ/2 wavelength.

3.8 Adding a Transmission Line in Parallel with an Impedance

In Chapter 2 we have seen that the open and short-circuited transmission lines could take on the equivalence of an inductor, capacitor, or series and parallel resonant circuits depending on the electrical length of the line. Therefore, the shunt transmission line will behave more like the lumped element movements on the Smith Chart.

Short Circuit Transmission Lines

At DC and low frequencies, a short circuit line is a very low inductance but this is not the case at higher RF and microwave frequencies.

Example 3-6: Three short-circuited transmission lines are given in Figures 3-23, 3-24, and 3-25. Figure 3-23 is a perfect short with 0 degree electrical length, Figure 3-24 is a line with 45 degree electrical length, and Figure 3-25 is a line with 90 degree electrical length. Display the impedance points on the Smith chart.

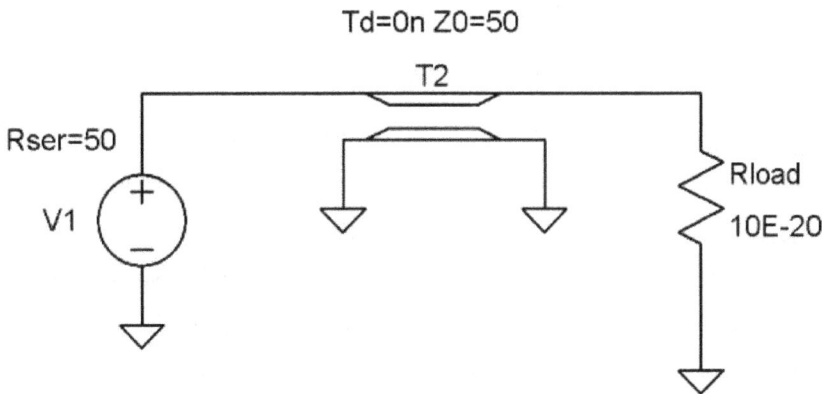

Figure 3-23 Short circuited transmission line 0 degree electrical length

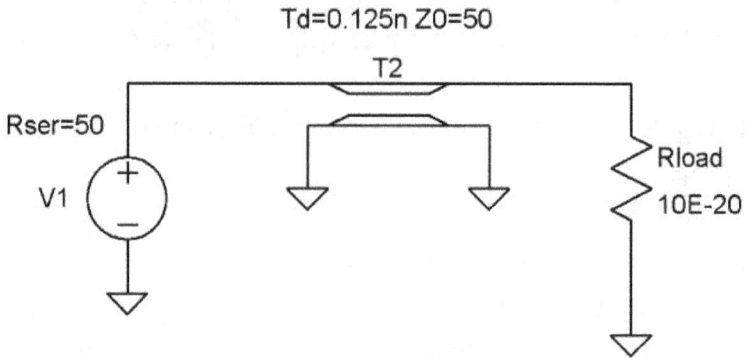

Figure 3-24 Short circuited transmission line 45 degree electrical length

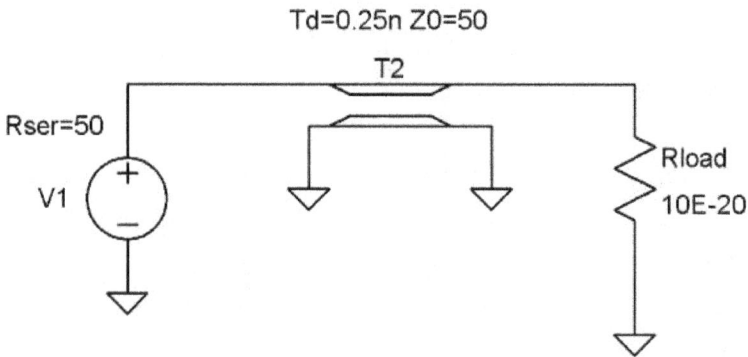

Figure 3-25 Short circuited transmission line 90 degree electrical length

Solution: Simulate the schematics and display the impedance on the Smith chart in Figure 3-26. Figure 3-26 point A shows that a short circuit transmission line with 0 degree length (perfect short) appears at the short circuit point on the Smith chart. If 45 degree of electrical length is added to the short circuit line, the impedance moves clockwise along the outer circumference of the Smith Chart to the position B shown at the top of Figure 3-26. As the line length is increased to 90 degree we see that the short circuit has been transformed to an open circuit at point C shown in Figure 3-26. At 180° line length the impedance will travel completely around the Smith Chart and appear as a short circuit again.

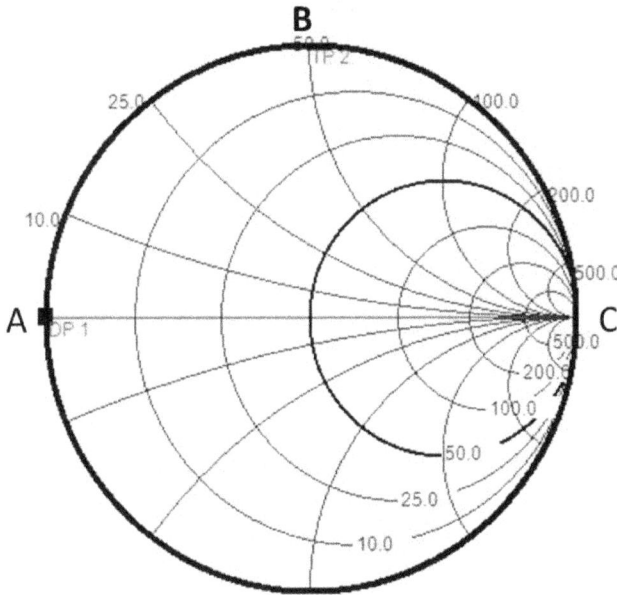

Figure 3-26 Short circuit transmission line impedance at electrical lengths:
A=0°, B=45°, and C=90°

The Impedance values at points A, B, and C are shown in Table 3-3.

Start	Point	Z	Q	Frequency
✔	DP 1	(0.000 + j0.000) Ω	Q=99999.000	1.0GHz
	TP 2	(0.000 + j50.000) Ω	Q=99999.000	1.0GHz
	TP 3	(-Infinity + jInfinity) Ω	Q=NaN	1.0GHz

Table 3-3 Data point values at points A, B, and C

Open Circuit Transmission Lines

Example 3-7: Three open-circuited schematics are given in Figures 3-27, 3-28, and 3-29. Figure 3-27 is an open circuit line with 0 degree electrical length, Figure 3-28 is an open circuit line with 45 degree electrical length, and Figure 3-29 is an open circuit line with 90 degree electrical length. Display the impedance points on the Smith chart.

Td=0n Z0=50

T2

Rser=50

V1

Rload
10E20

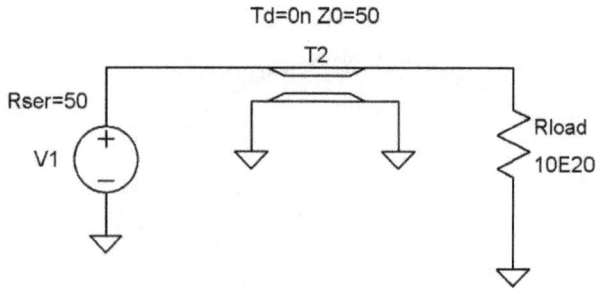

Figure 3-27 Open circuited transmission line 0 degree electrical lengths

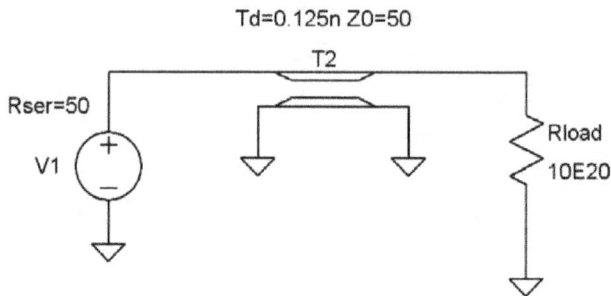

Td=0.125n Z0=50

T2

Rser=50

V1

Rload
10E20

Figure 3-28 Open circuited transmission line 45 degree electrical lengths

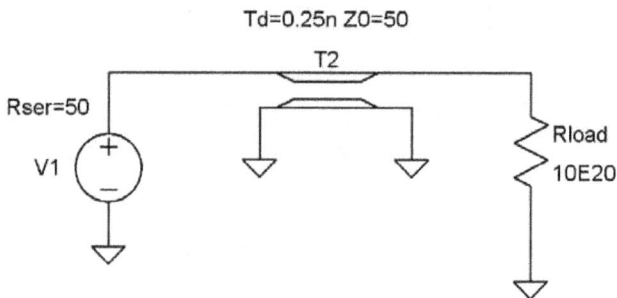

Td=0.25n Z0=50

T2

Rser=50

V1

Rload
10E20

Figure 3-29 Open circuited transmission line 90 degree electrical lengths

Solution: Figure 3-30 shows the characteristic of the open circuit transmission line. At 0 degree length it appears as a perfect open circuit. As the electrical length is increased, the impedance moves clockwise around the circumference to the 45 degree position at the bottom of the chart. At 90 degree electrical length the open circuit now appears as a short circuit.

This property of transforming open circuits to short circuits and vice versa is one that is used frequently throughout microwave circuit design.

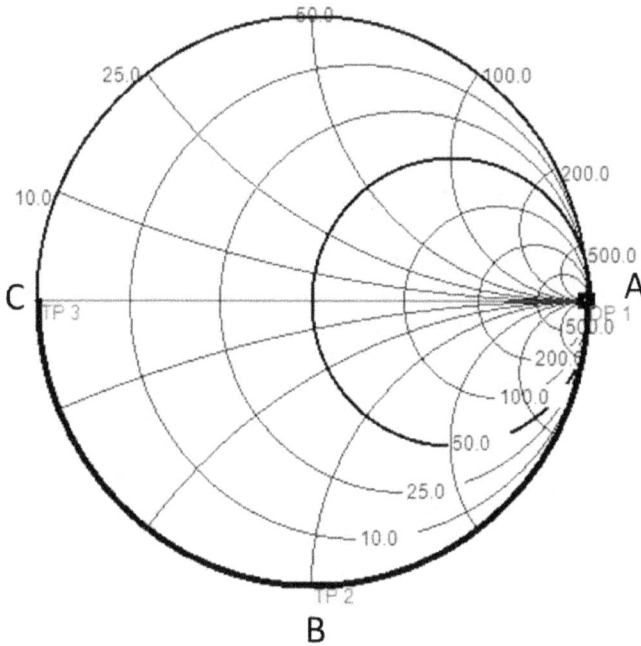

Figure 3-30 Open circuit transmission line impedance at electrical lengths: A=0°, B=45°, and C=90°

The Impedance values at points A, B and C are shown in the following Table.

Start	Point	Z	Q	Frequency
✔	DP 1	(999999950.526 + j0.000) Ω	Q=0.000	1.0GHz
	TP 2	(0.000 - j50.000) Ω	Q=99999.000	1.0GHz
	TP 3	(0.000 + j0.000) Ω	Q=0.000	1.0GHz

Table 3-4 Data point values at points A, B, and C

3.9 Open and Short Circuit Shunt Transmission Lines

For small fractional wavelength transmission lines the open circuit shunt transmission line acts as a shunt capacitor.

Example 3.8: Measure the electrical length of an open circuited shunt transmission line, or the amount of shunt capacitance, to move the impedance Z = 25 + j25 Ω from point A to the center of Smith Chart, as shown in Figure 3-33.

Solution: Plotting the impedance on the Smith Chart with the admittance circles shows that the impedance lies directly on the unit conductance circle, as shown in Figure 3-33. Therefore, a 50 Ohm transmission line of 43.38 electrical degrees move the impedance directly to to the center of Smith Chart.

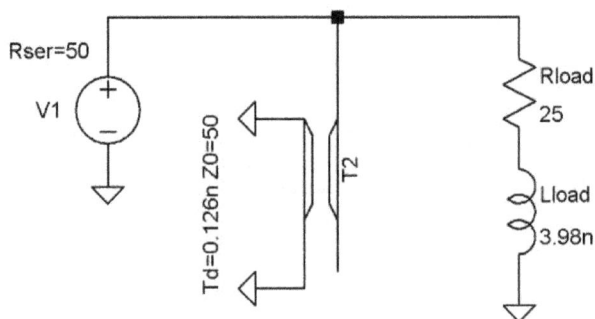

Figure 3-31 Open circuit shunt transmission line

The second schematic inset of Figure 3-32 shows that this circuit is equivalent to a 3.2 pF capacitor in shunt with the load impedance.

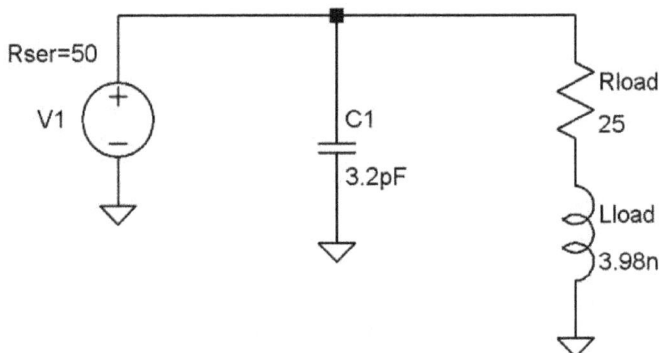

Figure 3-32 Open circuit shunt transmission line

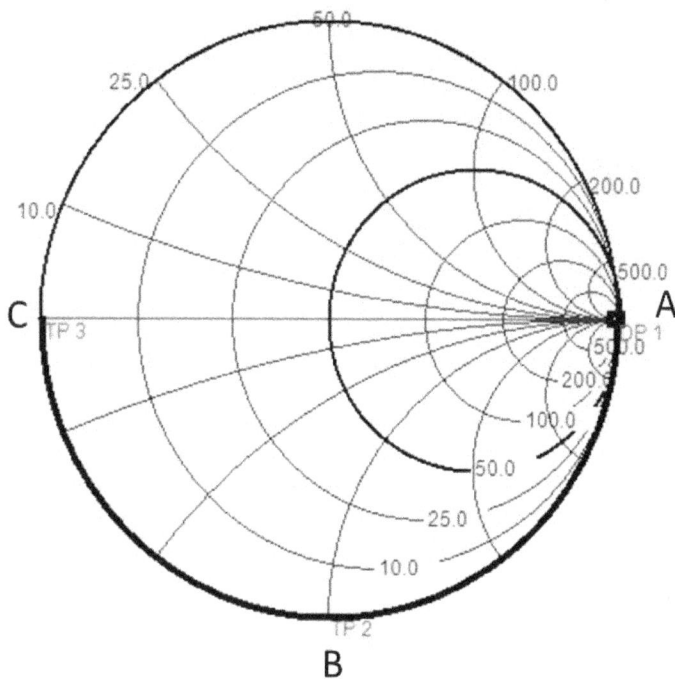

Figure 3-33 Moving from point to B

The Impedance values at points A and B are shown in Table 3-5.

Start	Point	Z	Q	Frequency
✔	DP 1	(25.000 + j25.000) Ω	Q=1.000	1.0GHz
	TP 2	(49.992 - j0.632) Ω	Q=0.013	1.0GHz

Table 3-5 Data point values at points A and B

For small fractional wavelength transmission lines the short circuit shunt transmission line acts as a shunt inductor. Consider the 4.3 –j14 Ω impedance as shown in Figure 3-34. This impedance lies on the unit conductance circle on the bottom half of the Smith Chart. A 50 Ω shunt transmission line added to the impedance moves along the constant conductance circle to the center of the Smith Chart.

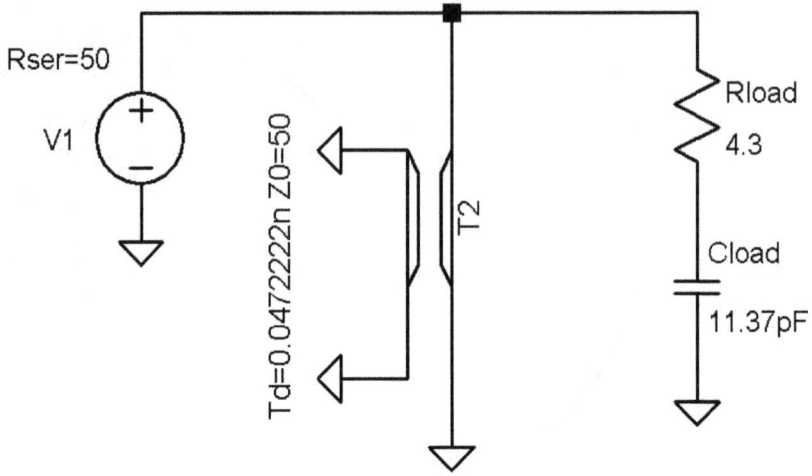

Figure 3-34 Short circuit shunt transmission line added to an impedance

Similarly a 2.4 nH shunt inductor has the same effect at a frequency of 1000 MHz. These movements form the basis for distributed network impedance matching which is covered in detail in Chapter 6.

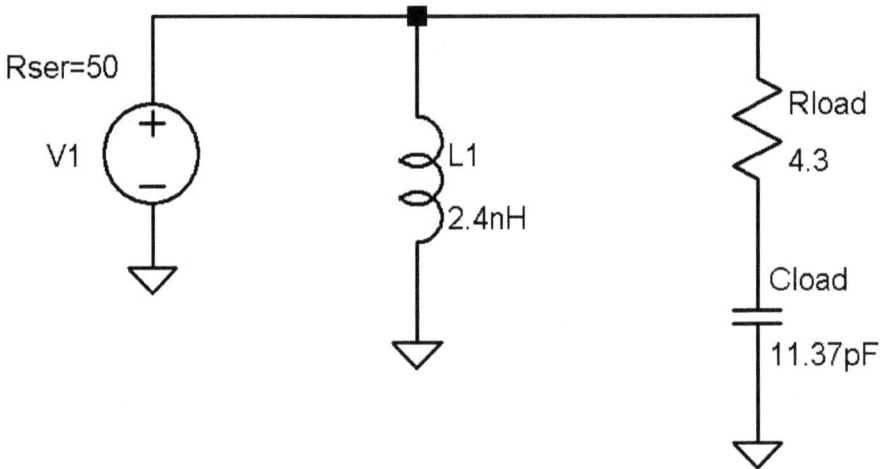

Figure 3-35 Shunt 2.4 nH inductor added to an impedance

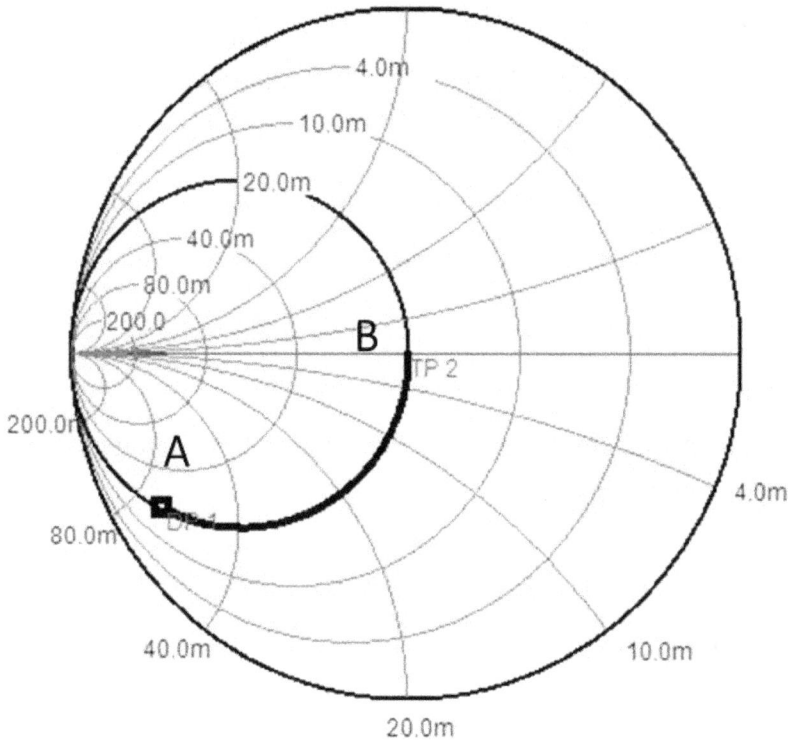

Figure 3-36 Moving Point A to B

The impedance values at points A and B are shown in Table 3-6.

Start	Point	Z	Q	Frequency
✔	DP 1	(4.300 - j14.000) Ω	Q=3.256	1.0GHz
	TP 2	(49.881 - j0.000) Ω	Q=0.000	1.0GHz

Table 3-6 Data point values at points A and B (Smith Chart Center)

References and Further Readings

[1]   David M. Pozar, *Microwave Engineering*, Third Edition, John Wiley and Sons, Inc., 2005.

[2]   Ali Behagi and Manou Ghanevati, Fundamentals of RF and Microwave Circuit Design, Practical Analysis and Design Tools, Techno Search, Ladera Ranch, California 2017

[3]   Ali Behagi, *RF and Microwave Circuit Design*, A Design Approach Using (**ADS**), Techno Search, Ladera Ranch, CA 2017.

[4]   Franklin F. Kuo, Network Analysis and Synthesis, John Wiley and Sons Inc., 1966

[5]   William Sinnema, Electronic Transmission Technology, Prentice-Hall, Inc, Englewood Cliffs, New Jersey 07632, 1979

[6]   Chris Bowick, RF Circuit Design, Second Edition, Newnes, Elsevier, 2008

Problems

3-1.   Place a 20 + j30 Ω impedance at point A on the Smith Chart. Add a series inductance to move the impedance along the constant resistance circle to point B having the impedance 20 + j50 Ω. Using a design frequency of 1000MHz, calculate the inductance that this reactance represents.

3-2.   Continuing with Problem 3-1 enable the admittance coordinates on the Smith Chart to add a shunt element. Add a shunt capacitance to move the impedance to the real axis. Measure the susceptance required to move to the real impedance axis by the difference between the intersecting susceptance lines.

3-3.   Calculate the magnitude of reflection coefficient for a desired VSWR = 2.

3-4.   Create a constant VSWR circle for a VSWR=20. Comment on the VSWR value required to place the VSWR circle on the circumference of the Smith Chart.

3.5   A load of 75 + j20 is connected to a 50 transmission line. Calculate the load admittance and the input impedance if the line is 0.2 wavelengths long.

3-6.   For the load impedance of 75 + j20, determine the reflection coefficient and the transmission coefficient.

3-7.   For the load impedance in Problem 3-4, determine the normalized value of the load impedance if impedance is normalized to 75 Ohm.

3-8.   A series RLC load, R = 100 Ω, L = 20 nH, C = 25 pF is connected to a 50 Ω transmission line. Calculate the VSWR and reflection coefficient at the load at 100 MHz.

3-9.   Using the RLC load impedance of problem 3-8 determine the impedance with a series transmission line of characteristic impedance of 50 Ω and electrical length of 180 degrees.

3-10.   Determine the input impedance of a network that has a reflection coefficient of 0.5 at an angle of 112°.

3-11.   Create a one-port S parameter text file with the impedance of Problem 3-10 at a frequency of 1 GHz. Plot the S parameter, S11, on the Smith Chart.

3-12.   Using the S parameter file of the C Band amplifier of Figure 3-8, plot the input return loss, S11, and output return loss, S22, on the Smith Chart. Determine the worst case input and output VSWR for this amplifier.

Chapter 4
Resonant Circuits and Filters

4.1 Introduction

The first half of this chapter examines resonant circuits. Lumped element resonant circuits and the lumped equivalent networks of mechanical and distributed resonators are considered. Resonant circuits are used in many applications, such as filters, oscillators, tuners, tuned amplifiers, and microwave communication networks. The analysis of basic lumped element series and parallel RLC resonant circuits is implemented software. The discussion turns to microwave resonators with an analysis of the Q factor measurement of transmission line resonators. Using the LTspice software a robust technique is demonstrated for the evaluation of Q factor from the measured S parameters of a resonant circuit. The second half of the chapter is an introduction to the vast subject of filter networks. The design of lumped element filters is introduced and followed by an introduction to distributed element filters. The chapter concludes with the conversion of lumped elements to distributed elements.

4.2 Resonant Circuits

Near resonance, RF and microwave resonant circuits can be represented either as a lumped element series or parallel RLC network.

Series Resonant Circuits

In this section we analyze the behavior of series resonant circuits.

Example 4.1: Consider the one port series resonator that is represented as a series RLC circuit of Figure 4-1. Analyze the circuit, with R = 10 Ω, L = 10 nH, and C = 10 pF.

Solution: Set up the schematic as shown in Figure 4-1.

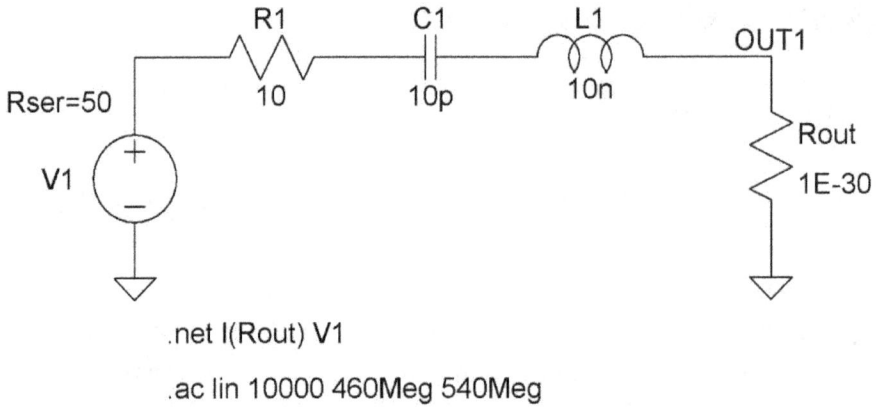

Figure 4-1 One-port series RLC resonator circuit

The plot of the resonator's input impedance in Figure 4-2 shows that the resonance frequency is about 503.3 MHz and the input impedance at resonance is 10 Ω, the value of the resistor in the network.

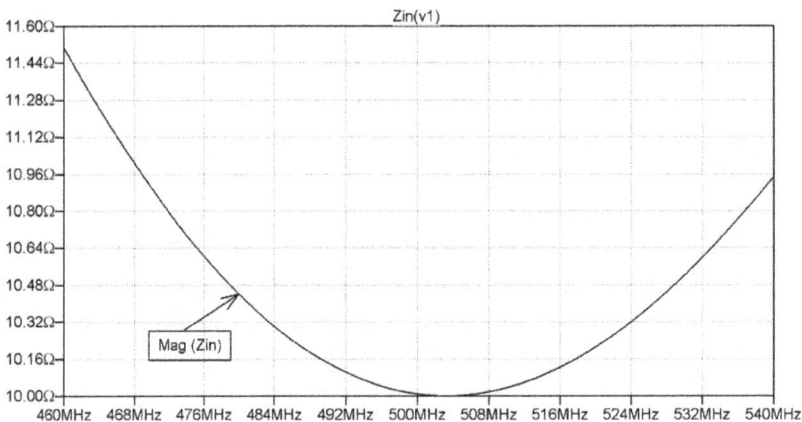

Figure 4-2 Input impedance plot showing the resonance frequency

The input impedance of the series RLC resonant circuit is given by,

$$Z_{in} = R + j\omega L - j\frac{1}{\omega C}$$

Where, $\omega = 2\pi f$ is the angular frequency in radian per second.

If the AC current flowing in the series resonant circuit is I, then the complex power delivered to the resonator is

$$P_{in} = \frac{|I|^2}{2} Z_{in} = \frac{|I|^2}{2}\left(R + j\omega L - j\frac{1}{\omega C}\right) \tag{4-1}$$

At resonance the reactive power of the inductor is equal to the reactive power of the capacitor. Therefore, the power delivered to the resonator is equal to the power dissipated in the resistor

$$P_{in} = \frac{|I|^2 R}{2} \tag{4-2}$$

Parallel Resonant Circuits

Example 4.2: Analyze a rearrangement of the RLC components of Figure 4-1 into the parallel configuration of Figure 4-3. The schematic of Figure 4-3 represents the lumped element representation of the parallel resonant circuit.

Solution: The one port parallel resonant circuit is shown in Figure 4-3.

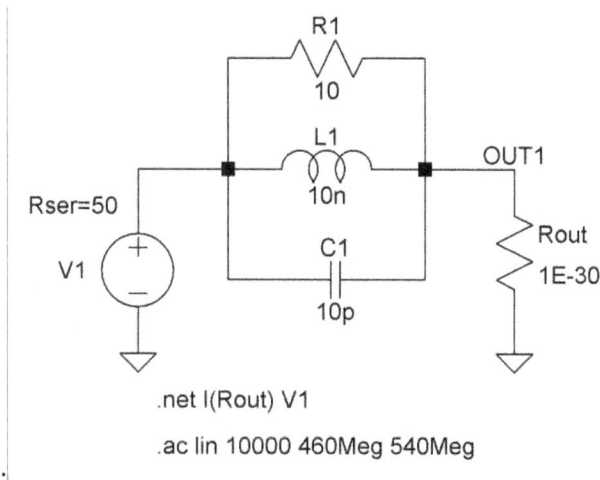

Figure 4-3 One-port parallel RLC resonant circuit

Simulate the schematic and display the input impedance in a rectangular plot. The plot of the magnitude of the input impedance shows that the resonance frequency is still 503.3 MHz where the input impedance is R = 10 Ω. Again this shows that the impedance of the inductor cancels the impedance of the capacitor at resonance. In other words, the reactance, X_L, is equal to the reactance, X_C, at the resonance frequency.

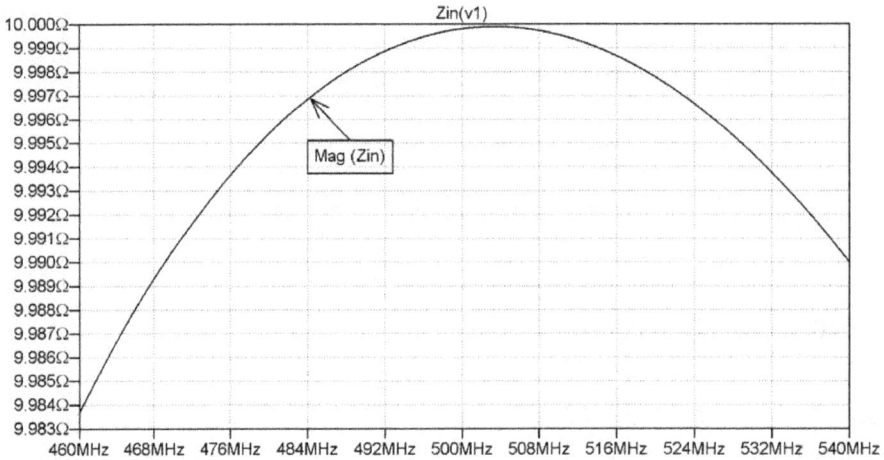

Figure 4-4 Input impedance of parallel RLC resonant circuit

The input admittance of the parallel resonant circuit is given by:

$$Y_{IN} = \frac{1}{R} + j\omega C - j\frac{1}{\omega L}$$

If the AC voltage across the parallel resonant circuit is V, then the complex power delivered to the resonator is

$$P_{in} = \frac{|V|^2}{2}Y_{in} = \frac{|V|^2}{2}\left(\frac{1}{R} + j\omega C - j\frac{1}{\omega L}\right) \qquad (4\text{-}3)$$

At resonance the reactive power of the inductor is equal to the reactive power of the capacitor. Therefore, the power delivered to the resonator is equal to the power dissipated in the resistor.

$$P_{in} = \frac{|V|^2}{2R} \qquad (4\text{-}4)$$

The resonance frequency for the parallel resonant circuit as well as the series resonant circuit is obtained by setting $\omega_0 C = \frac{1}{\omega_0 L}$ or:

$$\omega_o = 2\pi f_o = \frac{1}{\sqrt{LC}} \qquad (4\text{-}5)$$

Where, ω_0 is the angular frequency in radian per second and f_0 is equal to the frequency in Hertz.

Resonant Circuit Loss

In Figure 4-1 and 4-3 the resistor R1 represents the loss in the resonator. It includes the losses in the capacitor as well as the inductor. The Q factor can be shown to be a ratio of the energy stored in the inductor and capacitor to the power dissipated in the resistor as a function of frequency [6]. For the series resonant circuit of Figure 4-1 the unloaded Q factor is defined by:

$$Q_u = \frac{X}{R} = \frac{\omega_o L}{R} = \frac{1}{\omega_o RC} \qquad (4\text{-}6)$$

The unloaded Q factor of the parallel resonant circuit in Figure 4-3 is simply the inverse of the series resonant circuit.

$$Q_u = \frac{R}{X} = \frac{R}{\omega_o L} = \omega_o RC \qquad (4\text{-}7)$$

We can clearly see that as the resistance increases in the series resonant circuit, the Q factor decreases. Conversely as the resistance increases in the parallel resonant circuit, the Q factor increases. The Q factor is a measure of loss in the resonant circuit. Thus a higher Q corresponds to lower loss and a lower Q corresponds to a higher loss. It is usually desirable to achieve high Q factors in a resonator as it will lead to lower losses in filter applications or lower phase noise in oscillators. Note that the resonator Q of Equation (4-6) and (4-7) is defined as Q_u, the unloaded Q of the resonator. This means that the resonator is not connected to any source or load impedance and as such is unloaded. The measurement of Q_u requires that the resonator be attached (coupled) to a signal source or load of some finite impedance.

Equations (4-6) and (4-7) would then have to be modified to include the source and load resistance. We might also surmise that any reactance associated with the source or load impedance may alter the resonant frequency of the resonator. This leads to two additional definitions of Q factor, the loaded Q and external Q.

Loaded Q and External Q

Example 4.3: Analyze the parallel resonator that is attached to 50 Ω source and load as shown in Figure 4-5.

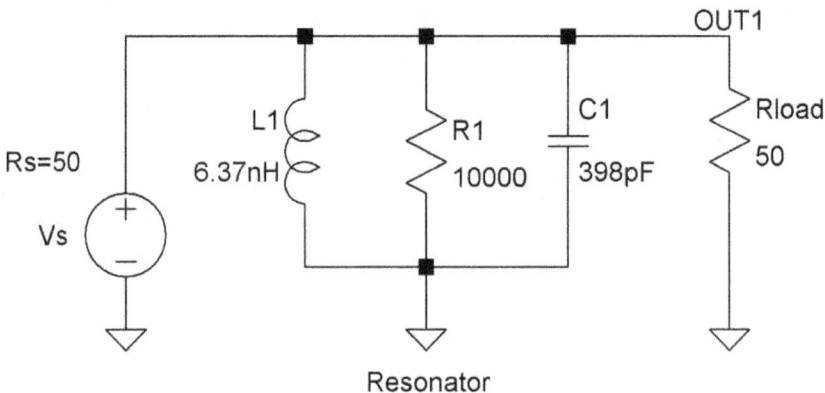

Figure 4-5 Parallel resonator with source and load impedance attached

Solution: Using Equation (4-7) to define the Q factor for the circuit requires that we include the source and load resistance which is 'loading' the resonator. This leads to the definition of the loaded Q, Q_L, for the parallel resonator as defined by Equation (4-8).

$$Q_L = \frac{R_S + R + R_L}{\omega_o L} \qquad (4\text{-}8)$$

Conversely we can define a Q factor in terms of only the external source and load resistance. This leads to the definition of the external Q, Q_E.

$$Q_E = \frac{R_S + R_L}{\omega_o L} \qquad (4\text{-}9)$$

The three Q factors are related by the inverse relationship of Equation (4-10).

$$\frac{1}{Q_L} = \frac{1}{Q_E} + \frac{1}{Q_U} \qquad (4\text{-}10)$$

At RF and microwave frequencies it is difficult to directly measure the Q_u of a resonator. We may be able to calculate the Q factor based on the physical properties of the individual inductors and capacitors as we seen in chapter 1. This is usually quite difficult and the Q factor is typically measured using a Vector Network Analyzer, VNA. Therefore, the measured Q factor is usually the loaded Q, Q_L. External Q is often used with oscillator circuits that are generating a signal. In this case the oscillator's load impedance is varied so that the external Q can be measured. The loaded Q of the network is then related to the fractional bandwidth by Equation (4-11).

$$Q_L = \frac{\sqrt{f_l f_h}}{BW_{-3dB}} \qquad (4\text{-}11)$$

Where, BW is the -3 dB bandwidth in Hz and higher frequencies at -3dB points. f_l and f_h are the lower and

4.3 Lumped Element Parallel Resonator Design

Example 4.4: In this example we design a lumped element parallel resonator at a frequency of 100 MHz. The resonator is intended to operate between a source resistance of 100 Ω and a load resistance of 400 Ω.

Solution: Best accuracy would be obtained by using S parameter files or Modelithics models for the inductor and capacitor. However a good first order model can be obtained by using the inductor and capacitor models that include the component Q factor. These models save us the work of calculating the equivalent resistive part of the inductor and capacitor model. Use the Q factors shown in the schematic of Figure 4-6.

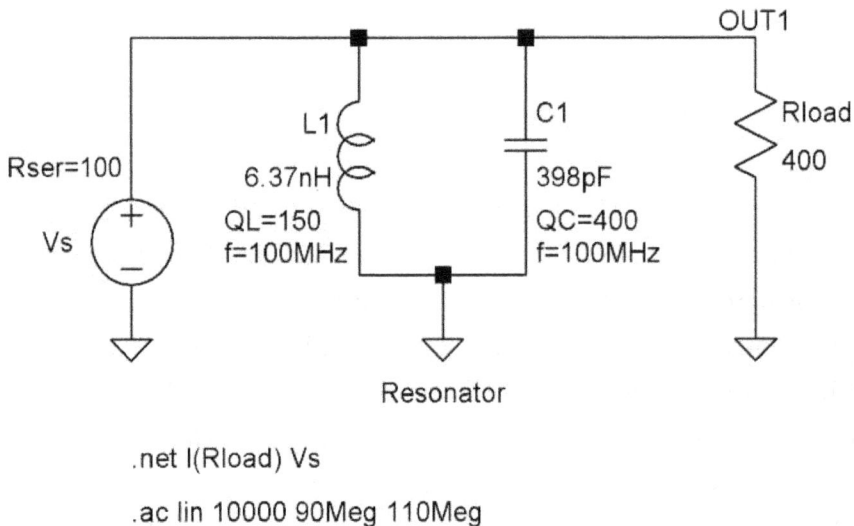

.net I(Rload) Vs

.ac lin 10000 90Meg 110Meg

Figure 4-6 Parallel resonator using inductor and capacitor with assigned Q_u values

Simulate the schematic and display the insertion loss, S21, in a rectangular plot.

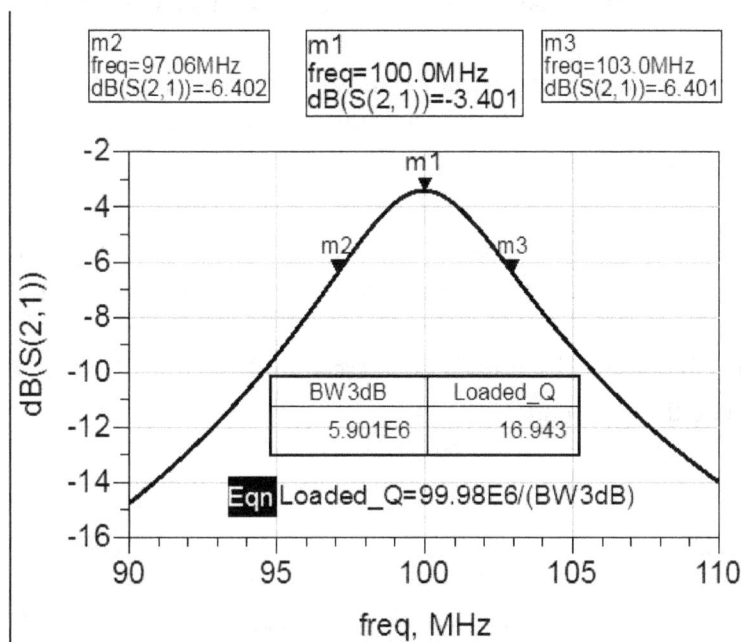

m2
freq=97.06MHz
dB(S(2,1))=-6.402

m1
freq=100.0MHz
dB(S(2,1))=-3.401

m3
freq=103.0MHz
dB(S(2,1))=-6.401

BW3dB	Loaded_Q
5.901E6	16.943

Eqn Loaded_Q=99.98E6/(BW3dB)

Figure 4-7 Insertion loss of Parallel resonator [3]

Note that markers [3] have been placed on the plot of the insertion loss, S21, that gives readout of -3 dB bandwidth. To place markers m2 and m3 manually at -3dB points, first place marker m1 at the peak of S21 trace. Next place marker m2 at a point below marker m1. Select marker m2 and press the up and down arrows to move the marker m2 as close to -3dB point as possible. Repeat the process for marker m3. The loaded Q_L is:

$$Q_L = \frac{\sqrt{f_l f_h}}{BW} = \frac{99.98 MHz}{5.95 MHz} = 16.80$$

The designer must use caution when sweeping resonant circuits. Particularly high Q band pass networks require a large number of discrete frequency steps in order to achieve the necessary resolution required to accurately measure the 3dB bandwidth. In this example the Linear Analysis is set up to sweep the circuit from 90 MHz to 110 MHz using 2000 points.

Effect of Load Resistance on Bandwidth and QL

In RF circuits and systems the impedances encountered are often quite low, ranging from 1 Ω to 50 Ω. It may not be practical to have a source impedance of 100 Ω and a load impedance of 400 Ω.

Example 4.5: Using the parallel LC example, change the load resistance from 400 Ω to 50 Ω and re-examine the circuit's 3 dB bandwidth and Q_L.

Solution: Change the load resistance from 400 Ω to 50 Ω in Figure 4-8 and simulate the schematic.

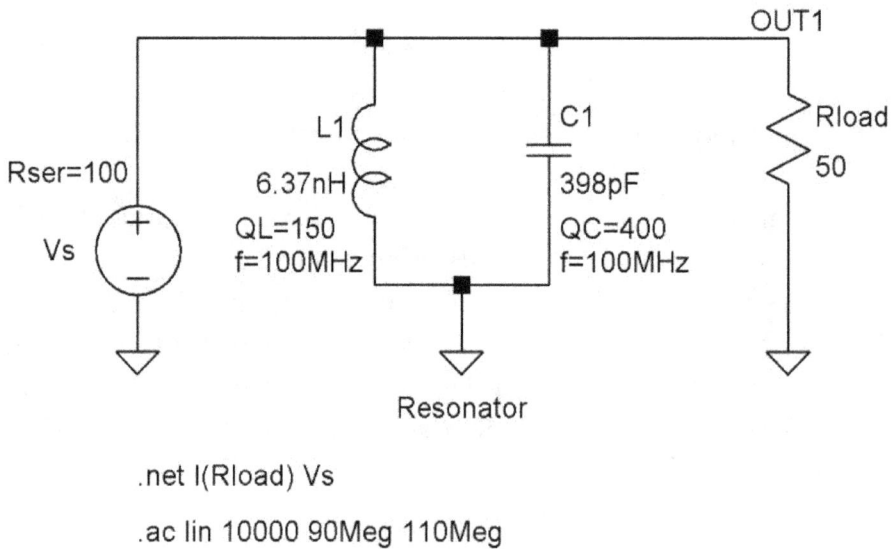

.net I(Rload) Vs

.ac lin 10000 90Meg 110Meg

Figure 4-8 A parallel resonance circuit

Simulate the schematic and display the insertion loss, S21, in a rectangular plot.

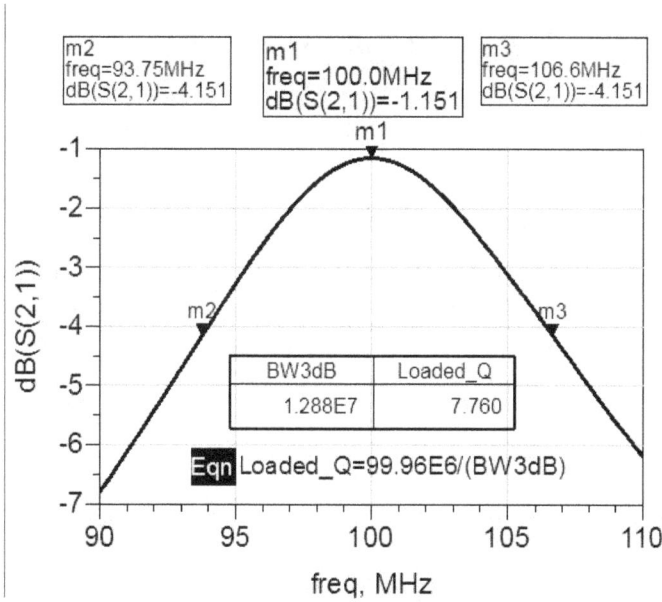

Figure 4-9 Parallel resonance circuit [3]

The 3 dB bandwidth [3] is 12.87 MHz resulting in a loaded Q factor of 7.766.

$$Q_L = \frac{\sqrt{f_l f_h}}{BW} = \frac{99.958 MHz}{12.87 MHz} = 7.766$$

The loaded Q factor has decreased by nearly half of the original value. We have increased the bandwidth or de-Q'd the resonator. This can also be thought of as tighter coupling of the resonator to the load.

4.4 Lumped Element Resonator Decoupling

To maintain the high Q of the resonator when attached to a load such as 50 Ω, it is necessary to transform the low impedance to high impedance presented to the load. The 50 Ω impedance can be transformed to the higher impedance of the parallel resonator thereby resulting in less loading of the resonator impedance. This is referred to as loosely coupling the resonator to the load. The tapped-capacitor and tapped-inductor networks can be used to accomplish this Q transformation in lumped element circuits.

Tapped Capacitor Resonator Design

Example 4.6: Consider rearranging the parallel LC network of Figure 4-8 with the tapped capacitor network shown in Figure 4-10. Re-examine the circuit's 3 dB bandwidth and Q_L.

Solution: The new capacitor values for C1 and C2 can be found by the simultaneous solution of the following equations.

$$C_T = \frac{C1 \cdot C2}{C1 + C2} \tag{4-12}$$

$$R_{L1} = R_L \left(1 + \frac{C2}{C1}\right)^2 \tag{4-13}$$

R_{L1} is the higher, transformed, load resistance. In this example substitute R_{L1} = 400 =, the original load resistance value. C_T is simply the original capacitance of 398 pf. The capacitor values are found to be: C1 = 616.1 pF and C2 = 1126.23 pF. The new resonator circuit is shown in Figure 4-10.

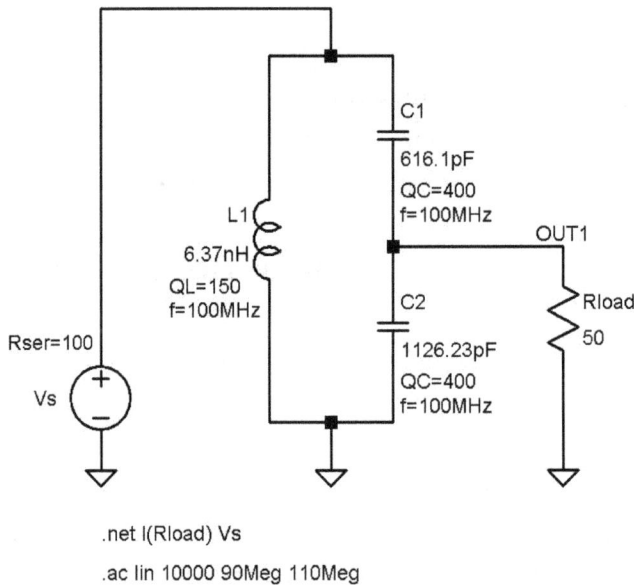

.net I(Rload) Vs

.ac lin 10000 90Meg 110Meg

Figure 4-10 Parallel LC resonator using tapped capacitor

Sweeping the circuit we see that the response has returned to the original performance of Figure 4-11[3].

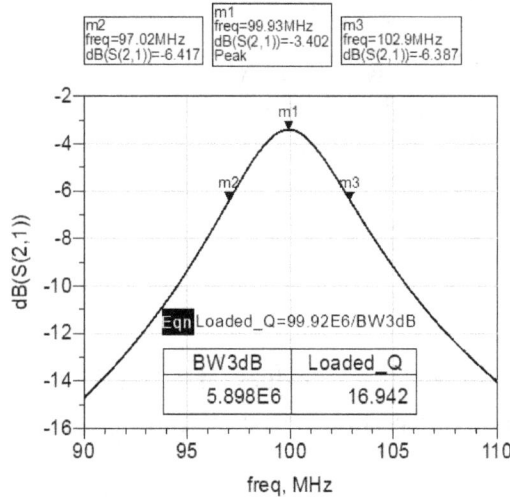

Figure 4-11 Tapped capacitor resonator response [3]

The 3 dB bandwidth has returned to 5.89 MHz making the Q_L equal to:

$$Q_L = \frac{\sqrt{f_l f_h}}{BW} = \frac{99.958\,MHz}{5.89\,MHz} = 16.97$$

The 50 Ω load resistor has been successfully decoupled from the resonator. The tapped capacitor and inductor resonators are popular methods of decoupling RF and lower microwave frequency resonators. It is frequently seen in RF oscillator topologies such as the Colpitts oscillator in the VHF frequency range.

Tapped Inductor Resonator Design

Example 4.7: Similarly design a tapped inductor network to decouple the 50 Ω source impedance from loading the resonator.

Solution: Replace the 100 Ω source impedance with a 50 Ω source and use a tapped inductor network to transform the new 50 Ω source to 100 Ω. Modify the circuit to split the 6.37 nH inductor, L_T, into two series inductors, L1 and L2. The inductor values can then be calculated by solving the following equation set simultaneously. R_{S1} is the higher, transformed, source resistance. In this example substitute $R_{S1} = 100\ \Omega$,

$$Rs1 = Rs\left(\frac{L_T}{L_1}\right)^2 \tag{4-14}$$

$$L_T = L_1 + L_2 \tag{4-15}$$

Solving the equation set results in values of L1=4.5 nH and L2=1.87 nH. The resulting schematic and response is shown in Figure 4-8.

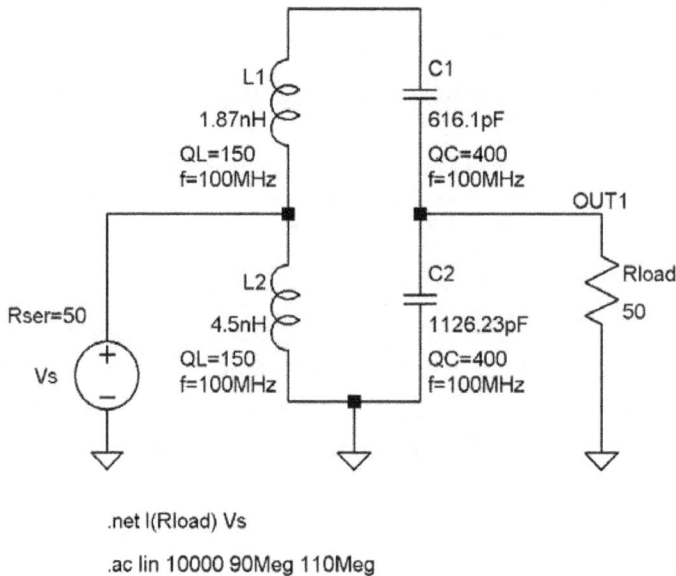

Figure 4-12 Tapped-inductor resonator (Courtesy Keysight ADS)

The new response in Figure 4-13 is very close to the plot of Figure 4-7 and 4-11. Therefore we now have a source and load resistance of 50 Ω and have not reduced the Q of the resonator from what we had with the original source resistance of 100 Ω and a load resistance of 400 Ω[3].

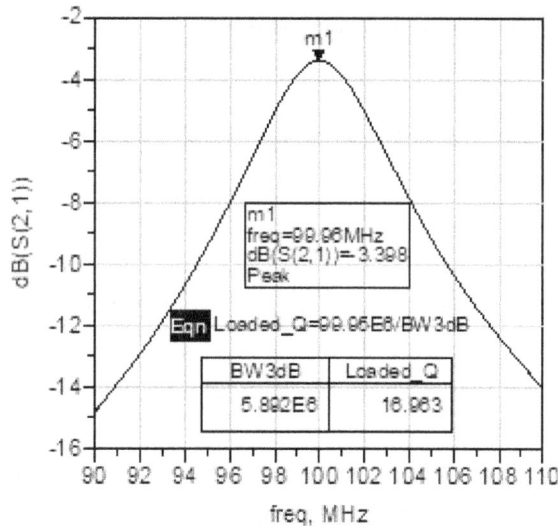

Figure 4-13 Tapped Inductor response [3]

4.5 Practical Microwave Resonators

At higher RF and microwave frequencies resonators are seldom realized with discrete lumped element RLC components. This is primarily due to the fact that the small values of inductance and capacitance are physically unrealizable. Even if the values could be physically realized we would see that the resulting Q factors would be unacceptably low for most applications. Microwave resonators are realized in a wide variety of physical forms. Resonators can be realized in all of the basic transmission line forms that were covered in Chapter 2. There are many specialized resonators such as ceramic dielectric resonator pucks that are coupled to a microstrip transmission line as well as Yittrium Iron Garnet spheres that are loop coupled to its load. These resonators are optimized for very high Q factors and may be tunable over a range of frequencies.

Figure 4-14 Ceramic dielectric resonator (puck) and coaxial resonator

Transmission Line Resonators

From Figure 2-18 in Chapter 2 we have seen that a quarter-wave short-circuited transmission line has a high input impedance and could be used as a parallel resonator . Similarly Figure 2-19 showed that a half-wave open circuited transmission line has a low input impedance and could be used as a parallel resonator too. Such parallel resonant circuits are often used as one port resonators. Near the resonant frequency, the one port resonator behaves as a parallel RLC network. As the frequency moves further from resonance the equivalent network becomes more complex. One port resonators are coupled to one another to form filter networks or directly to a transistor to form a microwave oscillator. Knowing the losses due to the physical and electrical parameters of the transmission line, one can calculate the Q_u of the transmission line resonator. The microstrip resonator Q_u is comprised of losses due to the conductor metal, the substrate dielectric, and radiation losses. The Q_u is often dominated by the conductor Q. Unfortunately it can be quite difficult to accurately determine the conductor losses in a microstrip resonator. T. C. Edwards has developed a set of simplified expressions for the conductor losses [5]. Equation (4-16) is an approximation of the conductor losses that treats the transmission line as a perfectly smooth surface.

$$ \alpha_c = 0.072 \frac{\sqrt{f}}{W_e Z_o} \lambda_g \qquad \text{dB/inch} \qquad (4\text{-}16) $$

Where,

  f = the frequency in GHz

  W_e = the effective conductor width (inches)

  Z_o = the characteristic impedance of the line

  α_c = Conductor loss in dB/inch

  λ_g = wavelength in dielectric in inches

A microstrip conductor is actually not perfectly smooth but exhibits a certain roughness. The surface roughness exists on the bottom of the microstrip conductor where it contacts the dielectric. This can be seen by magnifying the cross section of a microstrip line's contact with the dielectric material. The surface roughness is usually specified as an r.m.s. value.

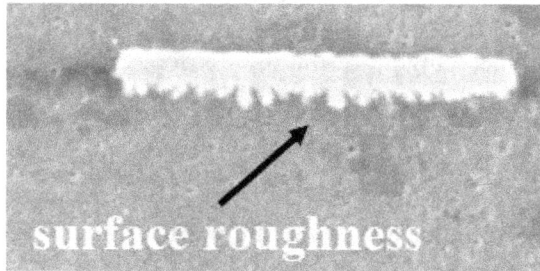

Figure 4-15 Cross section of microstrip line showing surface roughness at the conductor to dielectric interface (courtesy of Tektronix)

Edwards modified Equation (4-16) to include the effects of the surface roughness as given by Equation (4-17).

$$\alpha_c' = \alpha_c \left[1 + \frac{2}{\pi} \tan^{-1}\left(1.4 \left(\frac{\Delta}{\delta_s} \right)^2 \right) \right] \quad \text{dB/inch} \qquad (4\text{-}17)$$

Where, Δ is the surface roughness and δ_s is the conductor skin depth.

The corresponding Q factor related to the conductor is then given by:

$$Q_c = \frac{27.3\sqrt{\varepsilon_{eff}}}{\alpha_c \lambda_o} \qquad (4\text{-}18)$$

The dielectric loss is determined by the dielectric constant and loss tangent. It is calculated using Equation (4-19).

$$\alpha_d = 27.3 \frac{\varepsilon_r \left(\varepsilon_{eff} - 1 \right) \tan \delta}{\sqrt{\varepsilon_{reff}} \left(\varepsilon_r - 1 \right) \lambda_o} \quad \text{dB/inch} \qquad (4\text{-}19)$$

Where:

ε_r = the substrate dielectric constant

ε_{reff} = the effective dielectric constant

$\tan\delta$ = the loss tangent of the dielectric

λ_o = wavelength in inches

The corresponding Q factor due to the dielectric is then given by:

$$Q_d = 27.3 \frac{\sqrt{\varepsilon_{eff}}}{\alpha_d \lambda_o} \qquad (4\text{-}20)$$

We know that a microstrip line will also have some radiation of energy from the top side of the line. The open circuit stub will also experience some radiation effect from the open circuited end. On low dielectric constant substrates, $\varepsilon_r \leq 4.0$, the radiation losses are more significant for high impedance lines. Conversely for high dielectric constant substrates, $\varepsilon_r \geq 10$, low impedance lines experience more radiation loss. The radiation Q factor is presented as Equation (4-21).

$$Q_r = \frac{Z_o(f)}{480\pi\left(\dfrac{h}{\lambda_o}\right)^2 \left\{\left[\left(\dfrac{\varepsilon_{eff(f)}+1}{\varepsilon_{eff(f)}}\right)\right] - \left[\dfrac{\left(\varepsilon_{eff(f)}-1\right)^2}{2\left(\varepsilon_{eff(f)}\right)^{\frac{3}{2}}} \ln\left(\dfrac{\sqrt{\varepsilon_{eff(f)}}+1}{\sqrt{\varepsilon_{eff(f)}}-1}\right)\right]\right\}} \qquad (4\text{-}21)$$

Where, h = substrate thickness in cm.

Note that in Equation (4-21) the line impedance and effective dielectric constant are defined as functions of frequency. This includes the dispersion or frequency dependent effect of Z_o and ε_{eff}. Dispersion tends to slightly increase the ε_{eff} as the frequency increases. This dispersive Z_o and ε_{eff} are given in Equations (4-22) and (4-23).

$$\varepsilon_{eff(f)} = \cfrac{\varepsilon_r - \varepsilon_{eff}}{1 + \left[(0.6 + 0.009Z_o) \left(\cfrac{f}{Z_o / 8\pi(h - 2t)} \right)^2 \right]} \qquad (4\text{-}22)$$

Where,

h = substrate thickness in mils

t = conductor thickness in mils

$$Z_{o(f)} = Z_o \sqrt{\cfrac{\varepsilon_{eff}}{\varepsilon_{eff(f)}}} \qquad (4\text{-}23)$$

Finally the resultant overall unloaded Q factor, Q_u, of the microstrip line can be determined by the reciprocal relationship of Equation (4.24).

$$\frac{1}{Q_u} = \frac{1}{Q_c} + \frac{1}{Q_d} + \frac{1}{Q_r} \qquad (4\text{-}24)$$

Microstrip Resonators

Example 4.8: Consider a 5 GHz half wavelength open circuit microstrip resonator. The resonator is realized with a 50 Ω microstrip line on Roger's RO3003 dielectric. Calculate the unloaded Q factor of the resonator. The substrate parameters are defined as:

Dielectric constant	$\varepsilon_r = 3$
Substrate height	h = 0.030 in.
Conductor thickness	t = .0026 in.
Line Impedance	$Z_o = 50\ \Omega$
Conductor width	w = 0.075 in.
Loss tangent	$\tan\delta = 0.0013$

Solution: Using the simplified expression of Equation (4-16) for a smooth microstrip line the loss and conductor Q factor is calculated as:

$$\alpha_c = 0.072 \frac{\sqrt{5}}{0.077(50)} \cdot 1.52 = 0.063 \ dB/inch$$

$$Q_c = \frac{27.3\sqrt{2.41}}{(0.063)(2.36)} = 283.1$$

The dielectric loss and Q factor are then calculated from Equation (4-19) and (4-20).

$$\alpha_d = 27.3 \frac{(3)(2.41-1)(0.0013)}{\sqrt{2.41}(3-1)(2.36)} = 0.021 dB/inch$$

$$Q_d = 27.3 \frac{\sqrt{2.41}}{(0.021)(2.36)} = 875.2$$

For simplicity the radiation Q factor will be omitted. We will model the resonator using the linear simulator. Linear simulators often do not model the radiation effects of the microstrip line. Therefore the overall unloaded Q factors then becomes.

$$\frac{1}{Q_u} = \frac{1}{283.1} + \frac{1}{875.2} = \frac{1}{213.9}$$

$$Q_u = 213.9$$

Microstrip Resonator Model

The 5 GHz half wave open circuit microstrip resonator is modeled in Fig 4-16. Note that the source and load impedance has been increased to 5000 Ohm to avoid loading the impedance of the parallel resonant circuit.

Example 4-9: The 5 GHz half wave open-circuited microstrip resonator is shown in Figure 4-16. Simulate the resonator and measure the loaded Q factor and the group delay at 5 GHz.

Solution: Schematic of the microstrip half wave resonator is shown in Figure 4-16.

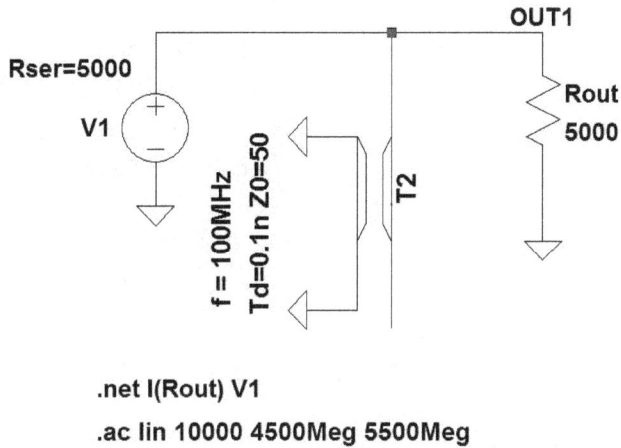

.net I(Rout) V1
.ac lin 10000 4500Meg 5500Meg

Figure 4-16 Half-wave open ended microstrip resonator

Simulate the schematic and display S21 in a rectangular plot, as shown in Figure 4-17.

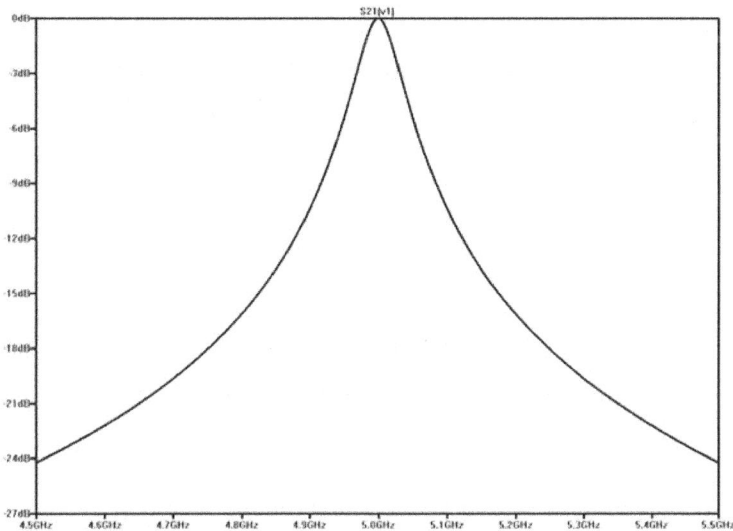

Figure 4-17 Response of the half-wave open circuit microstrip

For Q_L Calculation we need frequencies at 3 dB return loss.

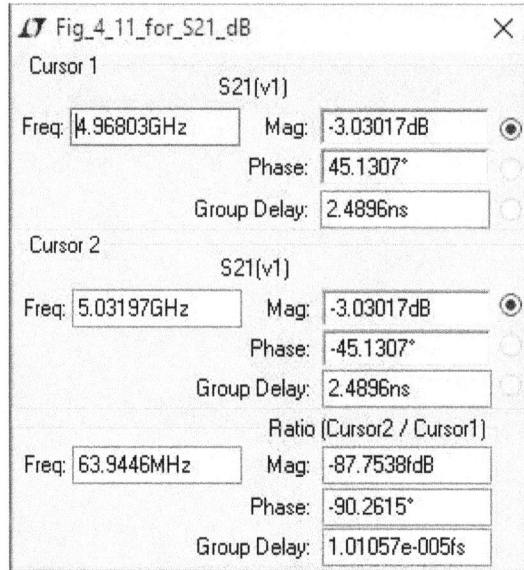

Figure 4-18: Data points at –3dB

Now we use Equation (4-11) to calculate the loaded Q factor of the half wave open-circuited microstrip resonator.

$$Q_L = (f_l * f_h)^{1/2} / BW = 4999/64 = 78.1$$

The insertion loss at the resonant frequency can be used to relate the loaded Q_L to the unloaded Q_u as shown by Equation (4-25).

$$InsertionLoss(dB) = 20\log\frac{Q_u}{Q_u - Q_L} \qquad (4\text{-}25)$$

It is also interesting to note that the loaded Q_L of the resonator is related to its group delay, as given in Equation (4-26).

$$Q_L = 2\pi f \left(\frac{t_d}{2}\right) \qquad (4\text{-}26)$$

Where:

  t_d is the group delay in seconds

  f is the frequency in Hz

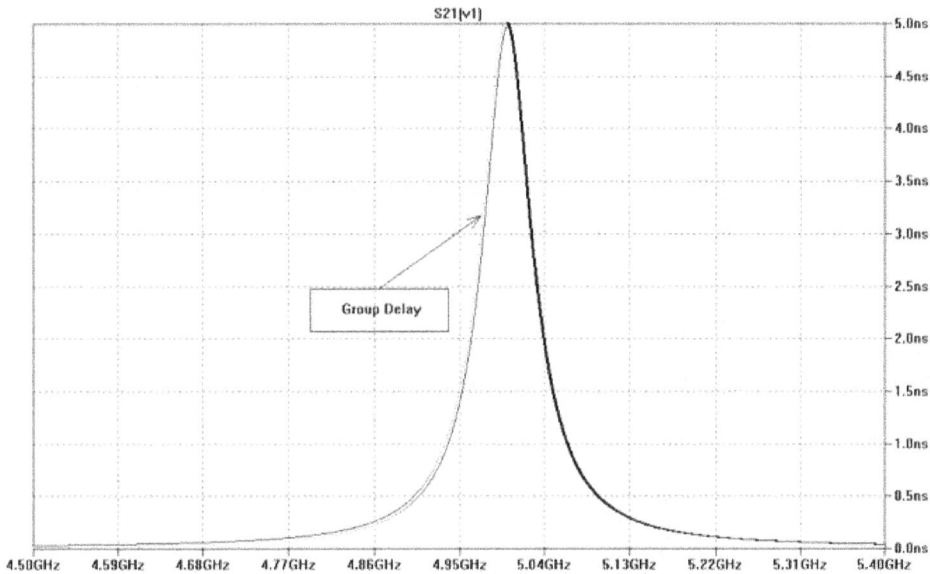

Figure 4-19 Group delay of the half-wave open circuited resonator

As Figure 4-19 shows the Group delay of the half-wave open-circuited microstrip resonator at 5 GHz is about 5 nanosecond.

4.6 Resonator Series Reactance Coupling

To reduce the loading on the half wave resonator of Figure 4-16, the source and load impedances of 5000 Ω were used. In practice the resonator is typically coupled to lower impedance circuits. If we attempt to examine the resonator on a network analyzer, most modern test equipment will have 50 Ω impedance levels. Such resonators are often coupled to the circuit by a highly reactive circuit element. This reactive element can be realized as a series capacitor or inductor. The resonator is then analyzed as a one port network.

Example 4-10: In the following capacitive coupled half-wave microstrip resonator change the value of the capacitor and find out when the resonator is critically coupled.

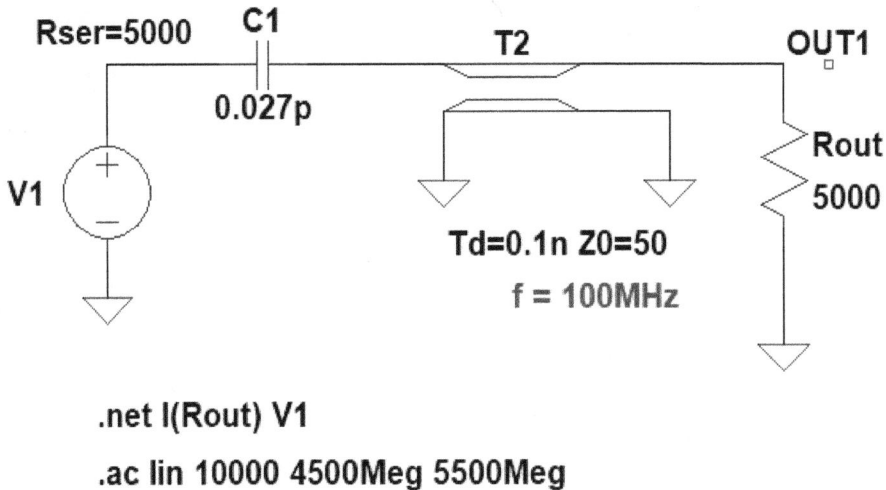

Rser=5000   C1   T2   OUT1

0.027p

V1   (+/-)

Rout
5000

Td=0.1n Z0=50
f = 100MHz

.net I(Rout) V1
.ac lin 10000 4500Meg 5500Meg

Figure 4-20 Capacitive coupled half-wave microstrip resonator

Solution: As the frequency is swept over a narrow frequency range around the resonant frequency, a circle is formed on the Smith Chart. This trace is known as the Q circle of the resonator. Figure 4-21 shows the scalar plot of S11 when the resonator is critically coupled The Smith Chart plot of Figure 4-22 shows three circles for different value of the coupling capacitance. We can see that as the coupling capacitance changes, the resonant frequency of the circuit also changes. The series capacitance acts to decrease the overall resonance frequency of the circuit. This new resonance frequency is known as the loaded resonance frequency. Because the series capacitance lowers the resonance frequency, the length of the resonator also decreases. With the coupling capacitance set at 0.027 pF the Q circle passes through the center of the Smith Chart at the resonant frequency. This is known as critical coupling and is characterized by having the lowest return loss on the scalar plot of Figure 4-21. The Smith Chart shows a larger Q circle which is a characteristic of an over coupled resonator. With the coupling capacitor is set to 0.027 pF the resonance frequency increases. The Smith Chart shows that the Q circle becomes much smaller thus under coupling the resonator.   As Figure 4-22 shows, the value of the coupling capacitor also has an impact on the size of the Q circle.

The diameter of the Q circle is dependent on the coupling of the resonator to the 50 = source.

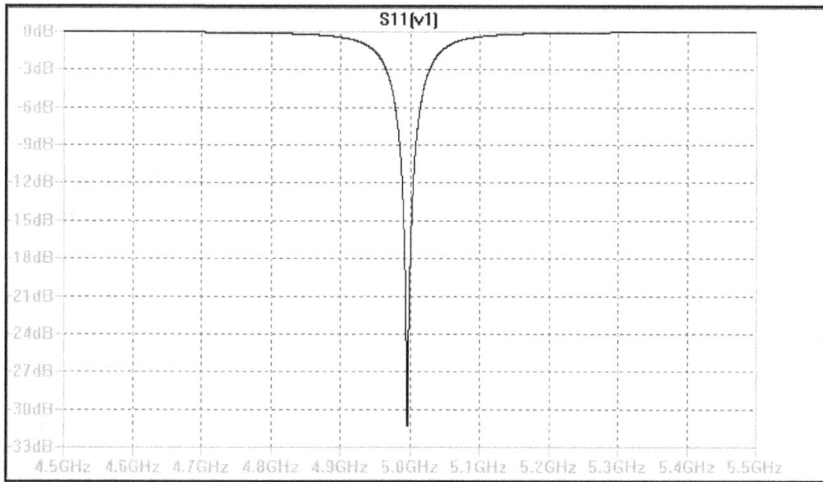

Figure 4-21 Capacitive critically coupled to half-wave microstrip resonator

The Smith chart plot at different capacitor value is shown in Figure 4-22,

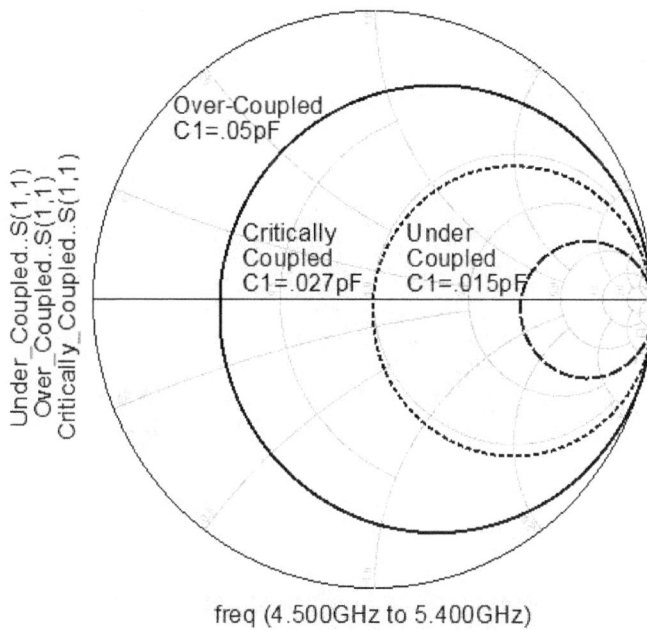

Figure 4-22 Capacitive coupled half-wave microstrip resonator

One Port Microwave Resonator Analysis

The microstrip half wave resonator was fairly easy to model and analyze in LTspice. Many microwave resonators are not as easy to model. High Q microwave resonators are often realized as metallic cavities or dielectric resonators for which there are no native models. The reactive coupling of the resonator to the circuit can be even more difficult to model. The coupling usually occurs by magnetic or electric coupling by a probe or loop inserted into the cavity. An E field probe coupled to a coaxial cavity resonator is shown in Figure 4-19. A dielectric resonator is coupled to a microstrip line by flux linkage in air as shown in Figure 4-20. Again there is no model in LTspice to directly model this coupling mechanism to the resonator. The designer is left to develop approximate models based on a lumped RLC equivalent models and couple the resonator to the circuit using an ideal transformer model. Linear simulation can still be of value in the design and evaluation process if we have a measured S parameter file of the resonator's reflection coefficient. Just as we have used S parameter models to represent capacitors and inductors we can also use the measured S parameters of a resonator. All modern vector network analyzers have the ability to save an S parameter data file for any measurement that can be made by the instrument. This section will show how we can analyze the S - parameter file of a microwave resonator.

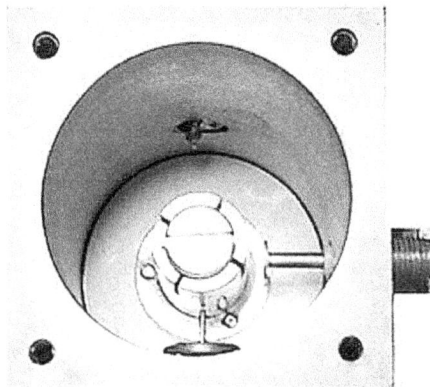

Figure 4-23: Cavity with E field probe coupled to center conductor

The coupling of microwave resonators is often characterized by a coupling coefficient k. The coupling coefficient is the ratio of the power dissipated in the load to the power dissipated in the resonator [5].

$$k = \frac{P_{load}}{P_{resonator}} = \frac{Q_o}{Q_{ext}} \qquad (4\text{-}27)$$

Where:

Q_o is the unloaded Q of the resonator

Q_{ext} is the external Q of the resonator

When P_{load} is equal to $P_{resonator}$, k = 1 and the critical coupling case exists. Substituting the reciprocal Q factor relationship of Equation (4-10) into Equation (4-27) we can relate the coupling coefficient to the loaded Q_L and unloaded Q_o of the resonator.

$$Q_L = \frac{Q_o}{1+k} \qquad (4\text{-}28)$$

Figure 4-24 Dielectric resonator coupled to microstrip transmission line

Smith Chart Q_O Measurement of Microstrip Resonator

Now the unloaded Q_o of the resonator can be calculated if the Q_L and k can be measured. Kajfez[8] has described a technique to extract the coupling coefficient k and Q_L values from the Q circle of the resonator. Consider the Q circle on the Smith Chart of Figure 4-25. A line that is projected from the center of the Smith Chart to intersect the Q circle with minimum length will intersect the circle at the loaded resonance frequency, f_L. The length of this vector is labeled as $|\Gamma_L|$. As the line projects along a path of the diameter of the circle it intersects the circle near the circumference of the Smith Chart at a point defined as $|\Gamma_d|$. The input reflection coefficient of the Q circle can be defined using the following empirical equation [5].

$$\Gamma_i = \Gamma_d \left| 1 - \frac{2k}{1+k} \cdot \frac{1}{1 + jQ_L 2 \dfrac{\omega - \omega_L}{\omega_o}} \right| \qquad (4\text{-}29)$$

Lines that are projected from Γ_d through the Q circle at the angles $\pm\phi$ are related to the loaded Q by Equation (4-30).

$$Q_L = \frac{f_L}{f_1 - f_2} \tan\phi \qquad (4\text{-}30)$$

If we set $\phi = 45°$ then Equation (4-30) reduces to the straightforward definition of Q_L given by Equation (4-31).

$$Q_L = \frac{f_L}{f_1 - f_2} \qquad (4\text{-}31)$$

In the previous section we have seen that the diameter of the Q circle was directly related to the coupling coefficient. The diameter of the circle can be easily measured from:

$$\left| \Gamma_d \right| - \left| \Gamma_L \right| = d \qquad (4\text{-}32)$$

The coupling coefficient is then derived from the diameter of the Q circle.

$$k = \frac{d}{2 - d} \qquad (4\text{-}33)$$

Finally the unloaded resonator Q_o is then calculated from Equation (4-28). We can also find the unloaded resonance frequency directly from the Q circle. Follow the reactive line on the Smith Chart that intersects the Q circle at Γ_d to the next Q circle intersection. The frequency at this Q circle intersection is the unloaded resonance frequency, f_o.

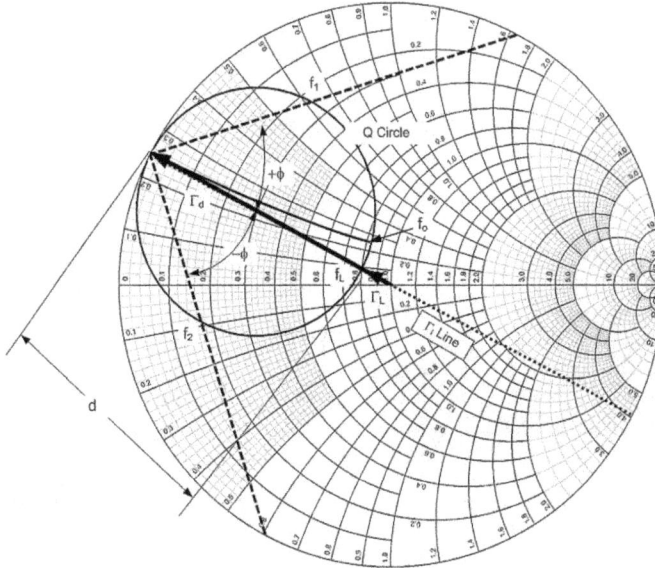

Figure 4-25 Resonator Q measurement from the resonator Q circle

4.7 Filter Design at RF and Microwave Frequency

In Section 4.3 we have seen that it is possible to change the shape of the frequency response of a parallel resonant circuit by choosing different source and load impedance values. Likewise multiple resonators can be coupled to one another and to the source and load to achieve various frequency shaping responses. These networks are referred to as filters.

Filter Topology

The subject of filter design is a complex topic and the subject of many dedicated texts. This section is intended to serve as a fundamental primer to this vast topic. It is also intended to set a foundation for successful filter design using the LTspice software. The four most popular filter types are: Low Pass, High Pass, Band Pass, and Band Stop. The basic transmission response of the filter types is shown in Figure 4-26.

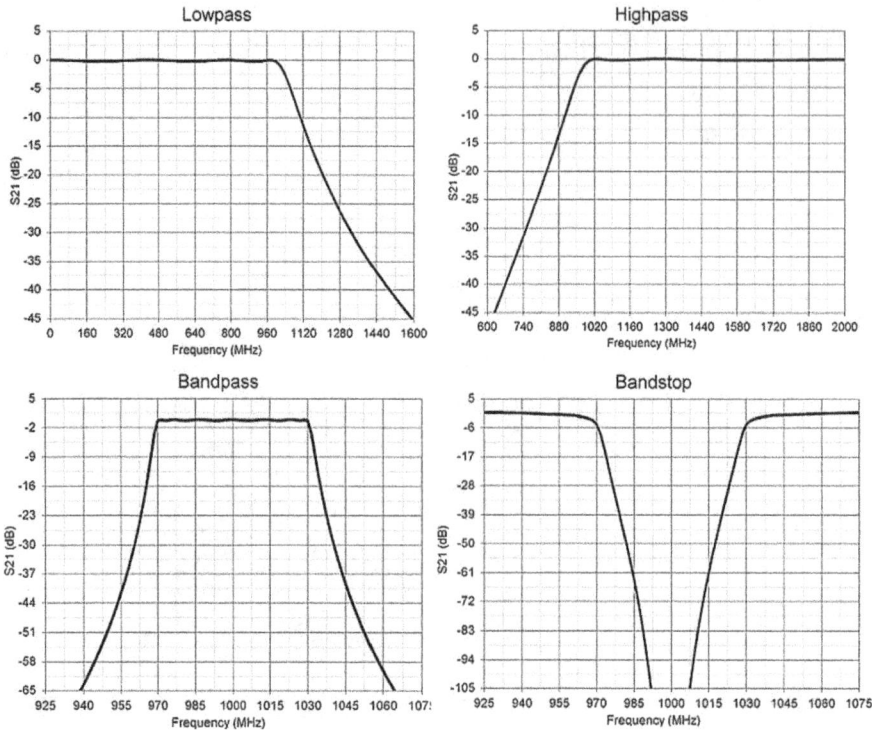

Figure 4-26 Transmission (S21) versus frequency characteristic

The filters allow RF energy to pass through their designed pass band. RF energy that is present outside of the pass band is reflected back toward the source and not transmitted to the load. The amount of energy present at the load is defined by the S21 response. The amount of energy reflected back to the source is characterized by the S11 response.

Filter Order

The design process for all of the major filter types is based on determination of the filter pass band, and the attenuation in the reject band. The attenuation in the reject band that is required by a filter largely determines the slope needed in the transmission frequency response. The slope of the filter's response is related to the order of the filter. The steeper the slope or 'skirt' of the filter; the higher is the order. The term order comes from the mathematical transfer function that describes a particular filter. The highest power of s in the denominator of the filter's Laplace transfer function is the order of the filter. For the simple low pass and high pass filters presented in this chapter the filter order is the same as the number of elements in the filter. However this is not the case for general filter networks. In more complex types of filters as well as bandpass and bandstop filters the filter order will not be equal to the number of elements in the filter. In general the filter order is the total of the number of transmission zeros at frequencies:

$$F = 0 \text{ (DC)}$$
$$F = \infty$$
$$0 < F < \infty \text{ (specific frequencies between DC and } \infty)$$

Transmission zeros block the transfer of energy from the source to the load. In fact the order of a filter network can be solved visually by adding up the number of transmission zeros that satisfy the above criteria. Figure 4-27 shows the relationship between the filter order and slope of the response for a Low Pass filter. Each filter of Figure 4-27 has the same cutoff frequency of 1000 MHz. The third order filter has an attenuation of about 16 dB at a rejection frequency of 2000 MHz. The fifth order filter shows an attenuation of 39 dB and the seventh order filter has more than 61 dB attenuation at 2000 MHz. It is therefore clear that the order of the filter is one of the first criteria to be determined in the filter design. It is dependent on the cutoff frequency of the pass band and the amount of attenuation desired at the rejection frequency.

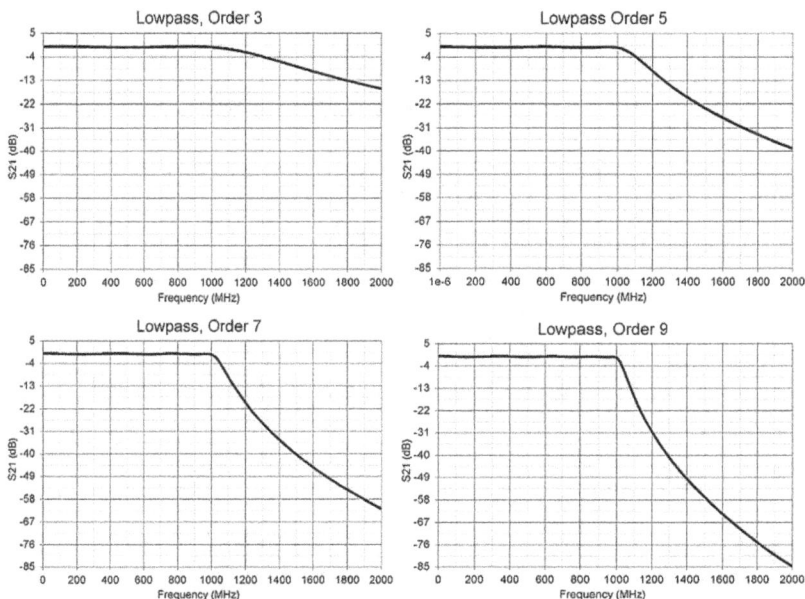

Figure 4-27 Relationship between filter order and the slope of S21

Filter Type

The shape of the filter passband and attenuation skirt can take on different shape relationships based on the coupling among the various reactive elements in the filter. Over the years several polynomial expressions have been developed for these shape relationships. Named after their inventors, some of the more popular passive filter types include: Bessel, Butterworth, Chebyshev, and Cauer. Figure 4-28 shows the general shape relationship among these filter types for a given seventh order filter. The Bessel filter type is a low Q filter and does not exhibit a steep roll off compared to its counterparts. The benefit of the Bessel filter is its linear phase or flat group delay response. This means that the Bessel filter can pass wideband signals while introducing little distortion. The Butterworth is a medium Q filter that has the flattest pass band of the group. The Chebyshev response is a higher Q filter and has a noticeably steeper skirt moving toward the reject band.

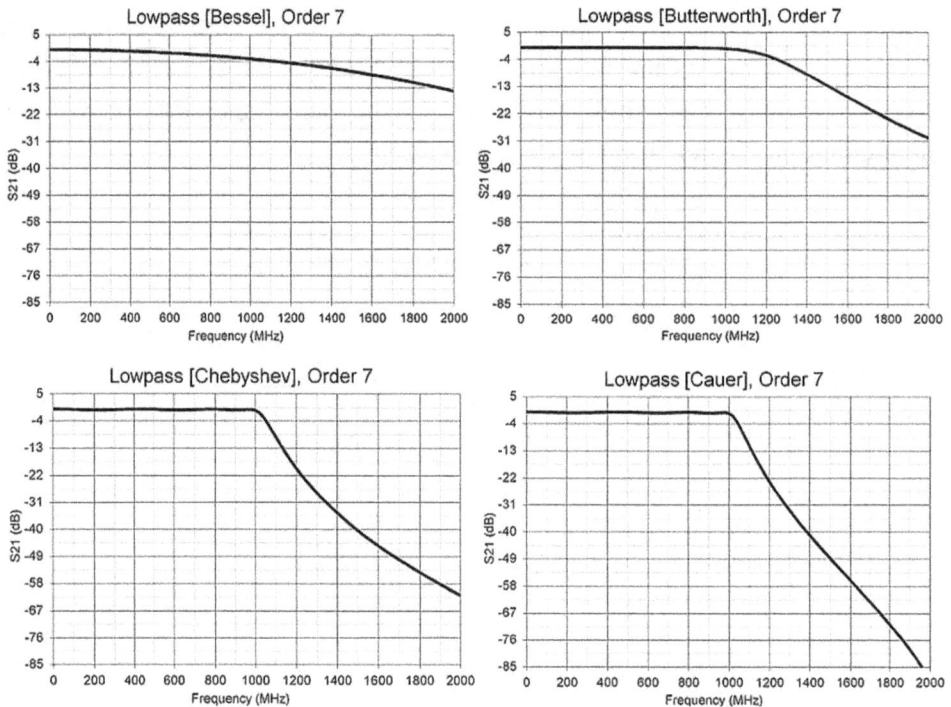

Figure 4-28 General characteristic shape of Bessel, Butterworth, Chebyshev, and Cauer filters

As a result it exhibits more transmission ripple in the pass band. The Cauer filter has the steepest slope of all of the four filter types. The Cauer filter is also known as an elliptic filter. Odd order Chebyshev and Cauer filters can be designed to have an equal source and load impedance. The even order Chebyshev and Cauer filters will have a different output impedance from the specified input impedance. Another interesting characteristic of the Cauer filter is that it has the same ripple in the rejection band as it has in the pass band. The Butterworth, Chebyshev and Cauer filters differ from the Bessel filter in their phase response. The phase response is very nonlinear across the pass band. This nonlinearity of the phase creates a varying group delay. The group delay introduces varying time delays to wideband signals which, in turn, can cause distortion to the signal. Group delay is simply the derivative, or slope, of the transmission phase and defined by Equation (4-34). Figure 4-29 shows the respective set of filter transmission characteristics with their corresponding group delay. Note the relative values of the group delay on the right hand axis.

$$\tau_g = -\frac{d\phi}{d\omega}$$
(4-34)

Where, ϕ is the phase shift in radians and ω is the frequency in radians per second.

From the group delay plots of Figure 4-29 it is evident that the group delay peaks near the corner frequency of the filter response. The sharper cutoff characteristic results in greater group delay at the band edge.

Filter Return Loss and Passband Ripple

The Bessel and Butterworth filters have a smooth transition between their cutoff frequency and rejection frequency. The forward transmission, S21, is very flat vs. frequency. The Chebyshev and Cauer filters have a more abrupt transition between their cutoff and rejection frequencies. This makes these filter types very popular for many filter applications encountered in RF and microwave engineering. It is important to note however that the steeper filter skirt results as a certain amount of impedance mismatch between the source and load impedance.

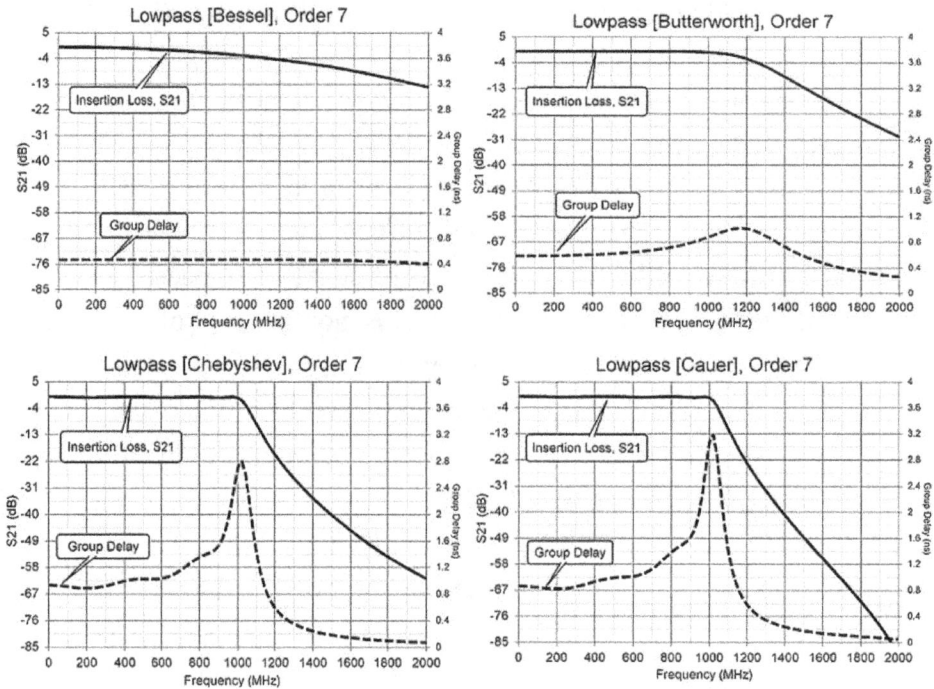

Figure 4-29 Group delay characteristic for various lowpass filter types

The Chebyshev and Cauer filter types have ripple in the forward transmission path, S21. The amount of ripple is caused by the degree of mismatch between the source and load impedance and thus the resulting return loss that is realized by these filter types. For a given Chebyshev or Cauer filter order, the roll off of the filter response is also steeper for greater values of passband ripple. The cutoff frequency of the filters that have passband ripple is then defined as the passband ripple value. For all-pole filters such as the Butterworth, the cutoff frequency is typically defined as the 3 dB rejection point. Figure 4-30 shows the passband ripple of a fifth order low pass filter for ripple values of 0.01, 0.1, 0.25, and 0.5 dB. Note that the ripple shown is produced by ideal circuit elements. In practice the finite unloaded Q or losses in the inductors and capacitors will tend to smooth out this ripple.

Lowpass [Chebyshev], Order 5, Ripple=0.01 dB

Lowpass [Chebyshev], Order 5, Ripple=0.1 dB

Lowpass [Chebyshev], Order 5, Ripple=0.25 dB

Lowpass [Chebyshev], Order 5, Ripple=0.5 dB

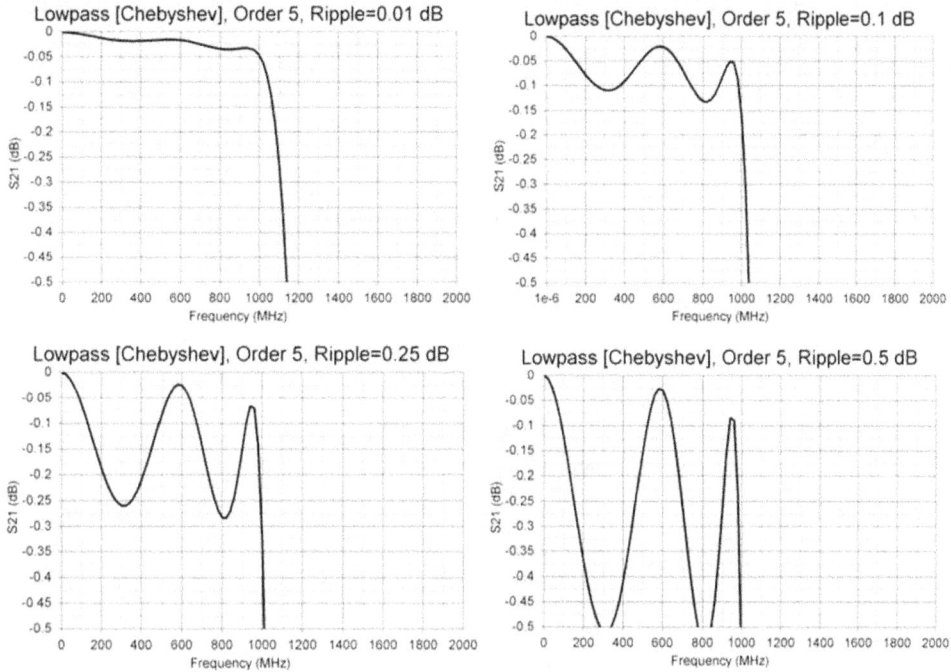

Figure 4-30 Passband ripple values in lowpass Chebyshev filter

In Chapter 2 the relationship for mismatch loss between a source and load was presented. For the Chebyshev and Cauer filters this mismatch loss is the passband ripple.

vswr ( )	ReturnLoss_dB ( )	Ripple_dB ( )
1.1	26.444	0.01
1.239	19.433	0.05
1.3	17.692	0.075
1.355	16.435	0.1
1.405	15.473	0.125
1.452	14.688	0.15
1.538	13.474	0.2
1.62	12.518	0.25
1.984	9.636	0.5

Table 4-1 Calculation of filter ripple versus VSWR

Figure 4-31 shows the same filters from Figure 4-30 with the return loss plotted along with the insertion loss, S21. We can see that for a given filter order, there is a tradeoff between filter rejection and the amount of ripple, or return loss, that can be tolerated in the passband. In most RF and microwave filter designs the 0.01 and 0.1 dB ripple values tend to be more popular. This is due to the trade-off between good impedance match and reasonable filter skirt slope. Figure 4-31 shows good correlation of the worst case return loss with that which is calculated in Figure 4-30. When tuning filters using modern network analyzers it is sometimes easier to see the larger changes in the return loss as opposed to the fine grain ripple as shown in Figure 4-29. For this reason it is common to tune the forward transmission of the filter by observing the level and response of the filter's return loss. Return loss is a very sensitive indicator of the filter's alignment and performance.

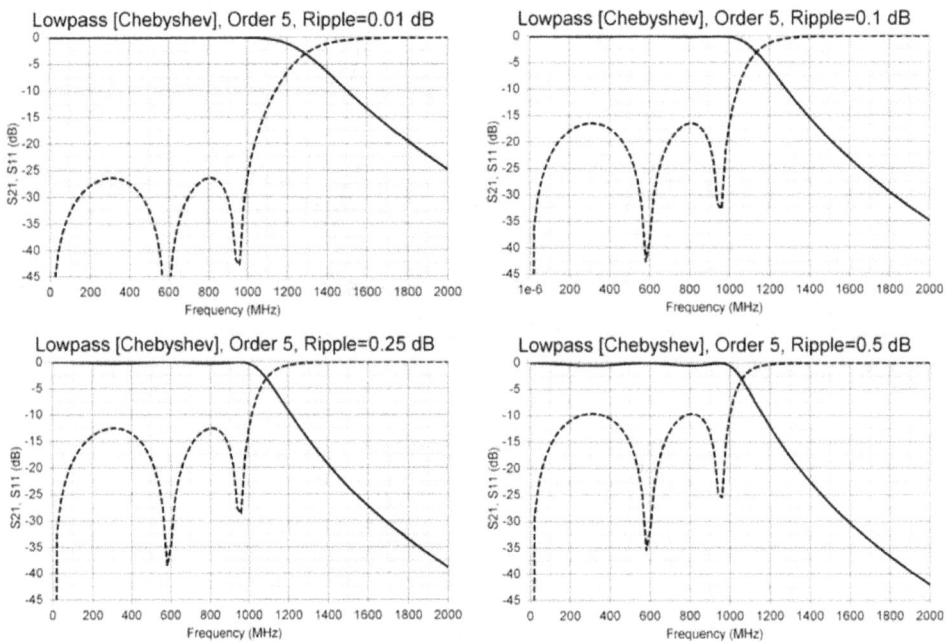

Figure 4-31 Lowpass Chebyshev filter rejection and return loss

4.8 Lumped Element Filter Design

Classical filter design is based on extracting a prototype frequency-normalized model from a myriad of tables for every filter type and order.

Fortunately these tables have been built into many filter synthesis software applications that are readily available. We will work through two practical filter examples, one low pass and one high pass filter.

Low Pass Filter Design

Example 4.11: As a practical filter design, consider a full duplex communication link (simultaneous reception and transmission) through a satellite with the following requirements:

- The uplink signal is around 145 MHz and the downlink is at 435 MHz.

- A 20 W power amplifier is used on the uplink with 25 dB gain.

- To have a low pass filter on the uplink to pass the 145 MHz uplink signal while rejecting any noise power in the 435 MHz band.

- It is necessary to provide a high pass filter on the downlink so that the 435 MHz downlink signal is received while rejecting any noise power at 145 MHz.

The transmitter and receiver antennas are on the same physical support boom so there is limited isolation between the transmitter and receiver. Even though the signals are at different frequencies, the broadband noise amplified by the power amplifier at 435 MHz will be received by the UHF antenna and sent to the sensitive receiver. Because the receiver is trying to detect very low signal levels, the received noise from the amplifier will interfere or 'de-sense' the received signals. Therefore it is necessary to design a 145 MHz Low Pass filter for this satellite link system. The specifications chosen for the filter design are selected as:

- Select a Chebyshev Response with 0.1 dB pass band ripple.

- Set the passband cutoff frequency (not the -3 dB frequency at 160 MHz

- The reject requirement is at least -40 dB rejection at 435 MHz.

Solution: Use the Passive Filter Synthesis Utility[3] to design the Low Pass filter. Select a filter of the Chebyshev type. On the Topology tab select a Lowpass filter type with a Chebyshev shape. Select the minimum capacitor subtype. On the Settings tab enter the cutoff frequency of 160

MHz and pass band ripple of 0.1 dB. Also set the cutoff frequency attenuation at 0.1 dB. The Filter Settings Tab is a great place to perform "what-if" analysis. The pass band ripple, filter order, cutoff frequency, and attenuation at cutoff can all be varied while observing their impact on the filter's characteristic. For the design example enter the parameters as shown in Fig. 4-30. The required filter order can be determined by increasing the order until the specification of -40 dBc attenuation at 435 MHz is achieved. As the filter response curve in Fig.4-30 shows, this Low Pass filter must be of fifth order to achieve the required attenuation. Along with the resulting attenuation and return loss the synthesis program creates the filter schematic with the synthesized component values.

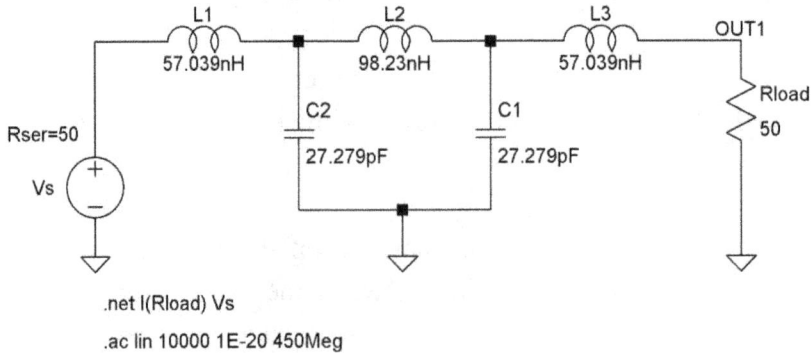

Figure 4-32 S parameter simulation of the passive filter

The response of the passive filter is shown in Figure 4-33.

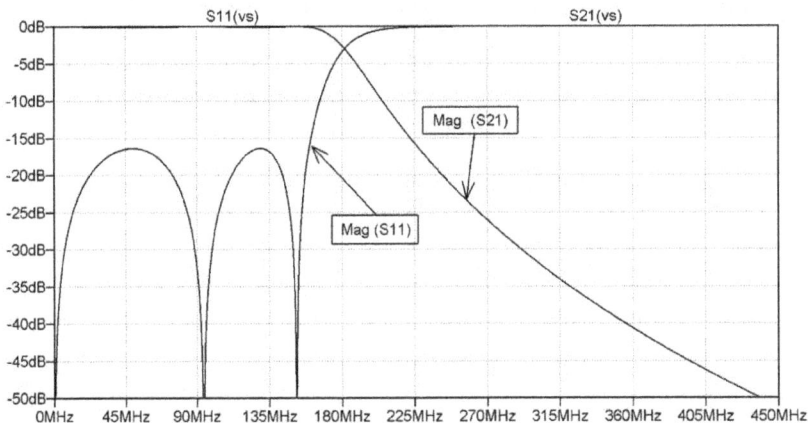

Figure 4-33 Response of the passive filter

Physical Model of the Low Pass Filter

The synthesized filter is an ideal design in the sense that ideal components have been used. To obtain a good 'real-world' simulation of the filter we need to use component models that have finite Q and parasitics such as multilayer chip capacitors. For power handling capability, choose the 700 series chip capacitors from ATC Corporation. We will use measured S parameter files to model the shunt capacitors. The measured S parameters will account for any package parasitic effects and the finite component Q factor. Most microwave chip capacitor manufacturers will supply S parameters for their products. ATC Corporation has a useful application for selection of chip capacitors called ATC Tech Select. This program is available from the ATC web site: www.atceramics.com Using the Tech Select program the engineer can access complete data sheets and other useful information including current and voltage handling capabilities of the various capacitors. Because the filter is passing relatively high power (20 W), we cannot use small surface mount style chip inductors. Instead we will use air wound coils to realize the series inductors. The inductors will be realized with AWG#16 wire nickel-tin plated copper wire. They will be wound on a 0.141 inch diameter form. Use the techniques covered in Chapter one, to design the inductors using the Air Wound inductor model. The filter model is then reconstructed using the S parameter files for the shunt capacitors and the physical inductor models for the series inductors. Make sure to model the substrate and the interconnecting printed circuit board traces as microstrip lines. Also model the ground connection of the shunt capacitors as a microstrip via hole. Although these PCB parasitic effects are normally more pronounced at frequencies above 2 GHz, it is often surprising the effect that these parasitics have at lower frequencies.

The final filter response is shown in Figure 4 -36. The response shows that the attenuation specification has been achieved. Because the circuit has physical models replacing the ideal lumped elements, the engineer can be confident that the filter can be assembled and will achieve the designed response. Figure 4-34 is a photo of the prototype low pass filter circuit with SMA coaxial connectors attached to the circuit board.

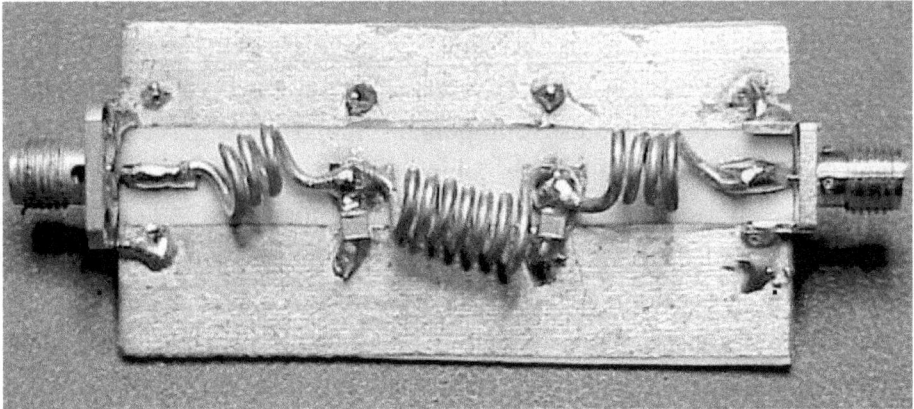

Figure 4-34 Physical prototype of the 146 MHz low pass filter
(Courtesy of BT Microwave)

High Pass Filter Design

Example 4.12: Design a high pass filter that passes frequencies in the 420 MHz to 450 MHz range. This filter could be placed in front of the preamp used in the downlink of the satellite system. This would help to keep out any of the transmit energy or noise power in the 146 MHz transmit frequency range. The High Pass Filter specifications are:

The pass band cutoff frequency (not the -3 dB point) is 420 MHz.

The filter has a Chebyshev response with 0.1 dB pass band ripple.

The reject requirement is at least -60 dB rejection at 146 MHz.

Solution: It can be shown that the 7th order high pass filter of Figure 4-35 will meet the requirements of Example 4.12.

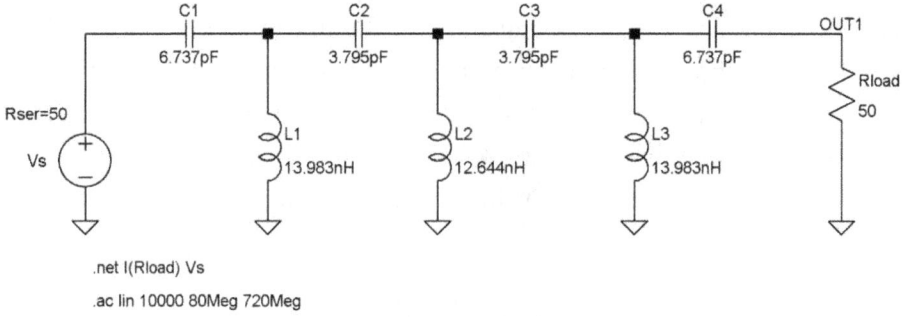

Figure 4-35 Passive model of the high pass filter

The simulated response of the passive filter is shown in Figure 4-36.

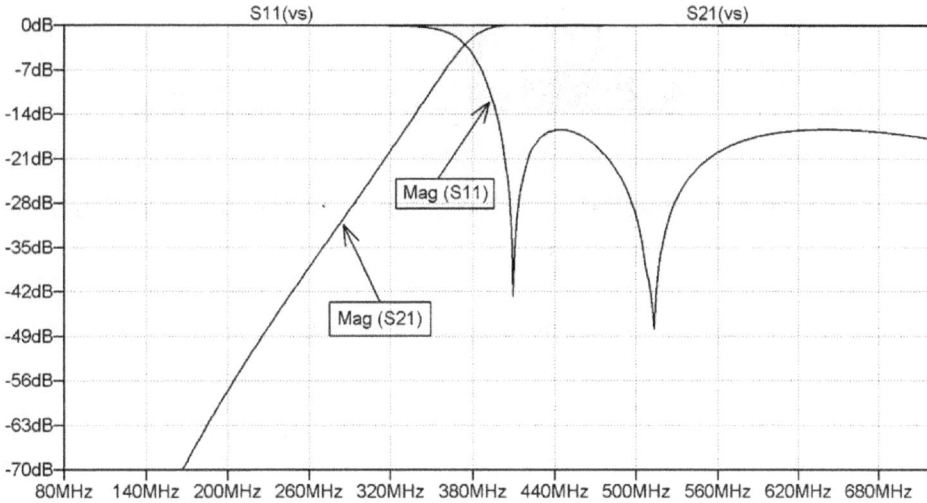

Figure 4-36 Response of the high pass filter

Physical Model of the High Pass Filter

Using the same techniques as described for the Low Pass Filter we can proceed with the High Pass Filter realization. The Passive Filter Design application calculated series capacitance values of 6.7 pF and 3.8 pF. Looking through the available ATC 700 series chip capacitors, the nearest values are 6.8 pF and 3.9 pF capacitors. We will select the S parameters files for these chip capacitors to use in the final filter model. The shunt inductors are realized using the Air Wound inductor model. Figures 4-35 and 4-36 show the final circuit model and filter response. From the response we can see that the required rejection specification of S21 < -60 dBc has been maintained. Figure 4-37 shows the completed assembly of the High Pass Filter on a printed circuit board with coaxial SMA connectors.

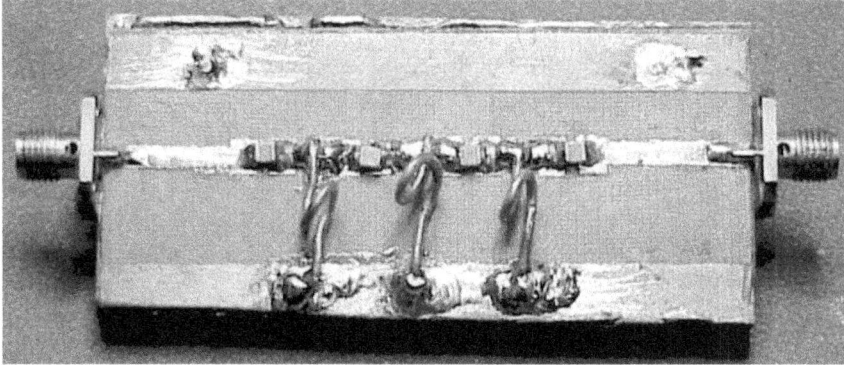

Figure 4-37 Physical prototype of the 420 MHz high pass filter
(Courtesy of BT Microwave)

Tuning the High Pass Filter Response

In many cases we would like to tune the component values to change or tweak the filter's characteristic to fit the desired response. The shunt inductors in the high pass filter are very easy to tune because they are modeled with the native inductor model. To tune the inductors simply check the tune box on the properties tab of the parameter to be tuned. Enable tuning of the inductor length of all three inductors. Using the Tune Window the length of each inductor can be increased or decreased by the step size or percentage selected. Each time the value is changed the analysis is run and the graph is updated. However the capacitors cannot be tuned because their physical model is based on an S parameter file that describes the capacitor's physical model. To change to another capacitor's S parameter file, we must edit the data file and browse to select a new S parameter file. Then we must sweep the circuit to observe the new response.

This process involves several steps and we lose the 'real time' sense of tuning the capacitor values and seeing the response change quickly.

4.9 Distributed Filter Design

Microstrip Stepped Impedance Low Pass Filter Design

In the microwave frequency region filters can be designed using distributed transmission lines. Series inductors and shunt capacitors can be realized with microstrip transmission lines. In the next section we will explore the conversion of a lumped element low pass filter to a design that is realized entirely in microstrip.

Example 4.13: Consider the lumped element 2 GHz low pass filter schematic and response shown in Figure 4-38. This low pass filter has a 3 dB bandwidth of approximately 2485 MHz. Use the microstrip equivalent models of series inductance and shunt capacitance to realize the filter in microstrip. The microstrip substrate is Rogers's 6010 material with a 0.025 dielectric thickness.

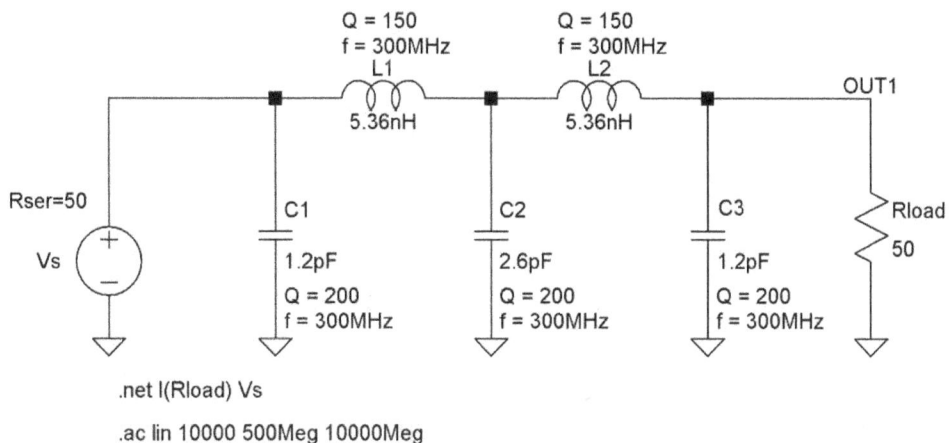

Figure 4-38 Lumped element 2.2 GHz low pass filter

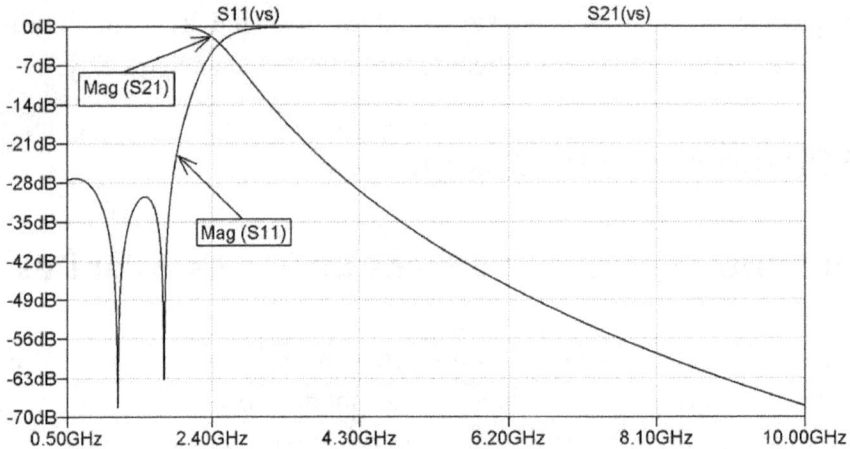

Figure 4-39 Lumped element 2.2 GHz low pass filter response

Lumped Element to Distributed Element Conversion

Example 4.14: Convert the ideal lumped low pass filter in Figure 4-38 to a realizable distributed element filter.

Solution: The series inductors will be realized as 80 Ω transmission lines of sufficient length to act as 5.36 nH inductors. The APCAD software program is used to calculate the microstrip line width for an 80 Ohm transmission line on the Rogers 0.025 inch RO3010 material. As Figure 4-41 shows the 80 Ω microstrip line has a line width of 6.26 mils with an effective dielectric constant of 6.079. To realize the required inductance a specific length of 80 Ω transmission line is required. Equation (2- 69) in chapter 2 is used to calculate the length of the microstrip line required to realize the inductance. The length of the required microstrip line for a 5.36 nH inductor is then found to be 321 mils. The shunt capacitors will be realized as 20 Ω transmission lines. Using APCAD the 20 Ω line width is shown in Figure 4-40 to be 102.8 mils with an effective dielectric constant of 7.902. Equation (2-70) in chapter 2 is used to find the length of 20 Ω transmission lines. The calculation shows that the required length for the 2.6 pF shunt capacitor is 217 mils. The microstrip line length for the 1.2 pF capacitors is then found to be 100 mils. Finally by converting the electrical lengths to time delays the schematic in Figure 4-42 is obtained.

Microstrip

W → 102.8
H
25
εr
T 0.7

Calculate Z0 [F4]

100
L

Z0 = 20.01 Ω

Elect Length = 0.024   λ
Elect Length = 8.6   degrees ▼
Elect Length = 281.102   mil (Air Line equiv.)
Delay = 23.816   ps
1.0 Wavelength = 4198.785   mil
Vp = 0.356   fraction of c
ε eff = 7.902
W/H = 4.112

Dielectric: ε r = 10.2

RT/duroid ® 6010LM ▼

Frequency: 1   GHz ▼

Length Units: mils ▼

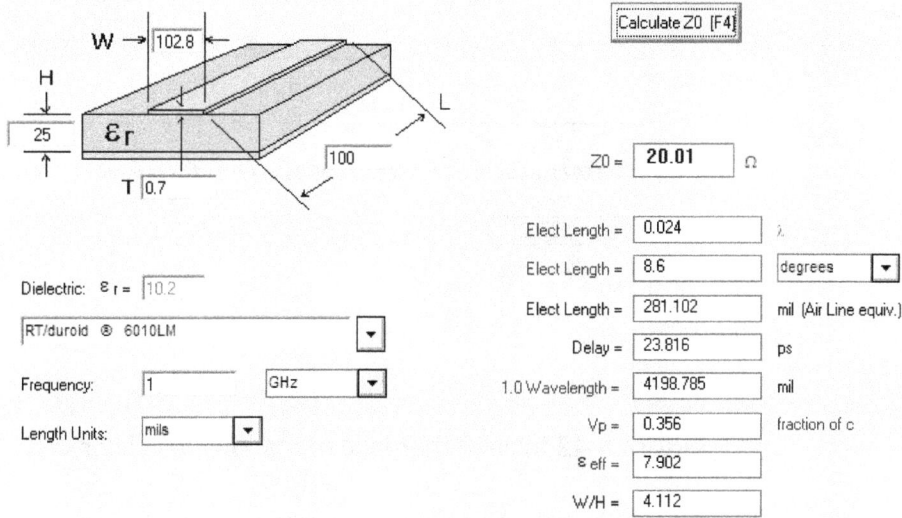

Figure 4-40 Calculations of 20 Ω microstrip line

Calculation of 80 Ohm microstrip width for the inductor is shown in Figure 4=41.

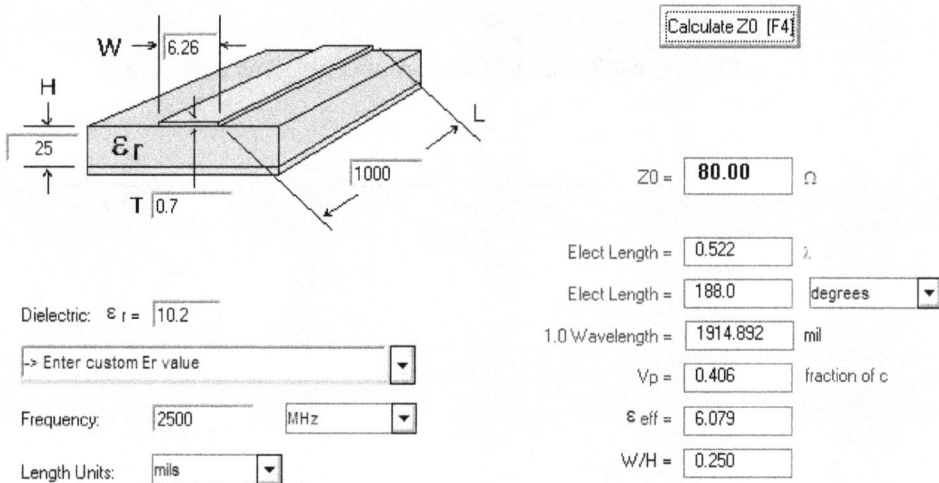

W → 6.26
H
25
εr
T 0.7

Calculate Z0 [F4]

1000
L

Z0 = 80.00 Ω

Elect Length = 0.522   λ
Elect Length = 188.0   degrees ▼
1.0 Wavelength = 1914.892   mil
Vp = 0.406   fraction of c
ε eff = 6.079
W/H = 0.250

Dielectric: ε r = 10.2

-> Enter custom Er value ▼

Frequency: 2500   MHz ▼

Length Units: mils ▼

Figure 4-41 Calculations of the 80 Ω microstrip line

Adding a short 50 Ω section to the input and output, the low pass filter response is shown in Figure 4-44.

Td=0.011n Z0=47.65 Td=.024n Z0=20.01 Td=0.06711n Z0=79.5 Td=0.052n Z0=20.01

T1   T2   T3   T4

Rser=50

V1

Td=0.06711n Z0=79.5 Td=0.024n Z0=20.01   Td=0.011n Z0=47.65

T5   T6   T7   OUT1

Rout
50

.net I(Rout) V1

.ac lin 10000 500Meg 10000Meg

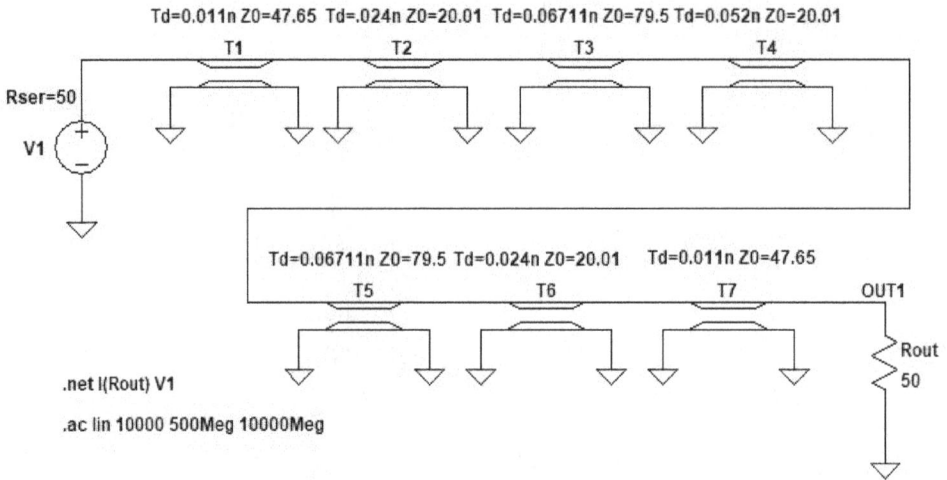

Figure 4-42 Initial Schematic of the low pass filter

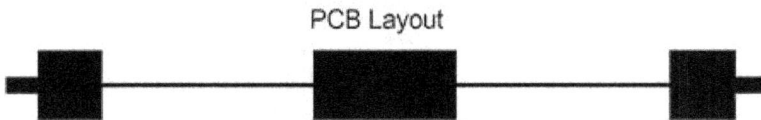

PCB Layout

Figure 4-43 Initial PCB layout of the low pass filter

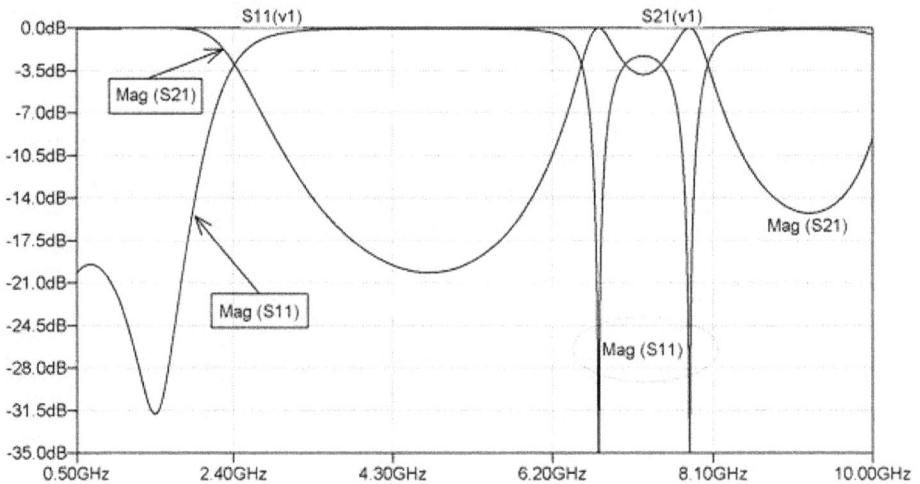

Figure 4-44 Initial response of the low pass filter

Examine the printed circuit board layout of the Low Pass Filter of Figure 4-43. Note the change in geometry as the impedance transitions from 50 Ω to 20 Ω and from 20 Ω to 80 Ω. These abrupt changes in geometry are known as discontinuities. Discontinuities in geometry result in fringing capacitance and parasitic inductance that will modify the frequency response of the circuit. At RF and lower microwave frequencies (up to about 2 GHz) the effects of discontinuities are minimal and sometimes neglected [7]. As the operation frequency increases, the effects of discontinuities can significantly alter the performance of a microstrip circuit.

References and Further Readings

[1]   Handbook of Filter Synthesis, Analtol I. Zverev, Wiley 1967

[2]   RF Circuit Design, Second Edition, Christopher Bowick, Elsevier 2008

[3]   Ali Behagi, *RF and Microwave Circuit Design*, A Design Approach Using (**ADS**), Techno Search, Ladera Ranch, CA 2017

[4]   Ali Behagi and Manou Ghanevati, *Fundamentals of RF and Microwave Circuit Design,* Practical Analysis and Design Tools, Techno Search, Ladera Ranch, California 2017

[5]   Foundations for Microstrip Circuit Design, T.C. Edwards, John Wiley & Sons, New York, 1981

[6]   Q Factor, Darko Kajfez, Vector Fields, Oxford Mississippi, 1994

[7]   David M. Pozar, *Microwave Engineering*, Second Edition, John Wiley and Sons, Inc. 1998

[8]   Q Factor Measurement with Network Analyzer, Darko Kajfez and Eugene Hwan, IEEE Transactions on Microwave Theory and Techniques, Vol. MTT-32, No. 7, July 1984.

[9]   High Frequency Techniques, Joseph F. White, John Wiley & Sons, Inc., 2005

[10]   Soft Substrates Conquer Hard Designs, James D. Woermbke, *Microwaves*, January 1982.

[11]   Principles of Microstrip Design, Alam Tam, RF Design, June 1988.

[12]   David M. Pozar, *Microwave Engineering*, Third Edition, John Wiley and Sons, 2005.

Problems

4-1.   Consider the one port resonator that is represented as a series RLC circuit. Analyze the circuit, with $R = 5\ \Omega$, $L = 5$ nH, and $C = 5$ pF. Plot the magnitude of the resonator input impedance and measure the resonance frequency.

4-2.   Consider the one port resonator that is represented as a parallel RLC circuit. Analyze the circuit, with $R = 500\ \Omega$, $L = 50$ nH, and $C = 50$ pF. Plot the magnitude of the resonator input impedance and measure the resonance frequency.

4-3.   Design a Butterworth lowpass filter having a passband of 2 GHz with an attenuation 20 dB at 4 GHz. Plot the insertion loss versus frequency from 0 to 5 GHz. The system impedance is $50\ \Omega$.

4-4.   Design a 5th order Chebyshev highpass filter having 0.2 dB equal ripples in the passband and cutoff frequency of 2 GHz. The system impedance is $75\ \Omega$. Plot the insertion loss versus frequency from 0 to 5 GHz.

4-5.   In a full duplex communication link, the uplink signal is around 200 MHz while the downlink is at 500 MHz. A 25 Watt power amplifier is used on the uplink with 20 dB gain.

(a) Design a low pass filter on the uplink to pass the 200 MHz uplink signal while rejecting any noise power in the 500 MHz band. Design the passband cutoff frequency is at 220 MHz, therefore, the filter should have a Chebyshev Response with 0.1 dB pass band ripple. The reject requirement is at least -40 dB rejections at 500 MHz.

(b) Design a High Pass Filter that passes frequencies in the 480 MHz to 520 MHz range. The High Pass Filter specifications are: The passband cutoff frequency is 480 MHz, therefore, the filter should have a Chebyshev response with 0.1 dB pass band ripple. The reject requirement is at least -60 dB losses at 190 MHz.

4-6.   Design a 75 Ω transmission line of sufficient length to act as a 10 nH inductor. Use the TLINE Transmission Line Synthesis Program to calculate the microstrip line width on the Rogers 0.025 inch RO3010 material.

4-7.   Using the microwave filter synthesis tool, design a stepped impedance low pass filter on RO3003 material that is 0.010 inches thick. Use a Chebyshev response with a 0.01dB ripple and a cutoff frequency of 4 GHz. Determine the worst case in band return loss and the rejection at 6 GHz.

4-8.   For the filter design of Problem 4-7 create an EM simulation using Momentum. Compare the EM simulation to a linear simulation. Comment on the rejection comparison at 6 GHz.

4-9.   For the filter design of Problem 4-7 determine the frequencies at which reentrant modes exist up through 20 GHz.

4-10.   Design a half wave microstrip resonator at 10 GHz using RO3003 substrate that is 0.020 thick. Initially design the resonator with a 50 W line impedance. Select a coupling capacitor to critically couple the resonator to the 50 Ω source. Then determine the resonator line impedance that results in the highest unloaded Q_o.

Chapter 5
Power Transfer and Impedance Matching

5.1 Introduction

Impedance matching is an integral part of RF and microwave circuit design and is used mainly for the efficient transfer of power from a source to the load. For example, in microwave amplifier design, the need for impedance matching arises when the amplifier must be properly terminated at both terminals in order to deliver maximum power from the source to the load. In narrowband applications impedance matching can be achieved, at a single frequency, with a two-element lossless network, known as L-network or L-section. In this chapter the basics of power transfer and the conditions for maximum power transfer are presented. The mathematical equations for the design of discrete L-networks are derived using MATLAB compatible expressions. This builds a solid foundation for the engineer to handle more complicated matching problems including broadband matching applications. In this chapter the analytical techniques for the design of matching arbitrary source and load impedances are developed. The analytical techniques are then used, in several examples, to design discrete narrowband and broadband impedance matching networks. For verification of the design the free software is used throughout.

5.2 Power Transfer Basics

At low frequencies, where the electrical wavelengths of the signals are much longer than the physical dimensions of the wires and lumped components, the phase of the voltage and current waveforms do not change significantly along the length of wires or components. Therefore, power is transmitted from the source to the load with little loss. At RF and microwave frequencies, where the signal wavelengths are equal to or shorter than the physical dimension of the network components, the amplitude and phase of the voltage and current waveforms change significantly as they travel from source to the load. In this case power is reflected at discontinuities and the maximum power will not reach the load. To achieve maximum power transfer we need to eliminate reflections at discontinuities by inserting proper impedance matching networks.

Maximum Power Transfer Conditions

For the network of Figure 5-1 where a voltage source, V_S, and the series impedance $Z_S = R_S + jX_S$ are connected to a network, having the input impedance $Z_{IN} = R_{IN} + jX_{IN}$, the power transferred to the network is given by Equation (5-1).

$$P_{Network} = \frac{\text{Re}\left[V_{IN} I_{IN}^*\right]}{2}$$

(5-1)

In Equation (5-1) Re denotes the real part and the symbol * denotes the conjugate value.

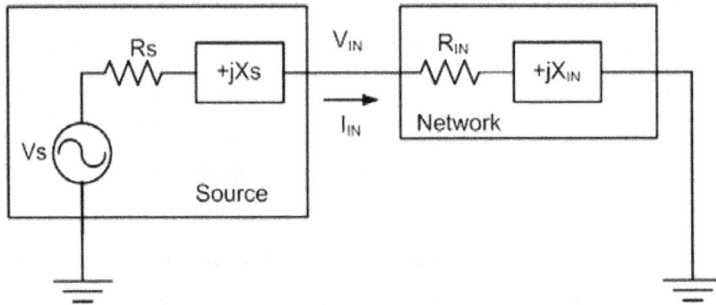

Figure 5-1 Voltage source connected to complex load impedance

In Figure 5-1, we assume V_S is sinusoidal steady state voltage source, R_S and R_{IN} are positive, and X_S and X_{IN} are real numbers. The input voltage and current to the network can be related to the source voltage as:

$$V_{IN} = \frac{V_S (R_{IN} + jX_{IN})}{(R_{IN} + R_S) + j(X_{IN} + X_S)}$$

And,

$$I_{IN}^* = \frac{V_S}{(R_{IN} + R_S) - j(X_{IN} + X_S)}$$

Therefore, multiplying V_{IN} by I_{IN}^*, Equation (5-1) can be written as:

$$P = \frac{\left(\dfrac{V_S^2 R_{IN}}{2}\right)}{\left(R_{IN} + R_S\right)^2 + \left(X_{IN} + X_S\right)^2} \qquad (5\text{-}2)$$

To maximize the power transfer, we differentiate Equation (5-2) with respect to R_{IN} and X_{IN} and set them equal to zero,

$$\frac{V_S^2}{2}\left[\left(R_{IN} + R_S\right)^2 + \left(X_{IN} + X_S\right)^2\right] - \frac{V_S^2 R_{IN}}{2}\left[2R_{IN} + 2R_S\right] = 0$$

And,

$$-\frac{V_S^2 R_{IN}}{2}\left(2X_{IN} + 2X_S\right) = 0$$

A simultaneous solution of these two equations for R_{IN} and X_{IN}, leads to:

$$R_{IN} = R_S \qquad (5\text{-}3)$$

And,

$$X_{IN} = -X_S \qquad (5\text{-}4)$$

Therefore, the maximum power transfer condition is that the load impedance be equal to the conjugate of the source impedance given in Equation (5-5).

$$Z_{IN} = Z_S^* \qquad (5\text{-}5)$$

This maximum power transfer condition between the source and load impedance can be divided into three cases.

Case 1:

If the source impedance is purely resistive the load impedance must also be purely resistive and equal to source resistance. In this case:

$$Z_{IN} = R_{IN} = R_S$$

Case 2:

If the source impedance is a resistor in series with a capacitor, $Z_S = R_S - jX_S$, the load impedance must be a resistor in series with an inductor such that:

$$Z_{IN} = R_S + jX_S$$

Case 3:

If the source impedance is a resistor in series with an inductor, $Z_S = R_S + jX_S$, the load impedance must be a resistor in series with a capacitor such that:

$$Z_{IN} = R_S - jX_S$$

Maximum Power Transfer with Purely Resistive Source and Load Impedance

This section explores in detail the three cases for maximum power transfer along with solutions using the LTspice software.

Example 5.1: Prove the maximum power transfer condition where the source and load impedance is purely resistive, as shown in Figure 5-2.

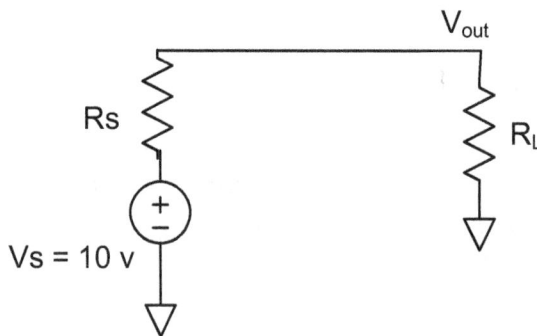

Figure 5-2 Network with purely resistive source and load impedance

Solution: When the source impedance is purely resistive and the load resistance is R_L, the maximum power is transferred to the load when $R_L = R_S$

Using the resistor voltage-divider principle we can determine the output voltage, V_{out}, for the following three possible cases:

Case I: $R_L = R_S$
Case II: $R_L < R_S$
Case III: $R_L > R_S$

Case I: If the input voltage is 10 VDC and $R_L = R_S = 50\ \Omega$, the output voltage is 5 Volts and the output power is 0.5 Watts. This is the maximum power that can be transferred.

$$V_{out} = V_S \frac{R_L}{(R_S + R_L)} = 10\frac{50}{(50+50)} = 5\ volts$$

$$P_L = \frac{V_{out}^2}{R_L} = \frac{5^2}{50} = 0.5\ Watts$$

Case II: If $R_L = 25\ \Omega$ and $R_S = 50\ \Omega$, the output voltage is 3.333 Volts and the output power is 0.444 Watts.

$$V_{out} = V_S \frac{R_L}{(R_S + R_L)} = 10\frac{25}{(50+25)} = 3.333\ volts$$

$$P_L = \frac{V_{out}^2}{R_L} = \frac{3.333^2}{25} = 0.444\ Watts$$

Case III: If $R_L = 100\ \Omega$ and $R_S = 50\ \Omega$, the output voltage is 6.666 Volts and the output power is 0.444 Watts.

$$V_{out} = V_S \frac{R_L}{(R_S + R_L)} = 10\frac{100}{(50+100)} = 6.666\ volts$$

$$P_L = \frac{V_{out}^2}{R_L} = \frac{6.666^2}{100} = 0.444\ Watts$$

Notice that the power output in Case I is greater than either Cases II or III. Therefore, the maximum power transfer is achieved only when $R_L = R_S$.

Example 5.2: Use the linear simulation techniques to demonstrate the maximum power transfer condition in a purely resistive system.

Solution: This is a very simple schematic where the voltage source is directly connected to 50 Ohm load, as shown in Figure 5-3. The input port is an RF signal source with 50 Ω source impedance and the output port is simply a 50 Ω resistor. Simulate the schematic from 450 to 550 MHz and add a rectangular graph with the insertion loss, S21, in dB. Figure 5- shows the plot indicating that S21 = 0 dB. Because there is zero insertion loss between the source and load, there is maximum power transfer.

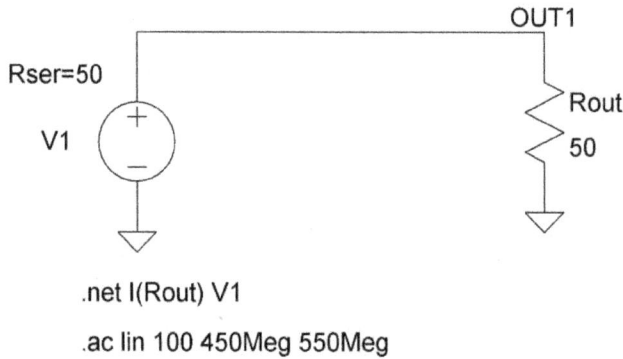

.net I(Rout) V1

.ac lin 100 450Meg 550Meg

Figure 5-3 Case I power transfer with $R_L = R_S$

Simulate the schematic and display the insertion loss, S21, in a rectangular plot as shown in Figure 5-4.

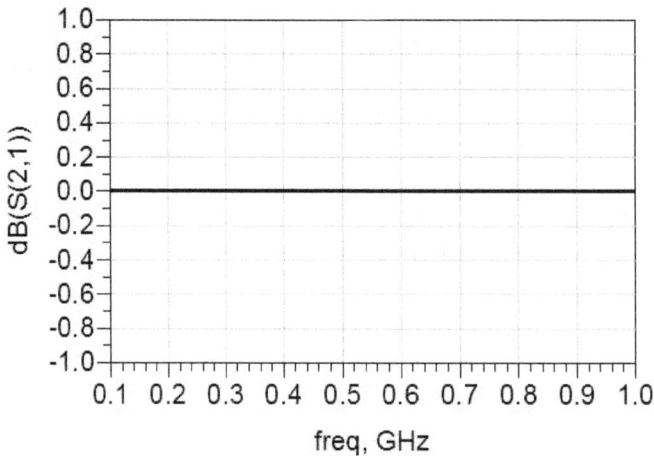

Figure 5-4 Insertion Loss with $R_L = R_S$

Notice that when the load impedance is equal to the source impedance the insertion loss is 0 dB indicating maximum power transfer.

Change the load impedance in Figure 5-3 to 25 Ω as shown in Figure 5-5.

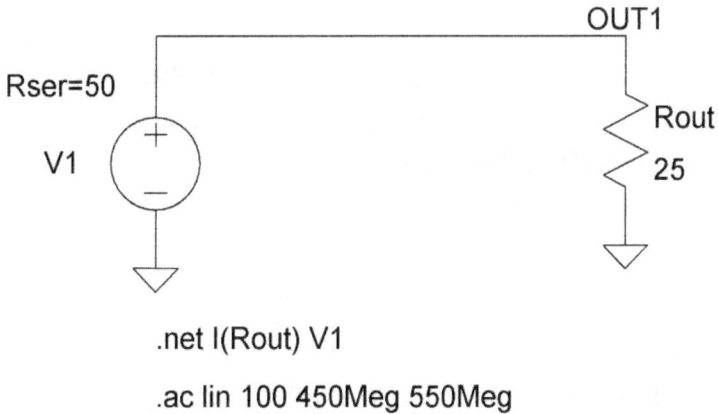

.net I(Rout) V1

.ac lin 100 450Meg 550Meg

Figure 5-5 Case II power transfer with $R_L < R_S$

Simulate the schematic and display the insertion loss, S21, in a rectangular plot as shown in Figure 5-6.

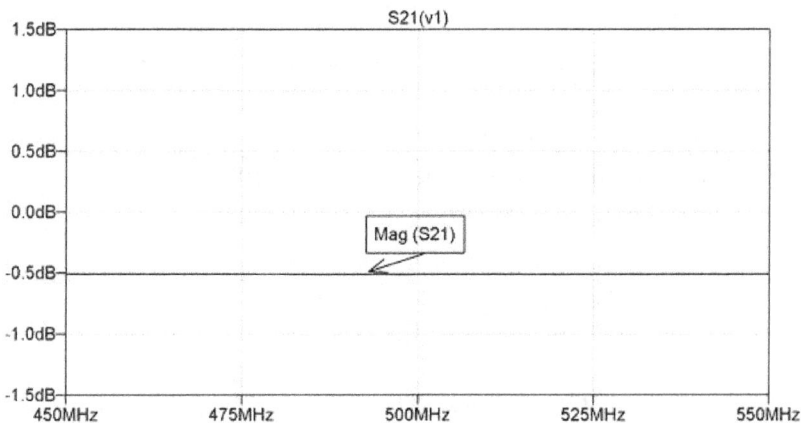

Figure 5-6 Insertion Loss with $R_L < R_S$

Notice that S21 = -0.512 dB indicating insertion loss. To calculate the amount of power loss, we know that the maximum power transfer is 0.5 Watts. To determine the amount of 0.512 dB power loss we first convert the

0.5 Watts to dBm which is the power in dB relative to 1 mW. Utilizing the conversion equation,

$$10 \log (\text{power in mW} = \text{power in dBm}.$$

$$10 \log (500 mW) \quad 26.98 \, dBm$$

The loss of 0.512 dB is subtracted from the 26.98 dBm to result in (26.98 dBm − 0.512 dB) = 26.47 dBm. Then converting from dBm back to mW we get 444 mW as calculated in Example 5.1.

$$10^{\left(\frac{26.47}{10}\right)} = 444 \ mW \quad or \quad 0.444 \ Watts$$

5.3 Maximum Power Transfer with Complex Load Impedance

According to Equation 5-5 the maximum power transfer occurs when $Z_S = Z_L{}^*$. Therefore, if $Z_L = R_L - jX_L$, then for maximum power transfer we must have $Z_S = R_L + jX_L$.

Example 5.3: If the load is 50 Ω in series with a 15 pF series capacitance, find the source impedance to have maximum power transfer at 500 MHz.

Solution: We can find a source inductance that cancels the reactance of the load at this frequency. The reactance of the 15 pF capacitor is:

$$X_C = \frac{1}{2\pi f C} = 21.231 \ \Omega$$

Therefore,

$$Z_L = 50 - j21.231 \ \Omega$$

For maximum power transfer the source impedance must be:

$$X_S = 50 + j21.231 \ \Omega$$

At 500 MHz the value of the source series inductor is:

$$L = \frac{21.231}{2\pi f} = 6.76 \quad nH$$

To demonstrate the maximum power transfer, create a schematic as shown in Figure 5-7. Place the specified 15 pF capacitance in series with the load and the 6.76 nH inductance in series with the source. Plot the S21 and VSWR in rectangular graphs as shown in Figures 5-8 and 5-9. Note that unlike the purely resistive source and load case, a complex conjugate match occurs at a single frequency. A perfect 1:1 VSWR is achieved at the conjugate match frequency of 500 MHz.

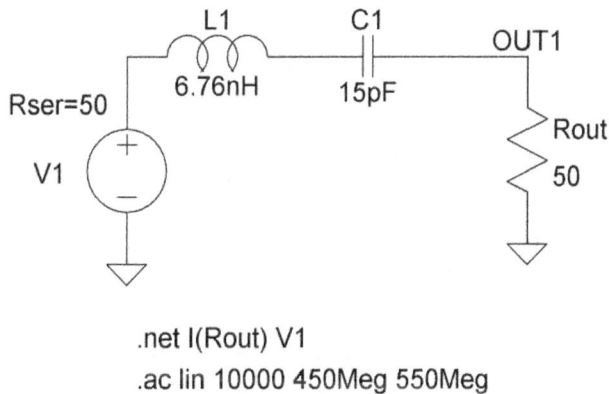

.net I(Rout) V1
.ac lin 10000 450Meg 550Meg

Figure 5-7 Maximum power transfer with complex impedance

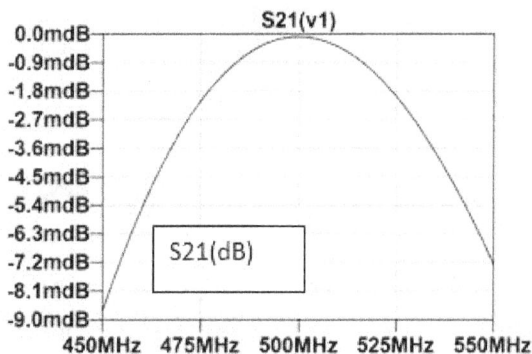

Figure 5-8 Maximum power transfer response

The Voltage Standing Wave Ratio is shown in Figure 5-9.

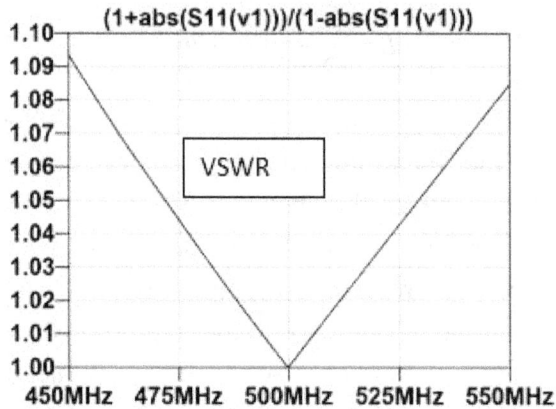

(1+abs(S11(v1)))/(1-abs(S11(v1)))

```
1.10
1.09
1.08
1.07
1.06          VSWR
1.05
1.04
1.03
1.02
1.01
1.00
    450MHz  475MHz  500MHz  525MHz  550MHz
```

Figure 5-9 Perfect 1:1 VSWR is achieved at 500 MHz

5.4 Analytical Design of the Impedance Matching Networks

One of the important tasks in RF and microwave engineering is the determination of how an arbitrary complex load impedance, $Z_L = R_L + jX_L$, is analytically matched to any complex source impedance, $Z_S = R_S + jX_S$, as shown in Figure 5-10. This problem arises mainly in the design of inter-stage matching networks between active devices or between an antenna and a transmitter.

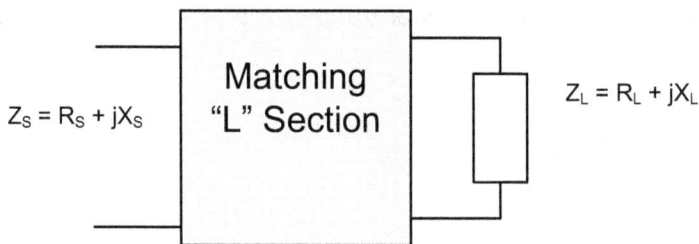

$$Z_S = R_S + jX_S \quad \boxed{\begin{array}{c} \text{Matching} \\ \text{"L" Section} \end{array}} \quad Z_L = R_L + jX_L$$

Figure 5-10 Complex impedance matching with an L-network

In Figure 5-10 the complex load impedance, $Z_L = R_L + j X_L$, is to be matched to the source impedance $Z_S = R_S + jX_S$. The only condition for impedance matching is that both R_S and R_L must be nonnegative while X_S and X_L could take any real value. In section 5.2 it was shown that maximum power would be transferred from the source to the load when the load impedance is the conjugate of the source impedance. For L-network matching there are only two configurations that can match an arbitrary load to arbitrary source impedances. In the first configuration the first element adjacent to the load is a series element as shown in Figure 5-11.

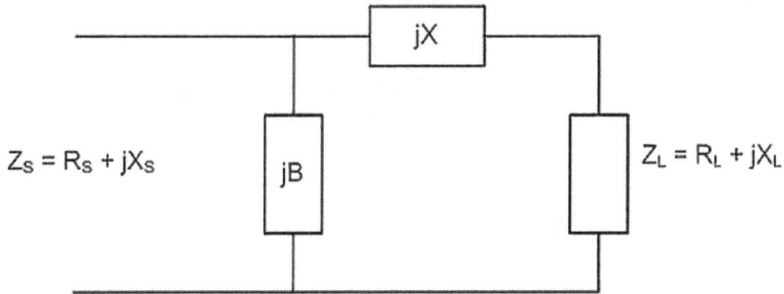

Figure 5-11 First impedance matching network configuration

In the second configuration the first element adjacent to the load is a shunt element as shown in Figure 5-12.

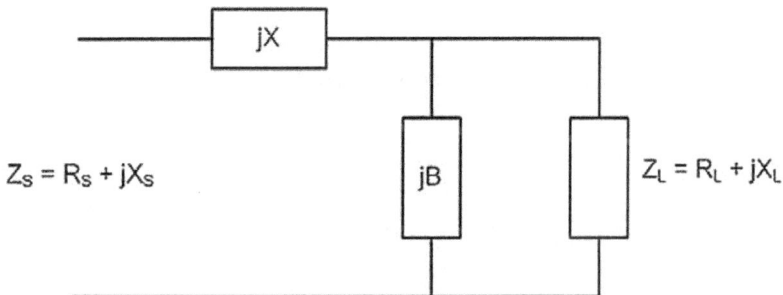

Figure 5-12 Second impedance matching network configuration

Matching a Complex Load to Complex Source Impedance

To match a complex load to any complex source impedance with a single L-network, either the first or the second configuration may be used. The choice of configurations depends on the conditions that source and load impedances dictate. Applying Equation $Z_S^* = Z_{IN}$ to the first matching configuration in Figure 5-11 we get:

$$R_S - jX_S = \cfrac{1}{jB + \left(\cfrac{1}{jX + R_L + jX_L}\right)} \qquad (5\text{-}6)$$

By separating the real and imaginary parts of Equation (5-6) we obtain two solutions for B and X as follows:

$$B_1 = \frac{R_L X_S + \sqrt{R_L R_S (R_S^{\,2} + X_S^{\,2} - R_L R_S)}}{R_L (R_S^{\,2} + X_S^{\,2})} \qquad (5\text{-}7)$$

$$X_1 = \frac{R_L X_S - R_S X_L}{R_S} + \frac{R_S - R_L}{B_1 R_S} \qquad (5\text{-}8)$$

And

$$B_2 = \frac{R_L X_S - \sqrt{R_L R_S (R_S^{\,2} + X_S^{\,2} - R_L R_S)}}{R_L (R_S^{\,2} + X_S^{\,2})} \qquad (5\text{-}9)$$

$$X_2 = \frac{R_L X_S - R_S X_L}{R_S} + \frac{R_S - R_L}{B_2 R_S} \qquad (5\text{-}10)$$

Similarly, applying the same procedure to the second configuration we have:

$$R_S - jX_S = jX + \cfrac{1}{jB + \left(\cfrac{1}{R_L + jX_L}\right)} \qquad (5\text{-}11)$$

Separating the real and imaginary parts of Equation (5-11), we also get two sets of solutions for B and X:

$$B_3 = \frac{R_S X_L + \sqrt{R_L R_S (R_L^{\,2} + X_L^{\,2} - R_L R_S)}}{R_S (R_L^{\,2} + X_L^{\,2})} \qquad (5\text{-}12)$$

$$X_3 = \frac{R_S X_L - R_L X_S}{R_L} + \frac{R_L - R_S}{B_3 R_L} \qquad (5\text{-}13)$$

And

$$B_4 = \frac{R_S X_L - \sqrt{R_L R_S (R_L{}^2 + X_L{}^2 - R_L R_S)}}{R_S (R_L{}^2 + X_L{}^2)} \qquad (5\text{-}14)$$

$$X_4 = \frac{R_S X_L - R_L X_S}{R_L} + \frac{R_L - R_S}{B_4 R_L} \qquad (5\text{-}15)$$

Conditions for the validity of solutions are that the arguments of the square roots in Equations (5-7), (5-9), (5-12) and (5-14) be positive or zero.

1. If $\; R_S{}^2 + X_S{}^2 - R_L R_S > 0 \qquad$ and $\qquad R_L{}^2 + X_L{}^2 - R_L R_S < 0$,

The two solutions obtained from Equations (5-7) through (5-10) are the only valid solutions.

2. If $\; R_S{}^2 + X_S{}^2 - R_L R_S < 0 \qquad$ and $\qquad R_L{}^2 + X_L{}^2 - R_L R_S > 0$,

The two solutions obtained from Equations (5-12) through (5-15) are the only valid solutions.

3. If $R_S{}^2 + X_S{}^2 - R_L R_S > 0 \qquad$ and $\qquad R_L{}^2 + X_L{}^2 - R_L R_S > 0$,

All four solutions obtained from Equations (5-7) through (5-10) and Equations (5-12) through (5-15) are the valid solutions.

Once the real values for B and X are calculated, the values of the matching elements are obtained from the following equations:

If B is positive, the matching element is a capacitor given by:

$$C = \frac{B}{2\pi f} \qquad (5\text{-}16)$$

If B is negative, the matching element is an inductor given by:

$$L = -\frac{1}{2\pi f B} \qquad (5\text{-}17)$$

If X is positive, the matching element is an inductor given by:

$$L = \frac{X}{2\pi f} \qquad (5\text{-}18)$$

If X is negative, the matching element is a capacitor given by:

$$C = -\frac{1}{2\pi f X} \qquad (5\text{-}19)$$

In the above equations, if the frequency is in Hz, the capacitors and inductors are in Farad and Henry, respectively.

In the next four examples we utilize Equations (5-7) through (5-15) to design the matching L-networks between two complex impedances.

Example 5.4: Design L-networks that match a complex load impedance, $Z_L = 10 - j15\ \Omega$, to a complex source impedance, $Z_S = 15 - j20\ \Omega$, at 2 GHz.

Note: Since $R_S^2 + X_S^2 - R_L R_S = 475 > 0$, and $R_L^2 + X_L^2 - R_L R_S = 175 > 0$ this Example has four solutions.

First Solution: Utilize Equations (5-7) and (5-8) in MATLAB Script to calculate the element values of the matching L-network. The procedure follows.

1. Enter design parameters

RS = 15; XS = -20; RL = 10; XL = -15; f = 2e9

2. Use Equations (5-7) and (5-8) to calculate the matching element values of the first L-network.

B1= ((RL*XS) + sqrt(RS*RL*(RS^2+XS^2-RS*RL)))/(RL*(RS^2+XS^2))

X1= (RL*XS-RS*XL)/RS + (RS-RL)/(B1*RS)

L1= X1/(2*pi*f)

C2= B1/(2*pi*f)

The MATLAB solutions are given in Table 5-1.

Name	Value
B1	0.011
C2	852.1e-15
L1	2.61e-9
RL	10
RS	15
X1	32.795
XL	-15
XS	-20
f	2e+9

Table 5-1: MATLAB Solutions

The calculation results in Table 5-1 show that B1 = 0.011 and X1 = 32.795, therefore, the series element is an inductor, L1 = 2.61 nH, and the shunt element is a capacitor, C2 = 0.852 pF.

To verify the solution, create a new schematic and add the matching elements as shown in Figure 5-13.

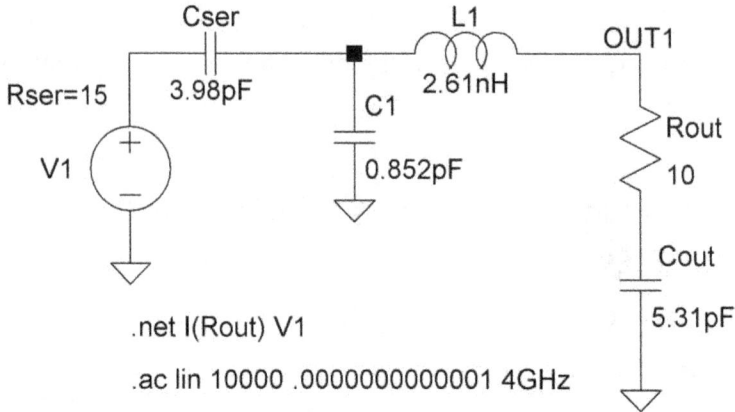

Figure 5-13 Schematic of the first matching L-network

Simulate the schematic and display the magnitude of the input reflection coefficient, Gamma, in a rectangular plot as shown in Figure 5-14.

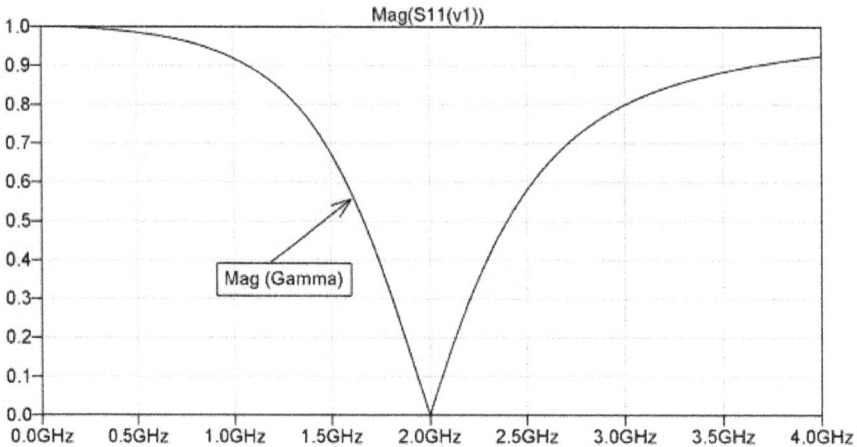

Figure 5-14 Magnitude of the input reflection coefficient

The magnitude of the input reflection coefficient in Figure 5-14 shows that the matching L-network perfectly matches the load impedance to the source impedance at a single frequency of 2 GHz.

Example 5.5: Design the second L-network that matches the complex load impedance, $Z_L = 10 - j15$ Ω, to a complex source impedance, $Z_S = 15 - j20$ Ω, at 2 GHz.

Second Solution: Use Equations (5-9) and (5-10) in MATLAB script to calculate the second solution. The procedure follows.

1. Enter Design Parameters

RS=15; XS=-20; RL=10; XL=-15; f=2e9

2. Use Equations (5-9) and (5-10) in MATLAB Script to calculate the matching element values

B2=((RL*XS)-sqrt(RS*RL*(RS^2+XS^2-RS*RL)))/(RL*(RS^2+XS^2))

X2=(RL*XS-RS*XL)/RS+(RS-RL)/(B2*RS)

C1=-1/(2*pi*f*X2)

L2=-1/(2*pi*f*B2)

The MATLAB solutions are given in Table 5-2.

Name	Value
B2	-0.075
C1	28.47e-12
L2	1.065e-9
RL	10
RS	15
X2	-2.795
XL	-15
XS	-20
f	2e+9

Table 5-2: MATLAB Solutions

The calculation results in Table 5-2 show that B2= -0.075 and X2 = -2.795, therefore, the series element is a capacitor, C1 = 28.47 pF and the shunt element is an inductor, L2 = 1.065 nH.

To plot the response of the matching network, create a new schematic and add the capacitor and inductor and connect the matching elements as shown in Figure 5-15.

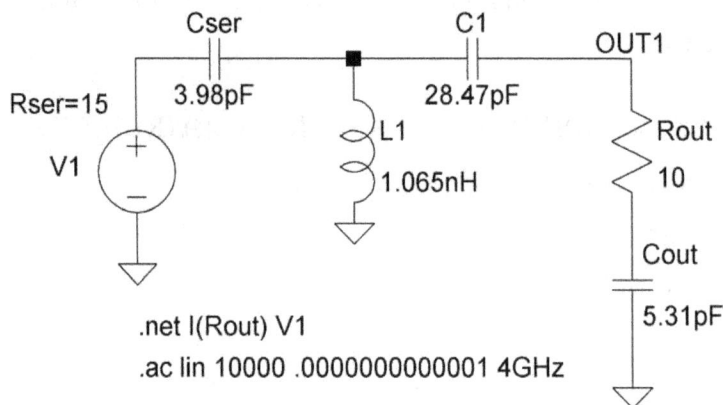

Figure 5-15 Schematic of the second matching L-network

Simulate the schematic and display the magnitude of the reflection coefficient, gamma, in a rectangular plot, as shown in Figure 5-16.

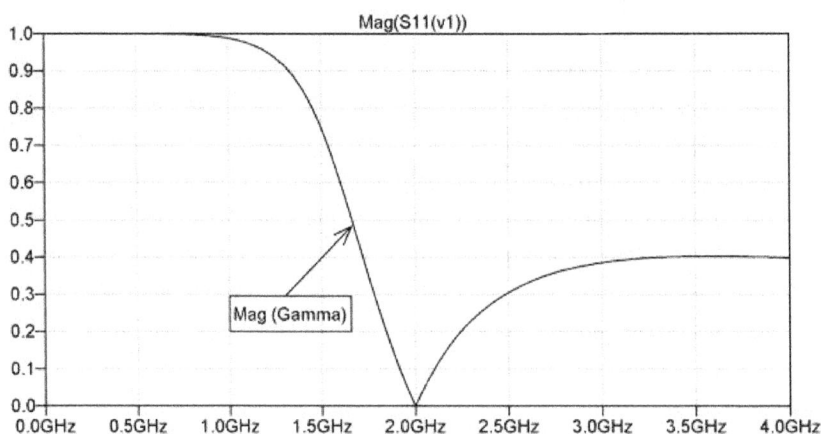

Figure 5-16 Magnitude of the input reflection coefficient

The simulated response in Figure 5-16 shows that the matching L-network perfectly matches the load impedance to the source impedance at a single frequency of 2 GHz.

Example 5.6: Design the third L-network that matches the complex load impedance, $Z_L = 10 - j15\ \Omega$, to a complex source impedance, $Z_S = 15 - j20\ \Omega$, at 2 GHz.

Third Solution: For the third solution use Equations (5-12) and (5-13) in MATLAB Script to calculate *B3* and *X3*. The procedure follows.

1. Enter design parameters

RS=15; XS=-20; RL=10; XL=-15; f=2e9

2. Use Equations (5-12) and (5-13) in MATLAB Script to calculate the matching element values

B3=((RS*XL)+sqrt(RS*RL*(RL^2+XL^2-RS*RL)))/(RS*(RL^2+XL^2))

X3=(RS*XL-RL*XS)/RL+(RL-RS)/(B3*RL)

L1=-1/(2*pi*f*B3)

L2=X3/(2*pi*f)

The MATLAB solutions are given in Table 5-3.

Name	Value
B3	-0.013
L1	6.16e-9
L2	2.881e-9
RL	10
RS	15
X3	36.202
XL	-15
XS	-20
f	2e+9

Table 5-3: MATLAB Solutions

The calculation results in Table 5-3 show that B3 = -0.013 and X3 = 36.202, therefore, the shunt element is an inductor L1= 6.16 nH and the series element is another inductor L2 = 2.881 nH, as shown in Figure 5-17.

To plot the response of the matching network, create a new schematic and add lumped elements as shown in Figure 5-17.

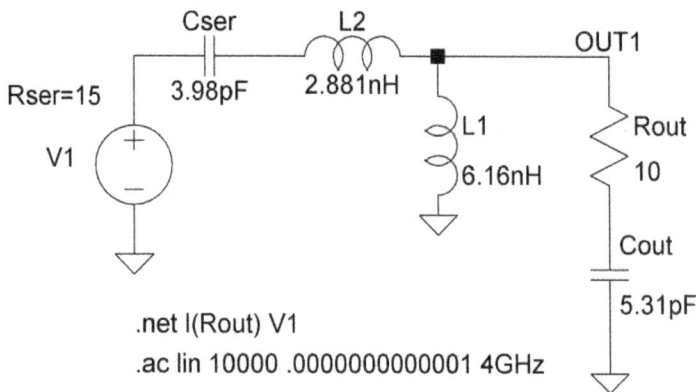

Figure 5-17: Schematic of the third matching L-network

Simulate the schematic and display S11 in a rectangular plot, as shown in Figure 5-18.

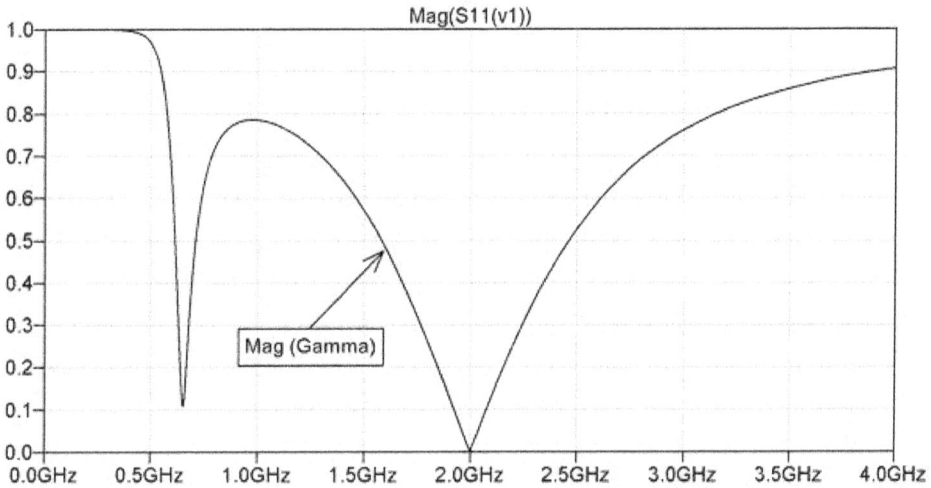

Figure 5-18 Magnitude of the input reflection coefficient

The simulated response in Figure 5-17 shows that the matching L-network perfectly matches the load impedance to the source impedance at a single frequency of 2 GHz.

Example 5.7: Design the fourth L-network that matches the complex load impedance, $Z_L = 10 - j15$ Ω, to a complex source impedance, $Z_S = 15 - j20$ Ω, at 2 GHz.

Fourth Solution: Use Equations (5-14) and (5-15) to calculate *B4* and *X4*.

1. Enter design parameters

RS=15; XS=-20; RL=10; XL=-15; f=2e9

2. Use Equations (5-14) and (5-15) in MATLAB script to calculate the matching element values

B4=((RS*XL)-sqrt(RS*RL*(RL^2+XL^2-RS*RL)))/(RS*(RL^2+XL^2))

X4=(RS*XL-RL*XS)/RL+(RL-RS)/(B4*RL)

The inductor values are found from L1 and L2.

L1=-1/(2*pi*f*B4)

L2=X4/(2*pi*f)

The MATLAB solutions are given in Table 5-4.

Name	Value
B4	-0.079
L1	1.002e-9
L2	302.2e-12
RL	10
RS	15
X4	3.798
XL	-15
XS	-20
f	2e+9

Table 5-4: MATLAB Solutions

The calculation results in Table 5-4 show that B4 = -0.079 and X4 = 3.798, therefore, the shunt element is an inductor L1=1.002 nH and the series element is another inductor L2=0.302 nH.

To plot the response of the matching network, create a new schematic and add the matching elements as shown in Figure 5-19.

Cser   L2   OUT1

3.98pF   0.302nH

Rser=15

V1

L1   Rout

1.002nH   10

Cout

.net I(Rout) V1   5.31pF

.ac lin 10000 .0000000000001 4GHz

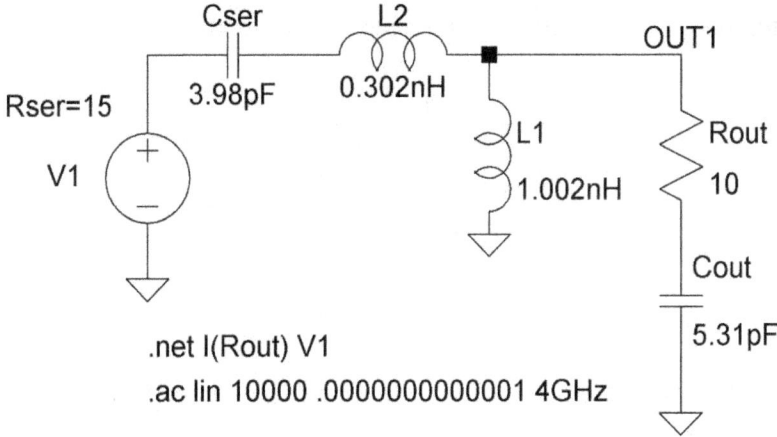

Figure 5-19 Schematic of the fourth matching L-network

Simulate the schematic and display S11 in a rectangular plot.

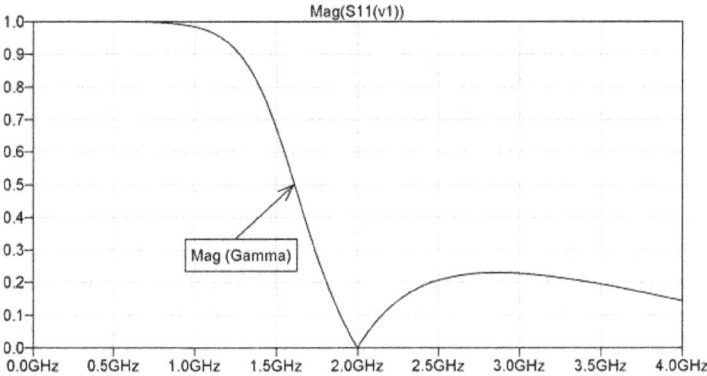

Mag(S11(v1))

Mag (Gamma)

Figure 5-20 Magnitude of the input reflection coefficient

The simulated response in Figure 5-20 shows that the matching L-network perfectly matches the load impedance to the source impedance at a single frequency of 2 GHz.

Matching a Complex Load to a Real Source Impedance

In amplifier design a common matching problem is the matching of a complex load impedance, $Z_L = R_L + jX_L$ to a real source impedance, $Z_S = R_S$. The complex impedance is usually the load and the real impedance is the characteristic impedance of transmission line connected to the source. In order to design the impedance matching networks we use both configurations of Figures 5-7 and 5-8. For the case of real source impedance the matching configuration of Figure 5-6 is redrawn in Figure 5-21.

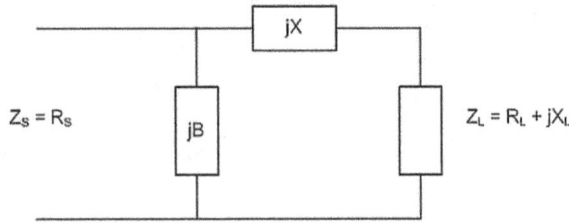

Figure 5-21 Matching complex load to resistive source (first configuration)

To derive the analytical expressions for B and X we utilize the maximum power transfer condition and set the conjugate of the source impedance equal to input impedance of the matching network followed by the load impedance, as given in Equation (5-20).

$$R_S = \cfrac{1}{jB + \left(\cfrac{1}{jX + R_L + jX_L} \right)} \qquad (5\text{-}20)$$

Note that for the real source resistor its conjugate is equal to itself. The solutions for B and X in Equation (5-20) can be obtained by substituting $X_S = 0$ in Equations (5-7) through (5-10).

$$B_1 = \frac{+\sqrt{R_S - R_L}}{R_S \sqrt{R_L}} \qquad (5\text{-}21)$$

$$X_1 = +\sqrt{R_L(R_S - R_L)} - X_L \qquad (5\text{-}22)$$

And,

$$B_2 = \frac{-\sqrt{R_S - R_L}}{R_S \sqrt{R_L}} \qquad (5\text{-}23)$$

$$X_2 = -\sqrt{R_L(R_S - R_L)} - X_L \qquad (5\text{-}24)$$

Note that the solutions given in Equations (5-21) through (5-24) are only valid if $R_L < R_S$. To calculate B and X, when $R_L > R_S$, we use the second matching configuration shown in Figure 5-122,

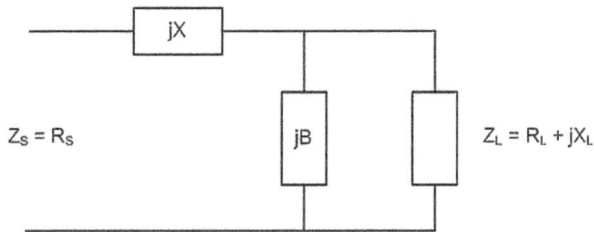

Figure 5-22 Matching complex load to resistive source (second configuration)

Applying the maximum power condition to the matching network in Figure 5-18, we have:

Applying the maximum power condition to the matching network in Figure 5-18, we have:

$$R_S = jX + \cfrac{1}{jB + \left(\cfrac{1}{R_L + jX_L}\right)} \qquad (5\text{-}25)$$

The solutions for B and X in Equation (5-25) can be obtained by reusing Equations (5-12) and (5-14) and substituting $X_S = 0$ in Equations (5-13) and (5-15).

$$B_3 = \frac{R_S X_L + \sqrt{R_L R_S (R_L{}^2 + X_L{}^2 - R_L R_S)}}{R_S (R_L{}^2 + X_L{}^2)} \qquad (5\text{-}26)$$

$$X_3 = \frac{R_S X_L}{R_L} + \frac{R_L - R_S}{B_3 R_L} \qquad (5\text{-}27)$$

and

$$B_4 = \frac{R_S X_L - \sqrt{R_L R_S (R_L{}^2 + X_L{}^2 - R_L R_S)}}{R_S (R_L{}^2 + X_L{}^2)} \qquad (5\text{-}28)$$

$$X_4 = \frac{R_S X_L}{R_L} + \frac{R_L - R_S}{B_4 R_L} \tag{5-29}$$

The conditions for the valid solutions are that the arguments of the square roots in Equations (5-26) through (5-29) be non-negative. Therefore, the two solutions obtained from Equations (5-26) through (5-29) are valid only if $R_L > R_S$. Combined conditions for valid solutions are summarized in Table 5-5.

Case #	First Condition	Second Condition	# of Solutions	Equations Used
1	$R_L < R_S$	$R_L^2 + X_L^2 - R_L R_S > 0$	4	(5-21) to (5-24) (5-26) to (5-29)
2	$R_L < R_S$	$R_L^2 + X_L^2 - R_L R_S < 0$	2	(5-21) to (5-24)
3	$R_L > R_S$	N/A	2	(5-26) to (5-29)

Table 5-5 Impedance matching conditions and the number of solutions

The solutions in Equations (5-26) through (5-29) can be simplified by normalizing the load impedance with respect to the source resistor. Therefore, if we let the source resistor be equal to Z_0, the normalized load resistance and reactance become,

$$r = \frac{R_L}{Z_0} \tag{5-30}$$

And

$$x = \frac{X_L}{Z_0} \tag{5-31}$$

The simplified equations are:

$$B_1 = \frac{\sqrt{\dfrac{(1-r)}{r}}}{Z_0} \tag{5-32}$$

$$X_1 = Z_0 \left[\sqrt{r(1-r)} - x \right] \tag{5-33}$$

$$B_2 = -\frac{\sqrt{(1-r)}}{Z_0} \qquad (5\text{-}34)$$

$$X_2 = -Z_0 \left[\sqrt{r(1-r)} + x \right] \qquad (5\text{-}35)$$

$$B_3 = \frac{x + \sqrt{r(r^2 + x^2 - r)}}{Z_0\,(r^2 + x^2)} \qquad (5\text{-}36)$$

$$X_3 = Z_0 \sqrt{\frac{(r^2 + x^2 - r)}{r}} \qquad (5\text{-}37)$$

$$B_4 = \frac{x - \sqrt{r\left(r^2 + x^2 - r\right)}}{Z_0\left(r^2 + x^2\right)} \qquad (5\text{-}38)$$

$$X_4 = -Z_0 \sqrt{\frac{\left(r^2 + x^2 - r\right)}{r}} \qquad (5\text{-}39)$$

Example 5.8: Design a single L-network that will match a real source impedance $Z_0 = 50\ \Omega$ to a complex load impedance, $Z_L = 7 - j22\ \Omega$, at a frequency of 1 GHz.

Notice that for this example $R_L < Z_0$ and $R_L{}^2 + X_L{}^2 - R_L\,Z_0 = 183 > 0$, therefore, the matching network has four solutions.

First Solution: We use Equations (5-32) and (5-33) in MATLAB script to calculate the element values of the matching network.
1. Enter design parameters and normalized load impedance

Z0=50; RL=7; XL=-22; f=1000e6

r=RL/Z0

x=XL/Z0

2. For the first solution use Equations (5-32) and (5-33) in MATLAB
Script to calculate the matching element values.

B1=sqrt((1-r)/r)/Z0

X1=Z0*sqrt(r*(1-r))-x*Z0

LS1=X1/(2*pi*f)

CP2=B1/(2*pi*f)

The MATLAB solutions are given in Table 5-6.

Name	Value
B1	0.05
CP2	7.889e-12
LS1	6.263e-9
RL	7
X1	39.349
XL	-22
Z0	50
f	1e+9
r	0.14
x	-0.44

Table 5-6: MATLAB Solutions

The calculation results in Table 5-6 show that B1 = 0.05 and X1 = 39.349,
therefore, the series element is an inductor L1= 6.263 nH and the shunt
element is a capacitor C2=7.889 pF, as shown in Figure 5-23.

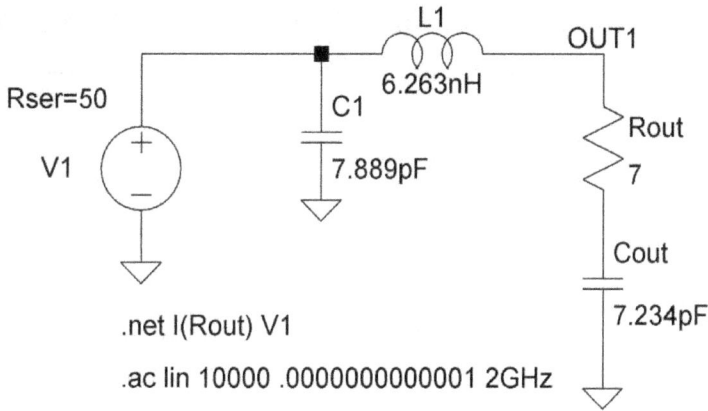

Figure 5-23: Schematic of the first matching L-network

Simulate the schematic and display S11 and S21 in a rectangular plot [2].

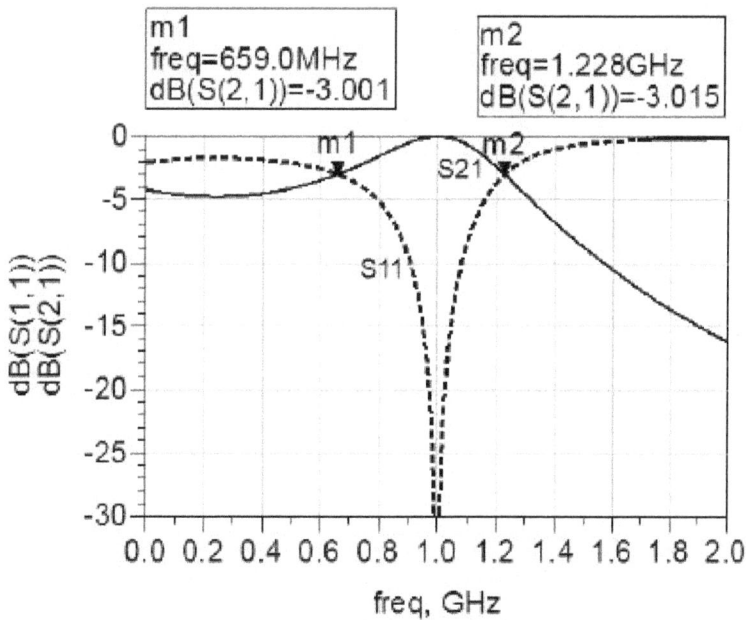

Figure 5-24 Magnitude of S11 and S21

The simulated response in Figure 5-24[2] shows that the matching L-networ
k perfectly matches the load impedance to the source impedance at 1 GHz.

Example 5.9: Design the second L-network that matches a real source
impedance, $Z_0 = 50\Omega$, to a complex load impedance, $Z_L = 7 - j22\Omega$, at a
frequency of 1 GHz.

Second Solution: Use Equations (5-34) and (5-35) to calculate the element values of the second matching network. The procedure follows.

1. Enter design parameters and normalized load impedance

Z0=50; RL=7; XL=-22; f=1000e6; r=RL/Z0; x=XL/Z0

2. Use Equations (5-34) and (5-35) in MATLAB script to calculate the matching element values

B2=-sqrt((1-r)/r)/Z0

X2=-Z0*sqrt(r*(1-r))-x*Z0

L1=X2/(2*pi*f)

L2=-1/(2*pi*f*B2)

The MATLAB solutions are given in Table 5-7.

Name	Value
B2	-0.05
L1	740.2e-12
L2	3.211e-9
RL	7
X2	4.651
XL	-22
Z0	50
f	1e+9
r	0.14
x	-0.44

Table 5-7: MATLAB Solutions

The calculation results in Table 5-7 show that B2 = -0.05 and X2 = 4.651, therefore, the series element is an inductor L1= 0.74 nH and the shunt element is another inductor L2= 3.211 nH, as shown in Figure 5-25.

To plot the response of the matching network, create a new schematic in LTspice and add the matching elements as shown in Figure 5-25.

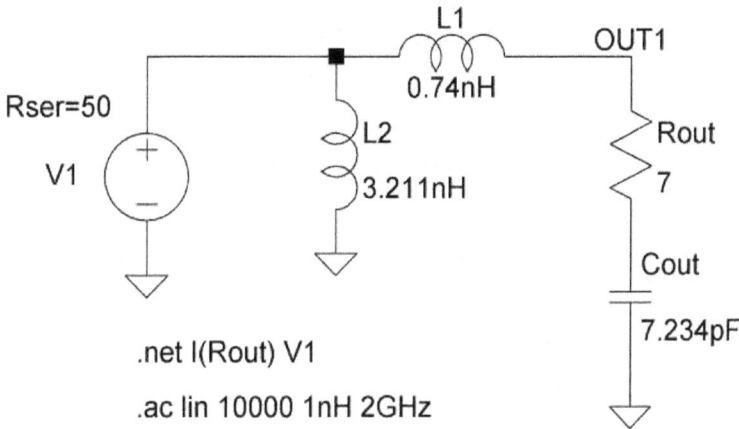

Figure 5-25 Schematic of the second matching L-network

Simulate the schematic and display S11 in a rectangular plot.

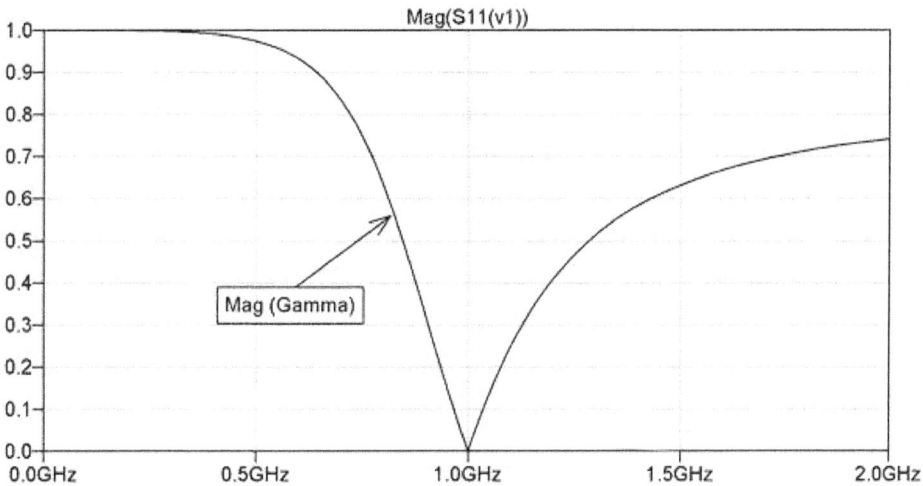

Figure 5-26 Magnitude of the input reflection coefficient

The simulated response in Figure 5-26 shows that the matching L-network perfectly matches the load impedance to the source impedance at 1 GHz.

Example 5.10: Design the third L-network that matches a real source impedance, $Z_0 = 50\ \Omega$, to a complex load impedance, $Z_L = 7 - j22\ \Omega$, at a frequency of 1 GHz.

Third Solution: We use Equations (5-36) and (5-37) to calculate the element values of the matching network. The procedure follows.

1. Enter design parameters and normalize the load impedance

Z0=50; RL=7; XL=-22; f=1000e6

r=RL/Z0

x=XL/Z0

2. Use Equations (5-36) and (5-37) in MATLAB Script to calculate the matching element values

B3=(x+sqrt(r*(r^2+x^2-r)))/(Z0*(r^2+x^2))

X3=Z0*sqrt((r^2+x^2-r)/r)

L2=X3/(2*pi*f)

L1=-1/(2*pi*f*B3)

The MATLAB solution are given in Table 5-8.

Name	Value
B3	-0.032
L1	5.008e-9
L2	5.754e-9
RL	7
X3	36.154
XL	-22
Z0	50
f	1e+9
r	0.14
x	-0.44

Table 5-8: MATLAB Solutions

The calculation results in Table 5-8 show that B3 = -0.032 and X3 = 36.154, therefore, the shunt element is an inductor L1= 5.008 nH and the series element is another inductor L2=5.754 nH, as shown in Figure 5-26.

To plot the response of the matching network, create a new workspace and open a new schematic window. Add capacitor and inductor and connect the matching elements as shown in Figure 5-27.

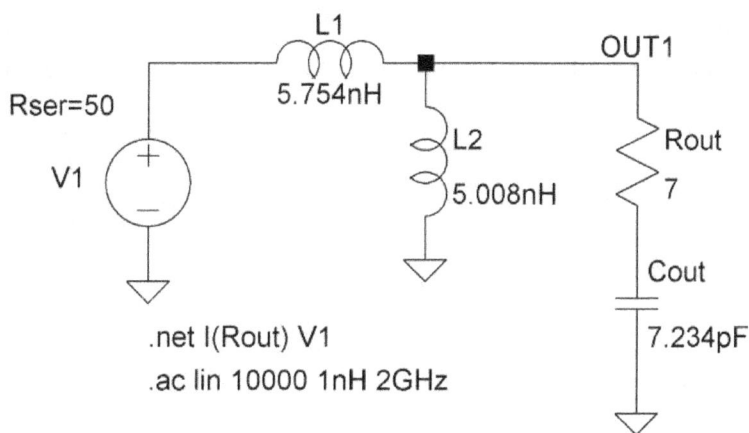

Figure 5-27 Schematic of the third matching L-network

Simulate the schematic from 500 to 1500 MHz and display the input return loss, S11, in a rectangular plot.

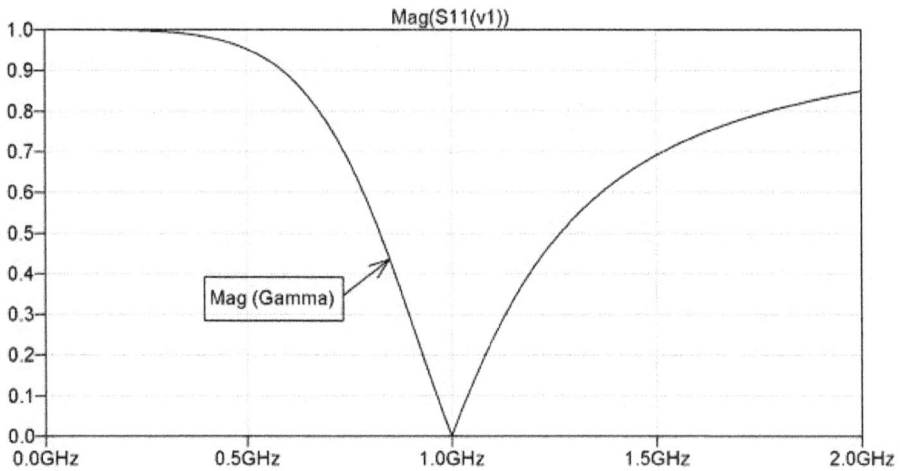

Figure 5-28 Magnitude of the input reflection coefficient

The simulated response in Figure 5-28 shows that the matching L-network perfectly matches the load impedance to the source impedance at 1 GHz.

Example 5.11: Design the fourth L-network that matches a real source impedance, $Z_0 = 50\ \Omega$, to a complex load impedance, $Z_L = 7 - j22$ Ω, at a frequency of 1 GHz.

Fourth Solution: For the fourth solution use Equations (5-38) and (5-39) to calculate the element values of the matching network. The procedure follows.

1. Enter design parameters and normalize the load impedance

Z0=50; RL=7; XL=-22; f=1000e6; r=RL/Z0; x=XL/Z0

2. Use Equations (5-38) and (5-39) to calculate the matching element values

B4=(x-sqrt(r*(r^2+x^2-r)))/(Z0*(r^2+x^2))

X4=-Z0*sqrt((r^2+x^2-r)/r)

C2=-1/(2*pi*f*X4)

L1=-1/(2*pi*f*B4)

The inductor and capacitor values of the matching network is given by the following Equations.

C2=-1/(2*pi*f*X4)
L1=-1/(2*pi*f*B4)

The MATLAB solutions are given in Table 5-9.

Name	Value
B4	-0.051
C2	4.402e-12
L1	3.135e-9
RL	7
X4	-36.154
XL	-22
Z0	50
f	1e+9
r	0.14
x	-0.44

Table 5-9: MATLAB Solutions

The solutions in Table 5-9 show that B4 = -0.051 and X1 = -36.154, therefore, the shunt element is an inductor L1= 3.135 nH and the series element is a capacitor C2=4.402 pF.

To plot the response of the matching network, create a new schematic and add the matching elements as shown in Figure 5-29.

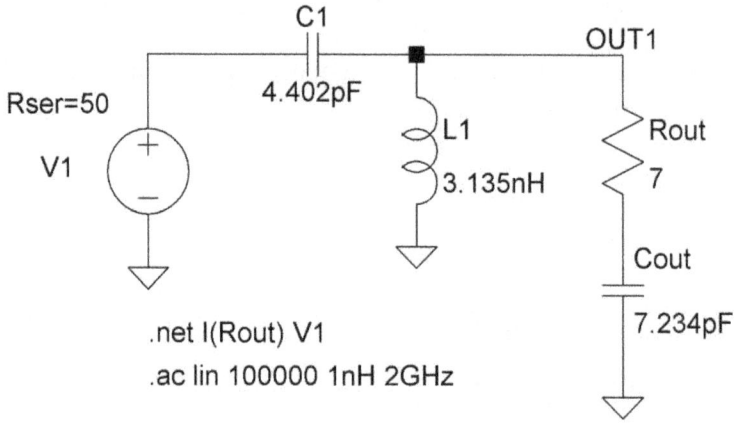

Figure 5-29 Schematic of the fourth matching L-network

Simulate the schematic and display S11 in a rectangular plot.

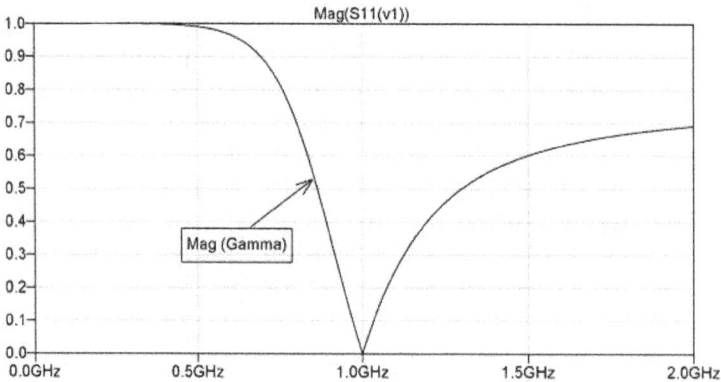

Figure 5-30 Magnitude of the input reflection coefficient

The simulated response in Figure 5-29 shows that the matching L-network perfectly matches the load impedance to the source impedance at 1 GHz.

Matching a Real Load to a Real Source Impedance

When source and load impedances are both real the first matching configuration of Figure 5-7 is redrawn in Figure 5-31.

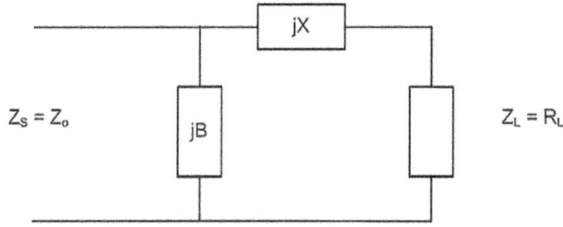

Figure 5-31 First matching configuration with $X_L = X_S = 0$

To design the matching network, apply the maximum power transfer condition and require that $Z_0^* = Z_{IN}$. Therefore,

$$Z_0 = \frac{1}{jB + \left(\dfrac{1}{jX + R_L}\right)} \tag{5-40}$$

Substituting $r = \dfrac{R_L}{Z_0}$ in Equation (5-40), we get:

$$Z_0 = \frac{1}{jB + \left(\dfrac{1}{jX + rZ_0}\right)} \tag{5-41}$$

The solutions for B and X in Equation (5-41) can be obtained by reusing Equations (5-32) and (5-34) and substituting x = 0 in Equations (5-33) and (5-35).

$$B_1 = \frac{\sqrt{(1-r)/r}}{Z_0} \tag{5-42}$$

$$X_1 = Z_0\sqrt{r(1-r)} \tag{5-43}$$

$$B_2 = -\frac{\sqrt{(1-r)/r}}{Z_0} \tag{5-44}$$

$$X_2 = -Z_0\sqrt{r(1-r)} \tag{5-45}$$

Note that the two solutions given by Equations (5-42) through (5-45) are only valid if *r* is less than 1 or $R_L < Z_0$. If r > 1, the second matching network configuration is used.

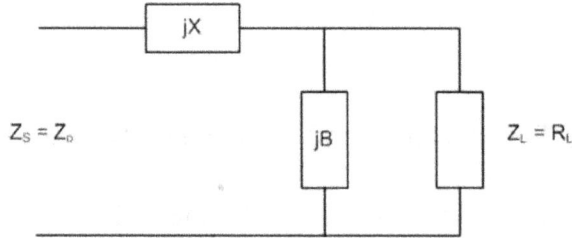

Figure 5-32 Second matching configuration with resistive load and source

To calculate the B and X values, we require that, $Z_0{}^* = Z_{IN}$, therefore,

$$Z_0 = jX + \cfrac{1}{jB + \left(\cfrac{1}{Z_0 r}\right)}$$

(5-46)

The solutions for B and X in Equation (5-46) can be obtained by substituting x = 0 in Equations (5-36) through (5-39).

$$B_3 = \frac{\sqrt{r-1}}{Z_0 r}$$

(5-47)

$$X_3 = Z_0\sqrt{r-1}$$

(5-48)

$$B_4 = \frac{-\sqrt{r-1}}{Z_0 r}$$

(5-49)

$$X_4 = -Z_0\sqrt{r-1}$$

(5-50)

Note that the two solutions given by Equations (5-47) through (5-50) are only valid if *r* is greater than 1 or $R_L > Z_0$. The combined conditions for the validity of solutions are given below and summarized in Table 5-10.

Case 1 If r is less than 1 there are two L-networks that match the two impedances. The two solutions are given by Equations (5-42) through (5-45).

Case 2 If r is greater than 1 there are two L-networks that match the two impedances. The two solutions are given by Equations (5-47) through (5-50).

Case No	Condition	Solutions	Equations
1	$r < 1$	2	(5-42) to (5-45)
2	$r > 1$	2	(5-47) to (5-50)

Table 5-10 Impedance matching conditions and the number of solutions

Example 5.12: Design an L-network to match a 10 Ω load to a 50 Ω source resistor at 500 MHz.

Because the load resistor is smaller than the source resistor, the example has two solutions given in Equations (5-42) through (5-45).

First Solution: For the first solution use Equations (5-42) and (5-43) to design the matching L-network. The procedure follows:

1. Enter design parameters and normalized load impedance

Z0=50; RL=10; f=500e6
r=RL/Z0

2. Use Equations (5-42) and (5-43) in MATLAB script to calculate the matching element values

B1=sqrt((1-r)/r)/Z0

X1=Z0*sqrt(r*(1-r))

L1=X1/(2*pi*f)

C2=B1/(2*pi*f)

The solution are given in Table 5-11.

Name	Value
B1	0.04
C2	12.73e-12
L1	6.366e-9
RL	10
X1	20
Z0	50
f	500e+6
r	0.2

Table 5-11: MATLAB Solutions

The calculation results in Table 5-11 show that B1 = 0.04 and X1 = 20, therefore, the series element is L1=6.366 nH and the shunt element is C2=12.73 pF.

To generate the schematic and plot the response of the matching network, create a new schematic and add the matching elements, as shown in Figure 5-33.

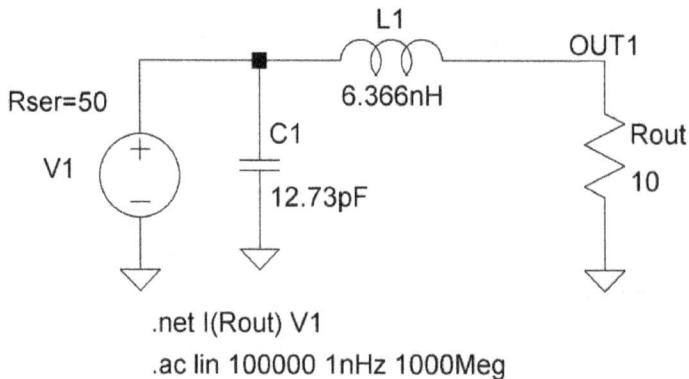

.net I(Rout) V1
.ac lin 100000 1nHz 1000Meg

Figure 5-33 Schematic of the first matching L-network

Simulate the schematic and display S11 and S21 in a rectangular plot.

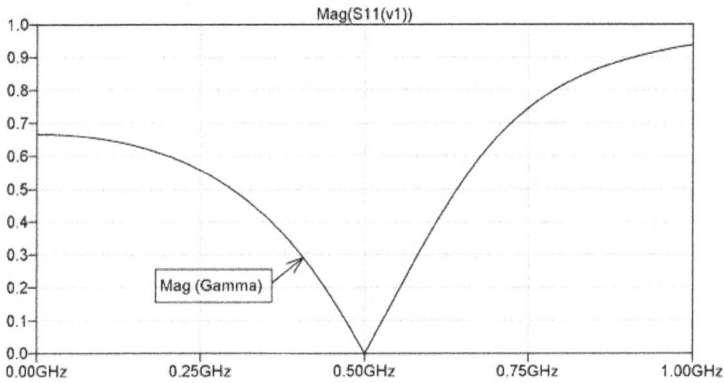

Figure 5-34 Magnitude of the input reflection coefficient

The simulated response in Figure 5-33 shows that the matching L-network perfectly matches the load impedance to the source impedance at 0.5 GHz.

Example 5.13: Design the second L-network to match a 10 Ω load to a 50 Ω source resistor at 500 MHz.

Second Solution: To design the matching L-network, use Equations (5-44) and (5-45) in Figure 5-31 and calculate the matching element values.

1. Enter design parameters and normalize the load impedance

Z0=50; RL=10; f=500e6

r=RL/Z0

2. Calculate matching element values

B2=-sqrt((1-r)/r)/Z0

X2=-Z0*sqrt(r*(1-r))

C1=-1/(2*pi*f*X2)

L2=-1/(2*pi*f*B2)

The MATLAB solutions are given in Table 5-12.

Name	Value
B2	-0.04
C1	15.92e-12
L2	7.958e-9
RL	10
X2	-20
Z0	50
f	500e+6
r	0.2

Table 5-12: MATLAB Solutions

The calculation results in Table 5-12 show that B2 = -0.04 and X2 = -20, therefore, the series element is a capacitor, C=15.92 pF, and the shun element is an inductor, L= 7.958 nH.

To generate the schematic and plot the response of the matching network, create a new schematic and add the matching elements as shown in Figure 5-35.

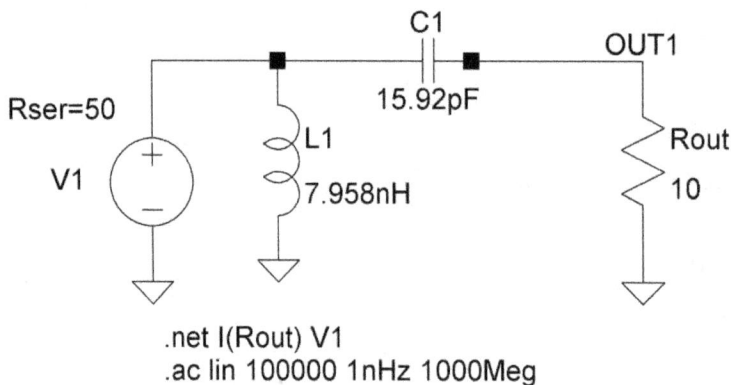

.net I(Rout) V1
.ac lin 100000 1nHz 1000Meg

Figure 5-35 Schematic of the second matching L-network

Simulate the schematic and display S11 in a rectangular plot.

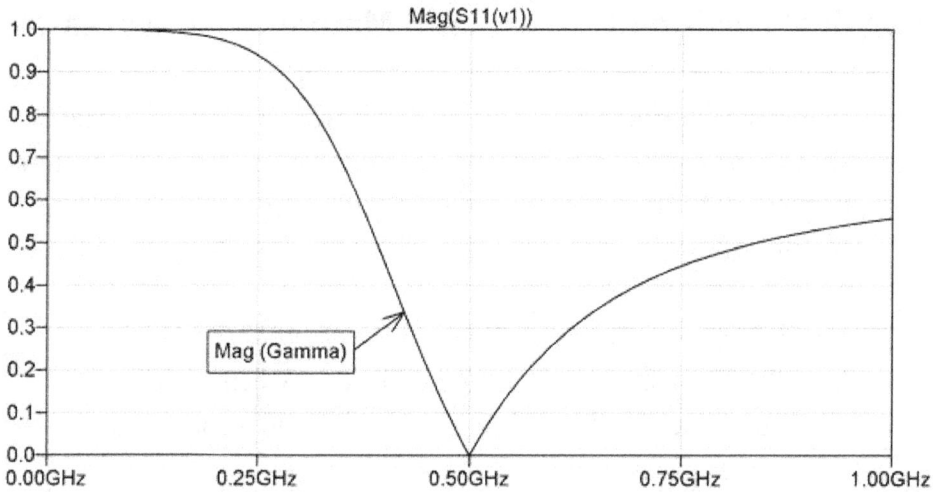

Figure 5-36 Magnitude of the input reflection coefficient

The simulated response in Figure 5-36 shows that the matching L-network perfectly matches the load impedance to the source impedance at 0.5 GHz.

5.5 Broadband Impedance Matching Networks

In the previous sections, the L-section matching networks achieved an impedance match at a fixed frequency capable of producing a 20 dB return loss over a narrow fractional bandwidth of less than 20%. Broadband networks are generally considered to have greater than 20% fractional bandwidths. In this section it is demonstrated that the bandwidth of a matching network can be increased by cascading L-networks. It is demonstrated that by cascading L-networks of equal Q factor, the bandwidth of a network can be increased. The design of equal-Q matching networks is based on the selection of intermediate, or virtual, resistors not necessarily 50 Ω, and then matching the load and source impedance to the virtual resistors. Successively adding additional L-networks of equal Q will continue to extend the bandwidth of the overall circuit.

Analytical Design of Broadband Matching Networks

This section demonstrates the importance of the proper selection of the intermediate network resistance that will result in the best broadband return loss. In Example 5.14 the complex source and load impedance are matched to one specific intermediate resistor thus creating two, equal-Q, L-networks.

The purpose is to show that this method provides a broader matching bandwidth at 20 dB return loss compared to the case when we chose a different resistor value. The Q of each L-network is the loaded Q factor defined by the source and load resistance ratio.

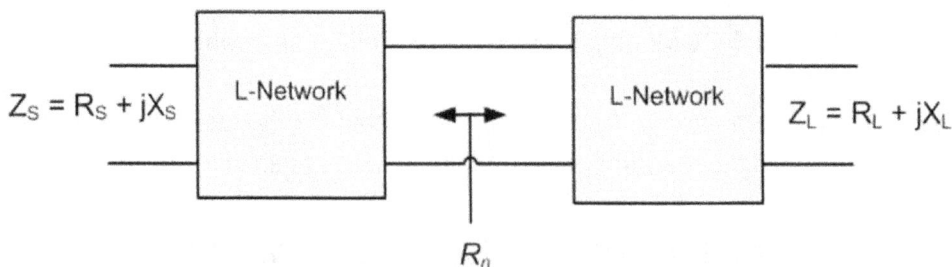

Figure 5-37 Cascaded L-networks with intermediate resistance, R_n

Example 5.14: A transmitter operates over a frequency range of 835 MHz to 1200 MHz. At its center frequency of 1 GHz the source impedance is $Z_S = 55 + j10$ Ω while the antenna input impedance is $Z_L = 20 + j15$ Ω. Design a cascade of two L-networks that matches the transmitter output impedance to the antenna input impedance. Calculate the Q factor for each L-network and show that the Q factors are the same.

Solution: The first step in the broadband solution is to calculate the intermediate resistor, R_n. The optimum intermediate resistor value is equal to square root of the product of the real part of the source and load impedance.

$$R_n = \sqrt{(55)(20)} = 33.166\,\Omega$$

Next we design two matching L-networks, one between the load and the intermediate resistor, and the other one between the source and intermediate resistor. Then we cascade the two L-networks to form the broadband matching L-networks between the source and load impedances.

Use Equations (5-32) and (5-33) to calculate the element values of the matching network between the load impedance and the optimum intermediate resistor. The procedure follows:

1. Enter design parameters and normalized load impedance

Z0=33.166; RL=20; XL=15; f=1000e6

r=RL/Z0

x=XL/Z0

2. Calculate matching element values

B1=sqrt((1-r)/r)/Z0

X1=Z0*sqrt(r*(1-r))-x*Z0

LS1=X1/(2*pi*f)

CP2=B1/(2*pi*f)

The MATLAB solutions are given in Table 5-13.

Name	Value
B1	0.024
B3	0.018
CP2	3,893e-12
LS1	195.3e-12
N	4
R1	39.764
R2	31.623
R3	35.149
RL	20
RS	50
X1	1.227
X3	28.008
XL	15
Z0	33.166
f	1e+9
r	0.603
x	0.452

Table 5-13: MATLAB Solutions

The calculation results in Table 5-13 show that the series element is an inductor, L1 = LS1 = 0.195 nH, and the shunt element is a capacitor, C1 = CP2 = 3.893 pF.

To plot the response of the first matching L-network, create a new schematic and add the capacitor and inductor, as shown in Figure 5-38.

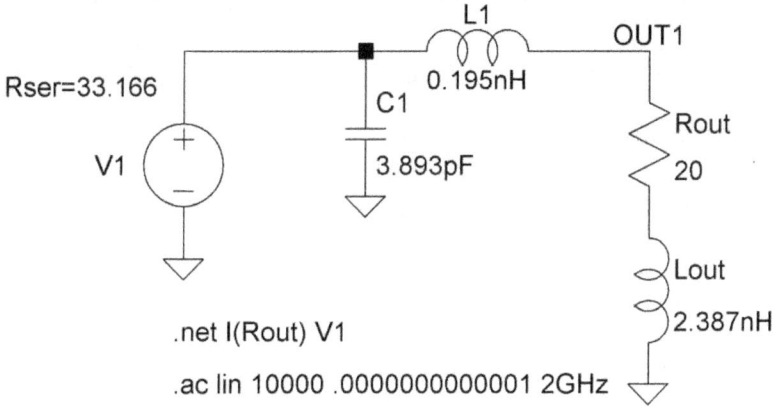

Figure 5-38 Schematic of the first matching L-network

Simulate the schematic and display S11 in a rectangular plot.

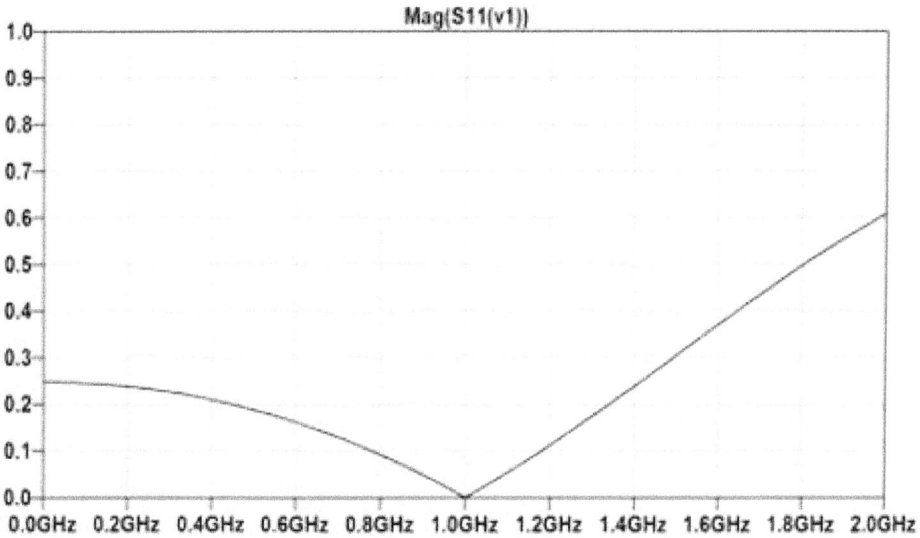

Figure 5-39 Magnitude of the input reflection coefficient

Figure 5-39 shows that the load impedance is matched to the optimum intermediate resistor at 1 GHz.

Then we use Equations (5-36) and (5-37) to design the matching L-network between the source impedance and the intermediate resistor. Procedure follows:

1. Enter design parameters and normalized load impedance for MATLAB calculations.

Z0=33.166; RL=55; XL=10; f=1000e6; r=RL/Z0; x=XL/Z0

2. Calculate matching element values

B3 = (x + sqrt(r*(r^2 + x^2-r)))/(Z0*(r^2+x^2))

X3 = Z0*sqrt((r^2 + x^2-r)/r)

LS1 = X3/(2*p

The MATLAB solutions are given in Table 5-14.

Name	Value
B1	0.024
B3	0.018
CP2	2.875e-12
LS1	4.458e-9
N	4
R1	39.764
R2	31.623
R3	25.149
RL	55
RS	50
X1	1.227
X3	28.008
XL	10
Z0	33.166
f	1e+9
r	1.658
x	0.302

Table 5-14: MATLAB Solutions

The calculation results in Table 5-14 show that the shunt element is a capacitor, C1= CP2 = 2.875 pF, and the series element is an inductor, L1= LS1 = 4.458 nH.

To plot the response of the matching network, create a new schematic and add capacitor and inductor as shown in Figure 5-40.

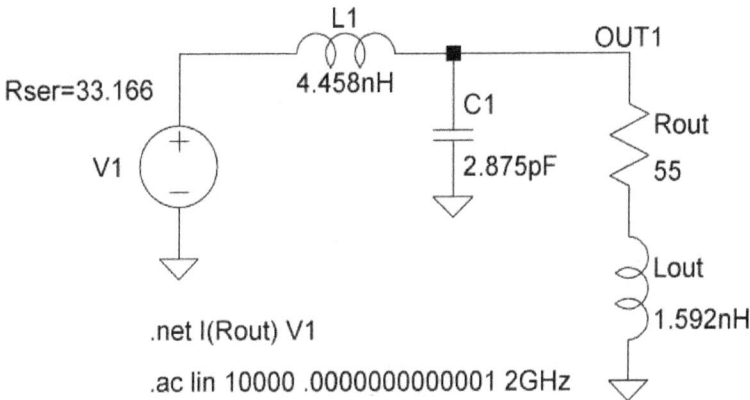

Figure 5-40 Schematic of the matching L-network

Simulate the schematic and display S11 in a rectangular plot.

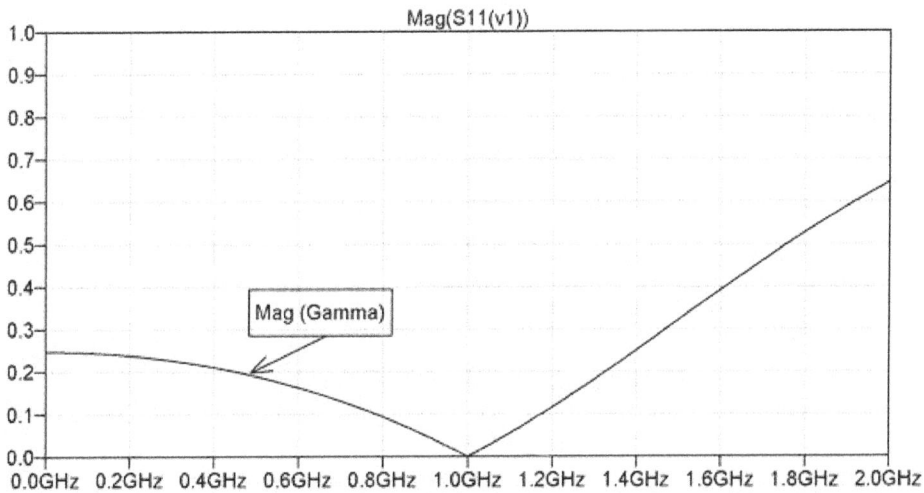

Figure 5-41 Magnitude of the input reflection coefficient

Figure 5-41 shows that the source impedance is matched to the optimum intermediate resistor at 1GHz.

Finally, cascade the two matching L-networks as shown in Figure 5-42.

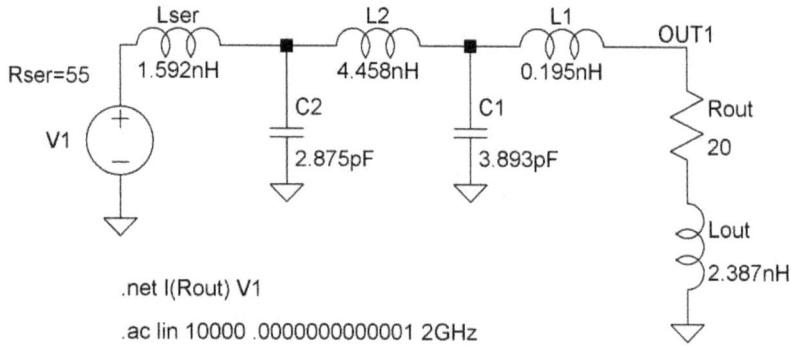

Figure 5-42 Schematic of the cascaded L-networks

Simulate the schematic in Figure 5-42 and display S11 in a rectangular plot.

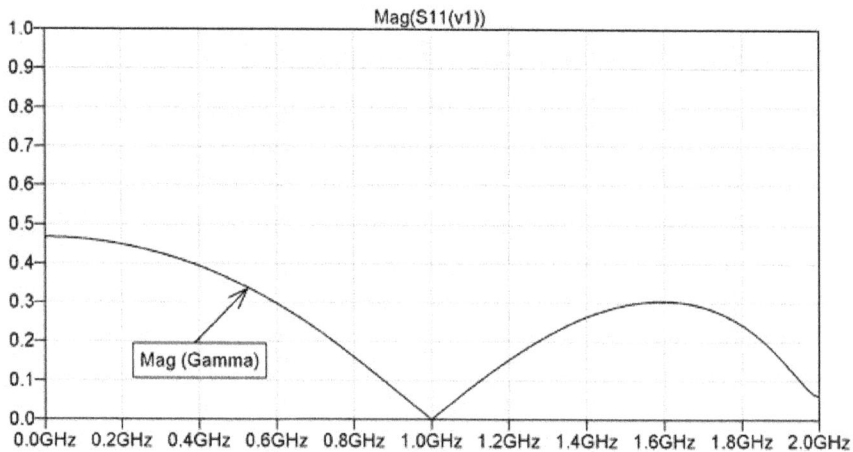

Figure 5-43 Magnitude of the input reflection coefficient

The simulated response in Figure 5-43 shows that the cascaded L-networks perfectly matches the complex load impedance to the complex source impedance at 1 GHz.

The Q factor of the each L-network is calculated by the following equation:

$$Q = \frac{1}{2}\sqrt{\frac{R_2}{R_1} - 1} \qquad\qquad R_2 > R_1$$

Where R_2 and R_1 are the input and output resistors of the L-networks.

For Example 5.14 the Q factors of the source and load are given by:

$$Q_{source} = Q_{load} = \frac{1}{2}\sqrt{\frac{33.166}{20} - 1} = \frac{1}{2}\sqrt{\frac{55}{33.166} - 1} = 0.405$$

Note that when the source and load matching networks have the same Q factor, the reflected power is zero and the maximum power would be transferred from the source to the load at the design frequency.

Example 5.15: Redesign the matching L-networks of Example 5.14 by selecting the intermediate resistance to be equal to 50 Ω instead of 33.166 Ω. Compare the Q factor values with the values for Example 5.14.

Solution: Change the intermediate resistor to 50 Ω and redesign the matching L-networks. Use the following equations to design the matching network between the source and the intermediate resistor.

1. Enter Design Parameters and Normalize Source Impedance
Z0=50; RL=55; XL=10; f=1000e6; r=RL/Z0; x=XL/Z0

2. Calculate B3 and X3 and the Matching Element Values
B3=(x+sqrt(r*(r^2+x^2-r)))/(Z0*(r^2+x^2))
X3=Z0*sqrt((r^2+x^2-r)/r)
LS2=X3/(2*pi*f)
CP1=B3/(2*pi*f)

Also use the following equations to design the matching network between the load and the intermediate resistor.

1. Enter Design Parameters and Normalize Load Impedance Z02=50;
RL2=20; XL2=15; f2=1000e6; r2=RL2/Z02; x2=XL2/Z02

2. Calculate Element Values
B12=sqrt((1-r2)/r2)/Z02
X12=Z02*sqrt(r2*(1-r2))-x2*Z0
LS12=X12/(2*pi*f)
CP22=B12/(2*pi*f)

Name	Value
B1	0.024
B12	0.024
B3	9.699e-3
CP1	1.544e-12
CP2	3.898e-12
CP22	3.898e-12
LS1	1.511e-9
LS12	1.511e-9
LS2	2.939e-9
RL	55
RL2	20
X1	9.945
X12	9.945
X3	18.464
XL	10
X12	15
Z0	50
Z02	50
f	1e+9
f2	1e+9
r	1.1
r2	0.4
x	0.2
x2	0.3

Table 5-15: Combined MATLAB Solutions

Notice that the matching element values in both schematics have changed due to the change in the intermediate resistor. The cascaded schematic of the matching L-networks is shown in Figure 5-44.

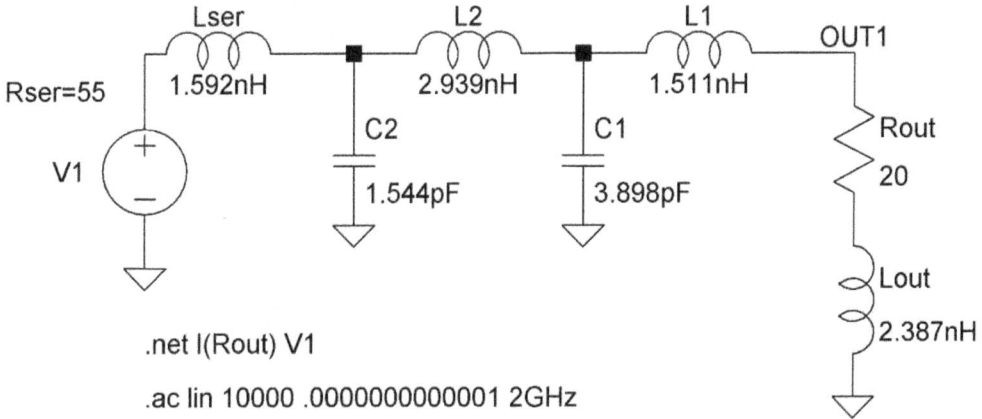

.net I(Rout) V1

.ac lin 10000 .0000000000001 2GHz

Figure 5-44 Schematic of the cascaded matching L-networks

Simulate the schematic in Figure 5-44 and display S11 magnitude in a rectangular plot.

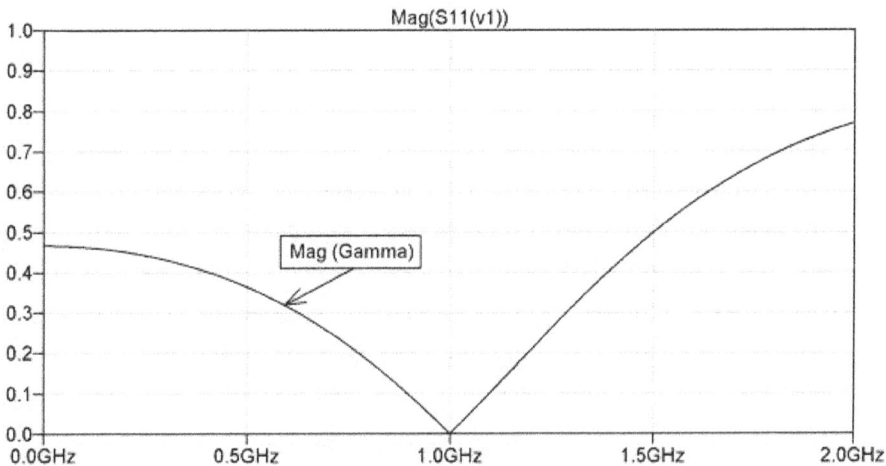

Figure 5-45 Magnitude of the input reflection coefficient

The simulated response in Figure 5-45 shows that the cascaded matching networks perfectly match the load to the source at 1 GHz.

A comparison of the fractional bandwidths for the two examples show that the matching L-networks in Figure 5-43 provides more matching bandwidth, at 20 dB return loss, than the matching L-networks in Figure 5-45. We can also show that the lesser bandwidth is due to the fact that the source and load matching L-networks in the second example do not have the same Q factors, as shown by the following calculations:

$$Q_{source} = \frac{1}{2}\sqrt{\frac{55}{50}-1} = 0.158$$

$$Q_{load} = \frac{1}{2}\sqrt{\frac{50}{20}-1} = 0.612$$

The equal Q factors in Example 5.14 causes the cascaded matching L-networks to transfer more power to the load while the unequal Q factors in example 5.15 cause the matching network to have some power reflected to the source.

Cascaded Broadband Impedance Matching Networks

In Example 5.14 we showed that cascading two equal-Q L-networks convert a narrowband matching L-network into a broadband matching network. Because the Q factor is related to the ratio of the source and load resistance on each L-network, the Q can be further reduced by applying multiple L-networks in cascade. In practice as the number of L-networks becomes greater than five, the element values become difficult to physically realize. In this section we show that by using a network of four equal-Q L-networks in cascade, we can lower the individual Q factor and provide an even greater matching bandwidth for the Example of 5.14.

Example 5.16: Redesign the matching network of Example 5.14 with four equal-Q L-networks. Compare the Q factor with Example 5.14.

Solution: This example uses the same steps developed for Example 5.14. The general equation for the calculation of any number of intermediate resistors is given by Equation (5-51).

$$R_n = R_S (r)^{n/N} \qquad\qquad n = 1, 2, 3, to\ N - 1 \qquad (5\text{-}51)$$

Where:

R_n is the intermediate resistor values

R_S is the source resistor value

r is the normalized load resistor, $r = R_L/R_S$

N is the number of cascading networks

Now we use Equation (5-51) to calculate intermediate resistors for $N = 4$.

1. Enter design parameters and normalize the load impedance

RS=55; RL=20;

r=RL/RS; N=4

2. Use MATLAB script to calculate intermediate resistor values

R1=RS*r^(1/N)

R2=RS*r^(2/N)

R3=RS*r^(3/N)

The calculation results show that R1 = 42.71 Ω, R2 = 33.166 Ω, and R3=25.755 Ω.

Now that the three intermediate resistors are calculated we start designing the four matching L-networks from the load impedance towards the source impedance. The procedure follows.

1. Enter design parameters and normalized load impedance

R3=25.755; RL=20; XL=15; f=1000e6; r=RL/R3; x=XL/R3

2. Use MATLAB script to calculate the matching element values

B1=sqrt((1-r)/r)/R3

X1=R3*sqrt(r*(1-r))-x*R3

CS1=-1/(2*pi*f*X1)

CP2=B1/(2*pi*f)

The MATLAB solutions for the first L-network is given in Table 5-16.

Name	Value
B1	0.021
B3	0.013
CP2	3.315e-12
CS1	37.26e-12
LS1	2.199e-9
N	4
R1	42.71
R2	33.166
R3	25.757
RL	20
RS	55
X1	-4.272
X3	24.547
XL	15
Z0	25.755
f	1e+9
r	0.777
.x	0,582

Table 5-16: MATLAB Solutions

The calculation results in Table 5-16 show that C1= CS1=37.26 pF and C2= 3.315 pF, as shown in Figure 5-46.

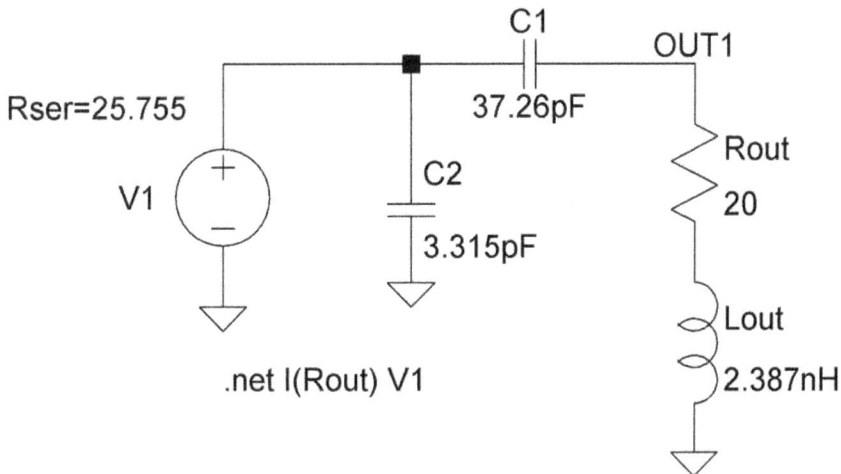

Figure 5-46 Schematic of the first matching L-network

Next design the matching L-network between R3 and R2 as follows. Here R3 is acting as a load.

1. Enter design parameters and normalized load impedance

R2=33.166; RL=25.755; f=1000e6

r=RL/R2

2. Calculate matching element values

B1 = sqrt((1 - r)/r)/R2

X1 = R2*sqrt(r*(1 - r))

LS1 = X1/(2*pi*f)

CP2 = B1/(2*pi*f)

The MATLAB solutions for the L-network between R2 and R3 is given in Table 5-17.

Name	Value
B1	0.016
B3	0.013
CP2	2.574e-12
CS1	37.26e-12
LS1	2.199e-9
N	4
R1	42.71
R2	33.166
R3	25.755
RL	25.755
RS	55
X1	13.816
X3	24.547
XL	15
Z0	33.166
f	1e+9
r	0.777
x	0,582

Table 5-17: MATLAB Solutions

The calculation results in Table 5-17 show that L= LS1=2.199 nH and C1=CP2= 2.574 pF, as shown in Figure 5-47.

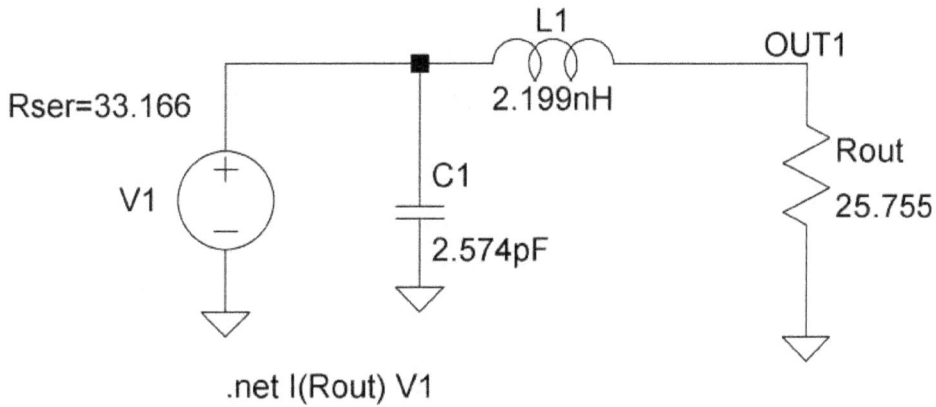

Figure 5-47 Schematic of the second matching L-network

Next design the matching L-network between R1 and R2 as follows. Here R2 is acting as a load.

1. Enter design parameters and normalized load impedance

R1 = 42.71; RL = 33.166; f = 1000e6
r = RL/R1

2. Calculate matching element values

B1 = sqrt((1 - r)/r)/R1
X1 = R1*sqrt(r*(1 - r))
LS1 = X1/(2*pi*f)
CP2 = B1/(2*pi*f)

The MATLAB solutions for the L-network between R1 and R2 is given in Table 5-18.

Name	Value
B1	0.013
B3	0.013
CP2	1.999e-12
CS1	37.26e-12
LS1	2.832e-9
N	4
R1	42.71
R2	33.166
R3	25.755
RL	33.166
RS	55
X1	17.791
X3	24.547
XL	15
Z0	42.71
f	1e+9
r	0.777
x	0.582

Table 5-18: MATLAB Solutions

The calculation results in Figure 5-18 show that L1= LS1=2.832 nH and C1=CP2=1.999 pF, as shown in Figure 5-48.

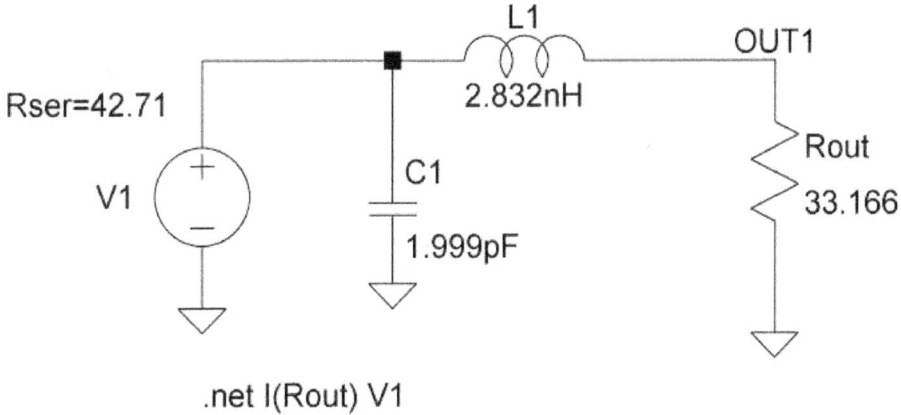

.net I(Rout) V1

Figure 5-48 Schematic of the third matching L-network

Finally, design the matching L-network between R1 and the source impedance. Here R1 is acting as a load.

1. Enter design parameters and normalized load impedance

R1 =Z_0= 42.71; RL = 55; XL = 10; f = 1000e6

r = RL/Z0

x = XL/Z0

2. Calculate matching element values

B3 = (x+sqrt(r*(r^2 + x^2 - r)))/(R1*(r^2 + x^2))

X3 = R1*sqrt((r^2 + x^2 - r)/r)

LS1 = X3/(2*pi*f)

CP2 = B3/(2*pi*f)

Part of the MATLAB solutions for the L-network between R1 and the source impedance is given in Table 5-19.

Name	Value
B1	0.013
B3	0.013
CP2	2.119e-12
CS1	37.26e-12
LS1	3.907e-9

Table 5-19: MATLAB Solutions

The calculation results in Table 5-19 show that L1=LS1=3.907 nH and C1= CP2=2.119 pF, as shown in Figure 5-49.

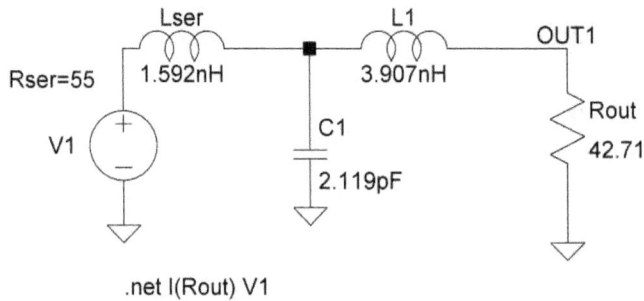

.net I(Rout) V1

Figure 5-49 Schematic of the fourth matching L-network

Cascade the four L-networks as shown in Figure 5-50.

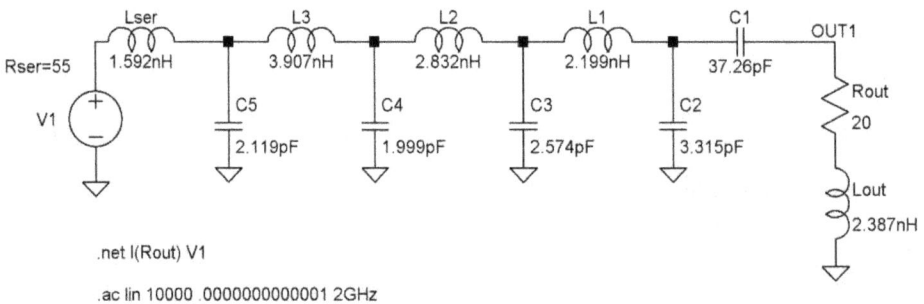

.net I(Rout) V1

.ac lin 10000 .0000000000001 2GHz

Figure 5-50 Schematic of the matching network using 4 L-networks

The simulated response of the schematic in Figure 5-50 is shown in Figure 5-51.

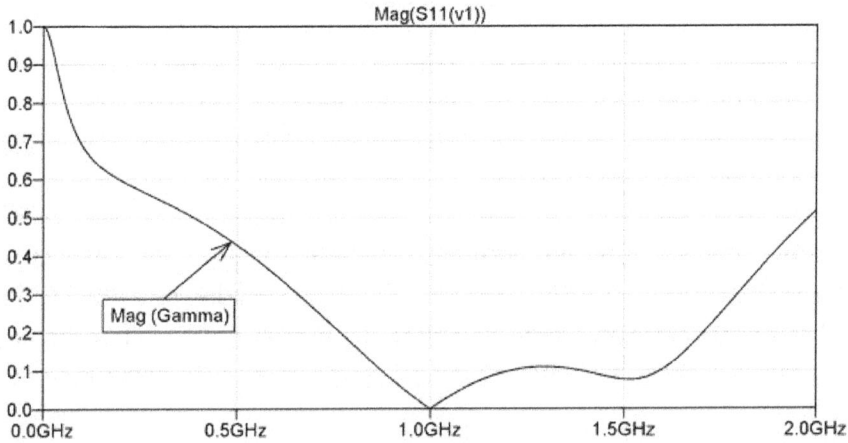

Figure 5-51 Magnitude of the input reflection coefficient

The S11 simulated response in Figure 5-51 shows that the four matching L-networks perfectly match the load impedance to the source impedance at 1 GHz.

As we stated earlier, the reason for wider matching bandwidth is that all four individual matching networks in Figure 5-50 have the same Q factor as shown in the following calculations.

$$Q_{4L} = \frac{1}{2}\sqrt{\frac{55}{42.71}-1} = \frac{1}{2}\sqrt{\frac{42.71}{33.166}-1} = \frac{1}{2}\sqrt{\frac{33.166}{25.755}-1} = \frac{1}{2}\sqrt{\frac{25.755}{20}-1} = 0.268$$

5.6 Derivation of Equations for Loaded Q and N

The loaded Q of a single L-network matching two real resistors R_1 and R_2 is given by:

$$Q_1 = \frac{1}{2}\sqrt{\frac{R_2}{R_1} - 1} \qquad\qquad R_2 > R_1$$

For a cascade of N equal-Q L-networks, the individual Q factor is given by:

$$Q_N = \frac{1}{2}\sqrt{\left(\frac{R_2}{R_1}\right)^{1/N} - 1} \quad N = 1,2,3 \qquad (5\text{-}52)$$

Notice that by increasing the number of L-networks the individual Q factor is decreased and, as a result, the matching bandwidth will be increased. Equation (5-52) can be used to calculate the number of L-networks needed for a given resistor ratio and Q factor. Equation (5-52) can be written as:

$$1 + 4Q_N^{\,2} = \left(\frac{R_2}{R_1}\right)^{1/N}$$

Or

$$\left[1 + 4Q_N^2\right]^N = \left(\frac{R_2}{R_1}\right)$$

Taking logarithm of both sides, the required number of cascaded L-networks, N, can be obtained from Equation (5-53).

$$N \geq \frac{\log\left(\dfrac{R_2}{R_1}\right)}{\log\left[1 + 4Q_N^2\right]} \qquad (5\text{-}53)$$

Notice that in Equation (5-53) the resistor R_2 must be greater than R_1 and the number of L-networks, N, must be a positive integer.

Example 5.17: For a load and source resistor ratio of 3, find the minimum number of cascaded L-networks to achieve a loaded Q of 0.3.

Solution: To calculate N substitute the resistor ratio and Q into Equation (5-53).

$$N \geq \frac{\log(3)}{\log\left[1+4(0.3)^2\right]} = \frac{0.477}{0.133} = 3.58$$

Therefore, a minimum of four L-networks is needed to achieve a Q of 0.3 or less.

For verification, substitute the resistor ratio and N into Equation (5-52) to calculate the Q factor.

$$Q_4 = \frac{1}{2}\sqrt{\left(\frac{R_2}{R_1}\right)^{1/4} - 1} = 0.28$$

5.7 Designing with Q-Curves on the Smith Chart

We have seen that increasing the number of matching L-networks increases the bandwidth of the matching networks. Each successive section is matched to an intermediate resistance, R_n. The overall Q of the network is determined by the ratio of the source to load resistor, as given by Equation (5-52). The Q factor is reduced as the number of L-networks is increased.

We have also shown that the bandwidth of such a cascade is optimized when the Q factor of all sections are the same. The cascade of successive L-networks of constant Q can also be accomplished on the Smith Chart. Curves of constant Q can be drawn on the Smith Chart at every point at which $Q = X / R$. Such a plot of Q curves is shown in Figure 5-52 for Q factor values of 1, 2, and 3.

A given Q curve is actually the external Q of a given L-network section. As such it does not account for loading on both ports of the network and is therefore two times the value of the Q as defined by Equation (5-52). The matching sequence progresses between the source impedance and the conjugate of the load impedance using the reactive element movements on the Smith Chart. Each L-network should intersect a point on the real axis of the Smith Chart.

The points on the real axis represent the intermediate resistance, R_n. The inductor and capacitor values must be chosen to stay within the boundary of the Q curve. Keeping each section equal distant to the Q curve results in a network in which each L-network has the same external Q factor. This is not to be confused with the overall loaded Q of the matching network.

The Q Curve selection can be visualized as a compromise between the desired bandwidth and the number of L-networks that may be required for a given impedance match. Using the smallest Q curve will result in the widest possible bandwidth. It can also be noted that for very high Q load impedances, it may not be possible to achieve a broadband match with reactive L-networks. For these types of loads it may be necessary to use transmission line transformer techniques [8]. Even in narrowband applications, engineers prefer to keep the matching network Q as low as possible in circuits that involve high power levels. High Q matching networks can create very high voltage or current peaks when excited with high RF power levels. The Q curve technique will be demonstrated by revisiting Example 5.14.

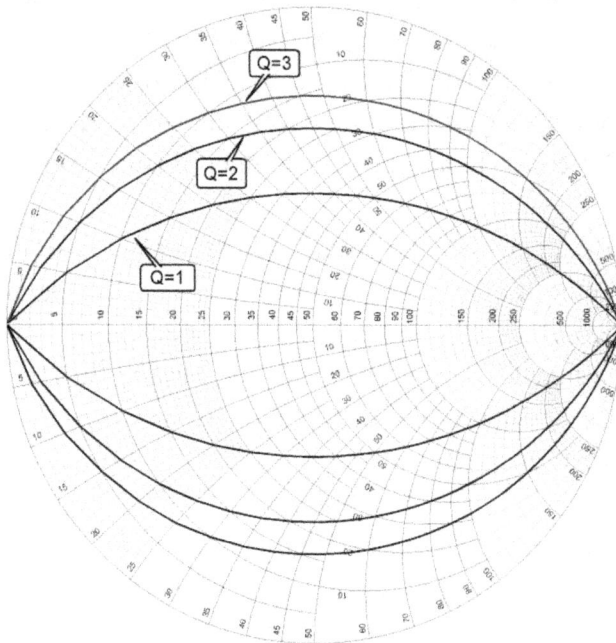

Figure 5-52 Smith Chart display of constant Q-curves

Q-Curve Impedance Matching Example

Example 5.18: A transmitter operates over a frequency range of 835 MHz to 1200 MHz. At its center frequency of 1 GHz the source impedance is Z_S = 55 + j10 Ω while the antenna input impedance is Z_L = 20 + j15 Ω. Use the constant Q curve method to implement the cascaded matching L-networks.

Solution: A separate schematic and analysis is created to generate the Q curve. Create a schematic that consists of a one port impedance element. Make the resistance and reactance values tunable variables. The S parameter simulation should be set at a fixed frequency in which any frequency can be used. Add a Parameter Sweep so that the resistance of the one port impedance can be swept over a range of 0.1 Ω to 4000 Ω in 5 Ω steps. Then write equations to calculate the impedance array that result in points of constant Q. These points can then be plotted on the Smith Chart to create the Q curve. The desired value of the Q Curve can be entered in the equations or in the Tune Window. The procedure follows:

Enter the desired Q Curve value twice the unloaded Q of Example 5.16.

$$Q=2(0.268) = 0.536$$

- Calculate R1 values using X1 from the Parameter Sweep

  - R1=X1./Q

- Define the complex impedance

  - Ztop=complex(R1,X1)

- Normalize the impedance to plot on Smith Chart

  - Ztopn = (Ztop-50)/(Ztop+50)

- Plot top circle

  - Setindep ("Ztopn","X1")

- Plot bottom half circle and define the complex impedance

  - Zbottom = complex (R1,-X1)

- Normalize the impedance to plot on Smith Chart

    - Zbottomn = (Zbottom-50)/(Zbottom+50)

- Plot bottom circle

    - Setindep ("Zbottomn","X1")

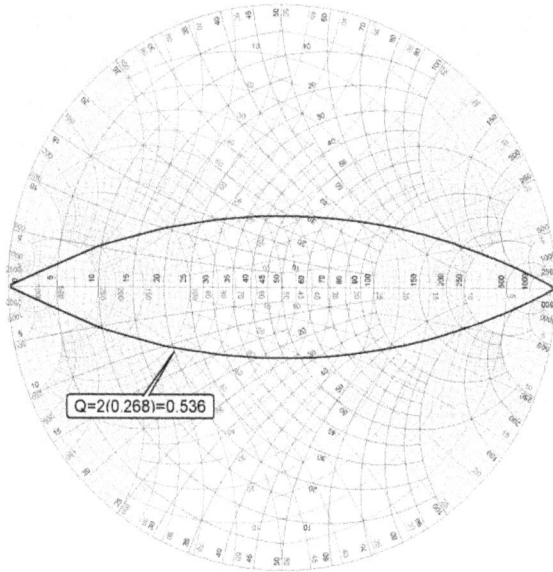

Figure 5-53 Q curve plot for Q=.536

The Q curve for a Q value of 0.536 is shown in Figure 5-53. Because Q-curve plots do not consider loading on both ports of the network, we need to double the individual Q factor in order to obtain the external Q used in plotting the Q curve. For the network in Figure 5-50 the individual Q factor was 0.268, therefore, for plotting the Q curve on the Smith Chart we must use $Q = 2 (0.268) = 0.536$ on the Smith Chart and in the equations.

Impedance Matching Progression: Using the lumped element movement techniques learned in Chapter 3 the matching network can be designed on the Smith Chart. On the Smith Chart we are matching from the load at Z_A to the conjugate of the source at Z_I. The Smith Chart of Figure 5-54 shows the lumped element movements on the constant conductance and resistance circles that maintain equal distant arcs within the specified Q curve. This ensures that each L-network maintains equal Q. The matching sequence between the source and load impedance progresses by using a series inductor from Z_A to intersect the Q curve at Z_B. Then add a shunt capacitor to move from the Q curve to the real axis at point Z_C. This process is continued until the last shunt capacitor intersects the impedance at Z_I.

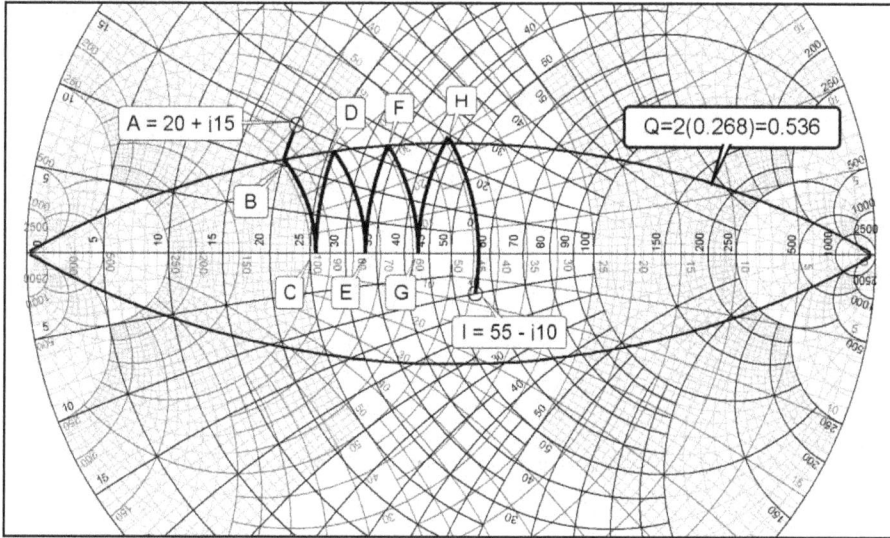

Figure 5-54 Q curve output for Example 5.4-3 (Q=0.536)

The Q curve plot of Figure 5-54 shows that all the node impedances fall on the same Q curve. To see the change in impedance as we move from the load impedance at 20 + j15 to the source impedance at 55 + j10 we have connected the points together indicating the addition of capacitors and inductors on the Smith Chart. The displayed node impedances on the Smith Chart show that three of the impedances fall on the real axis. These are the intermediate resistors. The intermediate virtual resistance values on the real axis are found to be:

$$Z_C = 25.777 \ \Omega \quad Z_E = 33.167 \Omega \quad Z_G = 42.715 \Omega$$

The impedances on the Q = 0.536 curve is measured as:

$$Z_B = 20 + j10.729 \ \Omega \qquad Z_D = 25.777 + j13.816 \Omega$$

$$Z_F = 33.167 + j17.797 \ \Omega \qquad Z_H = 42.715 + j24.55 \ \Omega$$

5.8 Limitations of the Broadband Matching Networks

From the previous example of section 5.5 we can deduce that the Q of the complex source and load impedance has an impact on the maximum bandwidth over which a good impedance match, low return loss, can be achieved. There is a finite limit on the achievable return loss for a given load impedance and circuit bandwidth known as Fano's limit [1]. Fano's Limit is the optimum reflection coefficient that can be achieved with a given load impedance. It is a theoretical limit that considers an infinite number of lossless matching elements. Fano's limit is defined by Equation (5-54).

$$|\Gamma| = e^{\frac{-\pi Q_L}{Q_{UL}}} \tag{5-54}$$

Where: Q_{UL} = the unloaded Q or the ratio of the reactance to resistance of the load and Q_L = the loaded Q of the network or the ratio of the center frequency of the network divided by the 3 dB bandwidth as defined by Equation (4-11) in chapter 4 section 4.2.

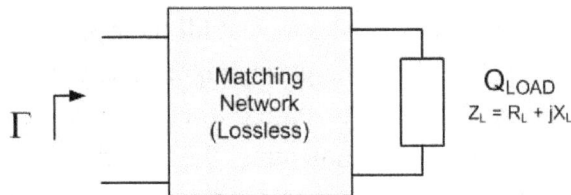

Figure 5-55 General configuration of impedance matching network

Example of Fano's Limit Calculation

In the following example we apply the Fano's limit to calculate the optimum reflection coefficient for a given network.

Example 5.19: Apply Fano's Limit to find the lowest reflection coefficient that can be achieved in the 3 dB range of the network shown in Figure 5-22.

Solution: From the network response in Figure 5-24 we notice that f_a=659 MHz and f_b=1228 MHz, therefore, the center frequency is $f_0 = \sqrt{(1228)(659)}$ = 899.5 MHz. The matching network loaded Q_L is defined as:

$$Q_L = \frac{f_0}{f_b - f_a} = \frac{899.5\,MHz}{1228\,MHz - 659\,MHz} = 1.580$$

For the load impedance, Z_L = 7- j22, the unloaded Q_{UL} is defined as:

$$Q_{UL} = \frac{X_L}{R_L} = \frac{22}{7} = 3.14$$

Fano's limit states that the best achievable Γ is given by:

$$|\Gamma| = e^{\frac{-\pi Q_L}{Q_{UL}}} = e^{\frac{-\pi(1.580)}{3.14}} = 0.20$$

From Table 2-3 we know that a reflection coefficient of 0.201 is equivalent to a return loss of 13.933dB or 1.503 VSWR. We must remember that Fano's limit gives a strictly theoretical result for return loss assuming lossless components and an infinite number of elements. The reflection coefficient defined by Fano's limit is essentially a rectangular return loss characteristic which is not possible to achieve in practice. It can be used however to give a quick estimate of the difficulty of a particular matching network, particularly for high Q load impedances.

5.9 Filter Characteristics of the L-networks

The matching L-networks that we have designed take the form of either a low pass or high pass characteristic. Based on the location of the inductor and capacitor the network acts as a low pass or a high pass filter. The low pass network consists of a series inductor and a shunt capacitor while the high pass network consists of a series capacitor and shunt inductor. Figures 5-56 and 5-57 show two low pass configurations that are used to match a source and load impedance based on the ratio of the real component of the source and load impedance. With the low pass configuration at very low frequencies, the inductive reactance is very small (short circuit) and the output is close to input. At very high frequencies the inductor has a high reactance and the output is near zero.

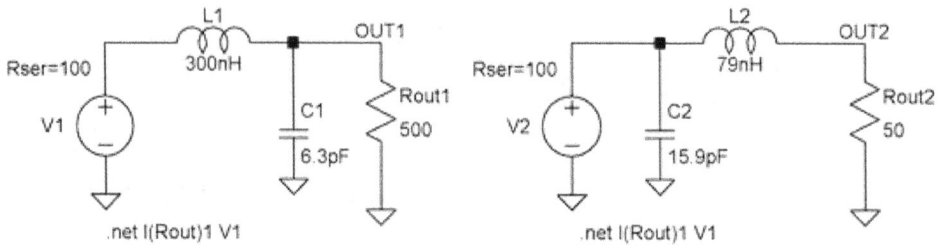

Figure 5-56 Low pass matching L-networks

The high pass network consists of a shunt inductor and a series capacitor. Figures 5-58 and 5-59 show two high pass configurations that are used to match a source and load impedance based on the ratio of the real component the source and load impedance. At very low frequencies the inductor has a very small reactance, (short circuit), and the output is close to zero. At high frequencies the inductor has a high reactance, (like an open), and the output is nearly equal to input. At very low frequencies the capacitor has a very high reactance, (open circuit), and the output is close to zero while at high frequencies the capacitor has a small reactance, (short circuit), and the output is nearly equal to the input.

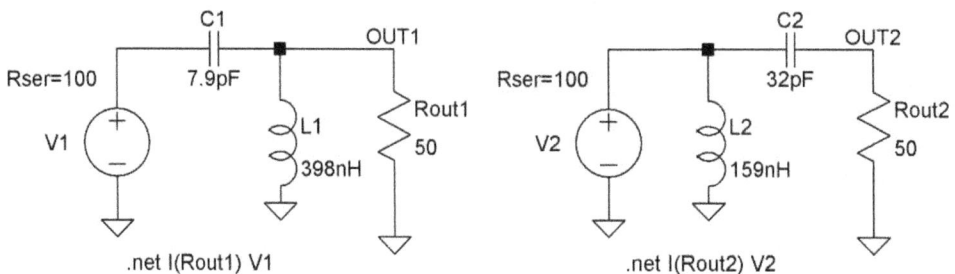

Figure 5-57 High pass matching L-networks

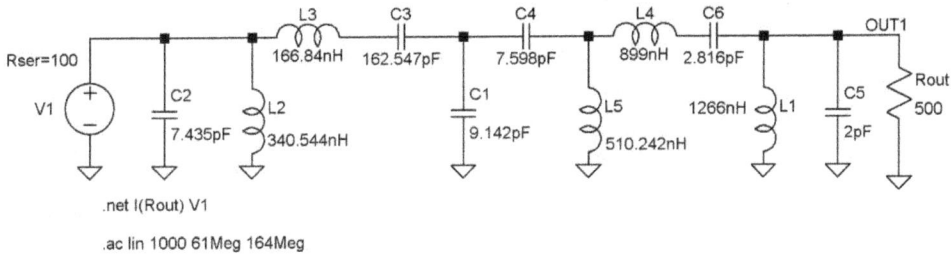

.net I(Rout) V1

.ac lin 1000 61Meg 164Meg

Figure 5-58 High pass matching L-networks

Effect of Finite Q on the Matching Networks

To complete the matching circuit design, the components need to be converted to their physical equivalents by accounting for the finite Q factor of the capacitors and inductors. This can be accomplished by using either S parameter or Modelithics models for the elements. A quick approach is to model each element with a Q factor that closely approximates the physical components.

Note that a Q factor can be specified for each element in the Matching network. In this design we'll demonstrate a technique to assign a common variable name representing the Q factor of each component. This allows the evaluation of the necessary Q factor of the components required to meet the specifications of the matching network.

The variable name, Qcap, is assigned to all capacitors in the matching network. Similarly the variable name, Qind, is assigned to all of the inductors. The frequency at which the Q factor is defined is also defined as a variable, Qf. This allows the Q factor of all inductors and all capacitors to be tuned simultaneously.

Figure 5-59 shows the revised schematic of Figure 5-58 with the Q factor defined for all components.

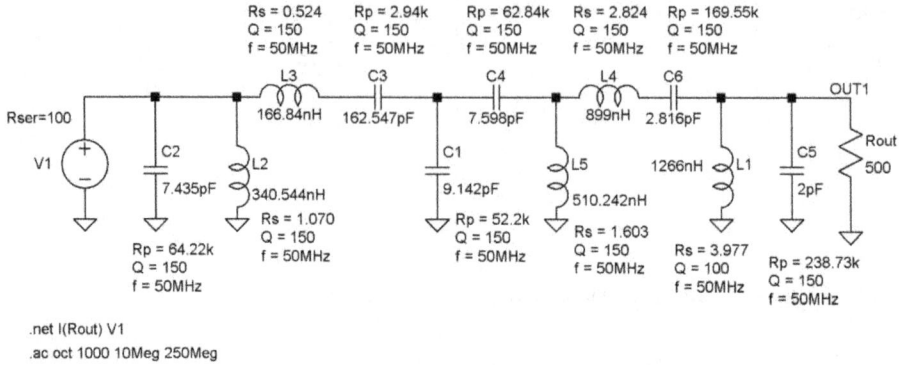

Figure 5-59 Fifth order matching schematic with Q factor variables

The network response with the initial Qcap and Qind values is shown in Figure 5-59. We can see that the insertion loss (>-1dB) and return loss (<20 dB) specifications have been met. It is not necessary to tune each individual capacitor Q in the matching network. As Figure 5-59 shows, a capacitor Q of 50 and an inductor Q of 40 results in an insertion loss that falls short of meeting the insertion loss specification. The response with Qcap=50 and Qind=40 is shown in Figure 5-60.

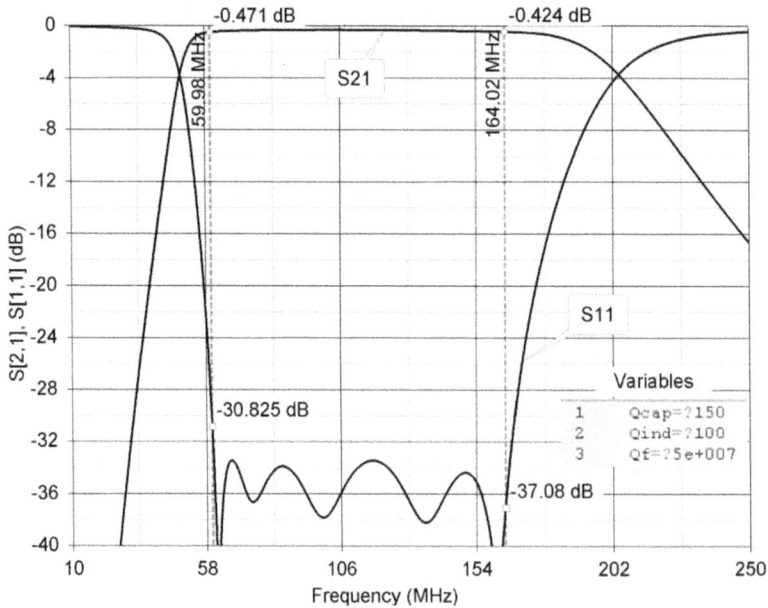

Figure 5-60 Response with Qcap=150 and Qind=100

References and Further Readings

[1] R. K. Feeney and D. R. Hertling, *RF/Wireless Principles & Practice*, Georgia Institute of Technology, 1998.

[2] Ali Behagi, *RF and Microwave Circuit Design*, A Design Approach Using (**ADS**), Techno Search, Ladera Ranch, CA 2017.

[3] Ali Behagi and Manou Ghanevati, *Fundamentals of RF and Microwave Circuit Design,* Practical Analysis and Design Tools, Techno Search, Ladera Ranch, California 2017

[4] Guillermo Gonzales, *Microwave Transistor Amplifiers – Analysis and Design*, Second Edition, Prentice Hall Inc., Upper Saddle River, NJ.

[5] Randy Rhea, *The Yin-Yang of Matching: Part 2 – Practical Matching Techniques*, High Frequency Electronics, March 2006

[6] Steve C. Cripps, *RF Power Amplifiers for Wireless Communications*, Artech House Publishers, Norwood, MA. 1999.

[7] David M. Pozar, *Microwave Engineering*, Third Edition, John Wiley & Sons, New York, 2005

[8] R. Ludwig, P. Bretchko, *RF Circuit Design*, Theory and Applications, Prentice Hall, Upper Saddle River, NJ, 2000

[9] Jerry Sevick, *Transmission Line Transformers*, The American Radio Relay League, Newington, CT., 1990.

Problems

5-1. For a load of 75 Ω with a 10 pF series capacitance we want to have maximum power transfer at 1 GHz. Find a source inductance that cancels the reactance of the load at this frequency.

5-2.   Design all the L-networks that will match a complex load impedance, $Z_L = 15 - j10\ \Omega$, to a complex source impedance, $Z_S = 25 + j20$ Ω, at a frequency of 1 GHz. Verify all the solutions by plotting the response of the matching networks.

5-3.   Design all the L-networks that will match a source impedance, $Z_S = 30\ \Omega$, to a complex load impedance, $Z_L = 25 + j20\ \Omega$, at a frequency of 1 GHz. Verify all the solutions by plotting the response of the matching networks.

5-4.   Design all the L-networks that will match a load impedance, $Z_L = 15$ Ω, to a source impedance, $Z_S = 75\ \Omega$, at a frequency of 1.5 GHz. Verify all the solutions by plotting the response of the matching networks.

5-5.   The output impedance of a transmitter operating at a frequency of 2 GHz is: $Z_S = 30 + j10\ \Omega$. Analytically design all the matching L-networks such that the maximum power is delivered to the antenna whose input impedance is $Z_L = 10 + j15\ \Omega$. Compare the matching bandwidth at 20 dB return loss with the method of using 50 Ω interconnecting line.

5-6.   Redesign the network of Problem 5-5 by matching the source and load impedances to a 50 Ω line. Compare fractional bandwidth and Q factors for examples 5.5 and 5.6.

5-7.   Redesign the matching network in Example 5.5 with four equal-Q L-networks. Compare the Q and bandwidth of the two networks.

5-8   For a load and source resistor ratio of 5, find the minimum number of cascaded L sections needed to achieve a loaded Q of 0.5.

5-9.   Use the A Impedance Matching Utility to solve a complex load matching example with $R_L = 1000\ \Omega$ in parallel with a 3 pF capacitance. For the source select $R_S = 50\ \Omega$, and for the load select the parallel RC with $R = 1000\ \Omega$ and $C = 3$ pF. Compare the Pi and Tee Matching Network response. Plot the insertion loss of both networks on the same plot.

Chapter 6
Distributed Impedance Matching Network Design

6.1 Introduction

Distributed networks are comprised of transmission line elements rather than discrete resistors, inductors, and capacitors. These transmission lines can take the form of the various transmission lines covered in chapter 2. At RF and microwave frequencies, where the wavelengths of the signals become comparable to the physical dimensions of the components, even lumped elements behave like distributed components. At microwave frequencies the distributed network is a more realizable form of a matching network than the lumped element versions in the Chapter 5. As a general rule, any electrical part larger than one tenth of the signal wavelength should be analyzed as distributed element. In an impedance matching network, operating at RF and microwave frequencies, reflections from short lengths of wire can create effects that are not predictable by the lumped element analysis. In this chapter both narrowband and broadband distributed matching networks are analyzed. Several examples are given to show how the matching networks are analytically and graphically designed. Distributed matching networks, discussed in this chapter, include quarter-wave and single-stub matching networks. For the quarter-wave matching networks the matching bandwidth, power loss and Q factor are analytically calculated.

6.2 Quarter-Wave Matching Networks

The quarter-wave transformer is a useful network for impedance matching between two resistors. At RF and microwave frequencies impedance matching between a resistive source and load impedance can easily be achieved by a 90 degree transmission line known as a Quarter-Wave Transformer or Quarter-Wave Network. In this section the quarter-wave matching network is defined and its properties are investigated.

Analysis of the Quarter-Wave Matching Networks

In chapter 2, Equation (2-41), we showed that the input impedance of a quarter-wave network with characteristic impedance Z_0 terminated in a resistive load R_L is given by:

$$Z_{IN} = \frac{Z_0^{\,2}}{R_L}$$

This equation can be written as:

$$Z_O = \sqrt{R_L Z_{IN}} \qquad\qquad (6\text{-}1)$$

In chapter 5 we also showed that maximum power transfer from a source with resistance R_S to a network terminated in a resistive load R_L is achieved only if the input impedance of the network is equal to the source resistance:

$$Z_{IN} = R_S$$

Therefore, the characteristic impedance of quarter-wave matching network, as shown in Figure 6-1, must satisfy the following equation.

$$Z_O = \sqrt{R_S R_L} \qquad\qquad (6\text{-}2)$$

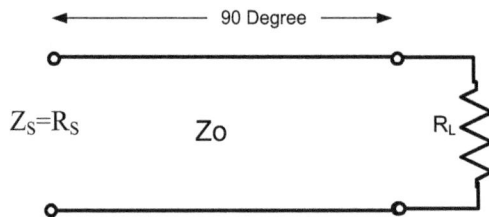

Figure 6-1 Quarter-wave network terminated in resistor R_L

Equation (6-2) states that the characteristic impedance of the quarter-wave network, matching R_L to R_S, must be equal to the square root of the product of source and load resistors.

If we normalize the load resistor R_L with respect to R_S,

$$r = \frac{R_L}{R_S} \qquad\qquad (6\text{-}3)$$

The characteristic impedance of the quarter-wave network becomes a function of R_S and r.

$$Z_O = R_S \sqrt{r} \qquad\qquad (6\text{-}4)$$

Notice that for $R_L > R_S$, the characteristic impedance Z_0 is greater than R_S while for $R_L < R_S$, the characteristic impedance Z_0 is less than R_S. The fractional bandwidth of a network, *FBW*, is defined in Equation (6-5) where f_H and f_L are the upper and lower frequencies of the bandwidth and the center frequency f_0 is equal to $\sqrt{f_H f_L}$, respectively.

$$FBW = \frac{f_H - f_L}{\sqrt{f_H f_L}} \qquad\qquad (6\text{-}5)$$

The fractional bandwidth of a quarter-wave matching network is given in Equation (6-6) [5]:

$$FBW = 2 - \frac{4}{\pi} \cdot \cos^{-1}\left(\frac{2\Gamma_m \sqrt{r}}{\sqrt{1 - \Gamma_m{}^2}\, |1 - r|} \right) \qquad\qquad (6\text{-}6)$$

Where, Γ_m is the magnitude of the reflection coefficient?

Equation (6-6) shows that the fractional bandwidth of a quarter-wave matching network depends upon the magnitude of the input reflection coefficient, Γ_m, and the mismatch ratio, r. The solutions to Equation (6-6) are only valid if,

$$\frac{2\Gamma_m \sqrt{r}}{\sqrt{1 - \Gamma_m{}^2}\, |1 - r|} \leq 1$$

At 3 dB return loss the reflection coefficient is $\Gamma_m = 0.707$, therefore, Equation (6-6) reduces to:

$$FBW_{3dB} = 2 - \frac{4}{\pi} \cdot \cos^{-1}\left(\frac{2\sqrt{r}}{|1-r|}\right) \qquad (6\text{-}7)$$

At 3 dB return loss Equation (6-7) has valid solutions only if,

$$\frac{2\sqrt{r}}{|1-r|} \leq 1 \quad or \quad r^2 - 6r + 1 \geq 0 \qquad (6\text{-}8)$$

Similarly, if we define the bandwidth at $\Gamma_m = 0.1$, corresponding to 20 dB return loss, as a good matching bandwidth it is insightful to evaluate the fractional bandwidth associated with this 20 dB return loss. From Equation (6-6) the fractional bandwidth at $\Gamma_{in} = 0.1$, is:

$$FBW_{20dB} = 2 - \frac{4}{\pi} \cdot \cos^{-1}\left(\frac{0.2\sqrt{r}}{\sqrt{0.99}\,|1-r|}\right) \qquad (6\text{-}9)$$

At 20 dB return loss Equation (6-9) has valid solutions only if,

$$\frac{0.2\sqrt{r}}{\sqrt{0.99}\,|1-r|} \leq 1 \quad or \quad 99r^2 - 202r + 99 \geq 0 \qquad (6\text{-}10)$$

The loaded quality factor, Q_L, of the quarter-wave matching network is defined as the inverse of the fractional bandwidth at 3 dB return loss; therefore, the loaded Q factor can be calculated from Equation (6-11).

$$Q_L = \frac{1}{FBW_{3dB}} = \frac{1}{2 - \frac{4}{\pi} \cdot \cos^{-1}\left(\frac{2\sqrt{r}}{|1-r|}\right)} \qquad (6\text{-}11)$$

Equation (6-11) shows that the validity condition in Equation (6-8) for the 3 dB fractional bandwidth is the same for the Q factor except that whenever the 3 dB fractional bandwidth tends towards infinity the Q factor tends towards zero. The Q factor given in Equation (6-11) is not to be confused with the unloaded transmission line Q factor as covered in chapter 4. This is actually an external Q factor as it relates to the loaded Q of the overall network. If f is the center (design) frequency, the bandwidth of the circuit is then calculated from Equation (6-12).

$$BW = (f) \cdot (FBW) \qquad (6-12)$$

Quarter-Wave Matching Network Design

In this subsection, based on the equations developed in section 6.2.1, two quarter-wave matching networks for $R_L = 2 \ \Omega$ and $R_L = 150 \ \Omega$ are designed. For each case the loaded Q factor and bandwidth is calculated at 3 and 20 dB return loss.

Example 6.1 Design a quarter-wave network intended to match a 50 Ω source to a 2 Ω load at 100 MHz. Calculate the Q factor and the fractional bandwidths at 3 and 20 dB return loss.

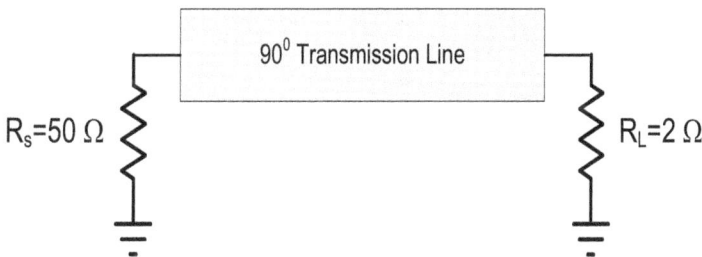

Figure 6-2 Matching quarter-wave transformer ($R_L < R_s$)

Solution: First we solve the problem analytically. Using the equations developed earlier, the characteristic impedance, loaded Q factor, and the bandwidths of the matching network are calculated as follows.

1. Enter design parameters and normalize the load impedance

RS=50; RL=2; f=100e6

r=RL/RS

2. Calculate Z1, BW, FBW, and Q

Z1=RS*sqrt(r)
FBW20dB = 2 - (4/PI)*acos((0.2*sqrt(r))/(sqrt(0.99)*(abs(1 - r))))

BW20dB = (f*FBW20dB)

FBW3dB = 2 - (4/PI)*acos((2*sqrt(r))/(abs(1 - r)))

BW3dB = (f*FBW3dB)

Q = 1/FBW3dB

The MATLAB solutions are given in Table 6-1.

Name	Value
Arg	1.009
BW20dB	5.333e+6
BW3dB	54.72e+6
FBW20dB	0.053
FBW3dB	0.547
Qe	1.827
RL	2
RS	50
Z1	10
f	100e+6
r	0.04

Table 6-1: MATLAB Solutions

The calculation results in Table 6-1 show that the Z1 = 10 Ω, Q_e factor is 1.827, and bandwidths at 3 and 20dB return loss are 54.72 MHz and 5.333 MHz, respectively.

The fractional bandwidth at 3 dB is 54.7 % and at 20 dB return loss is only 5.333 % indicating a narrowband matching network. This is characteristic of a narrowband matching network with a load resistor that is much smaller than the source resistor.

To measure the same parameters in LTspice, create a new schematic and place a transmission line between the source and load resistors. Calculate the time delay of a quarter-wave (90 degree) lossless transmission line at 100 MHz by using the following equation.

$$Td1 = L1/(360*f) = 90/(360*100e^6) = 2.5 \text{ ns}$$

The schematic is shown in Figure 6-3.

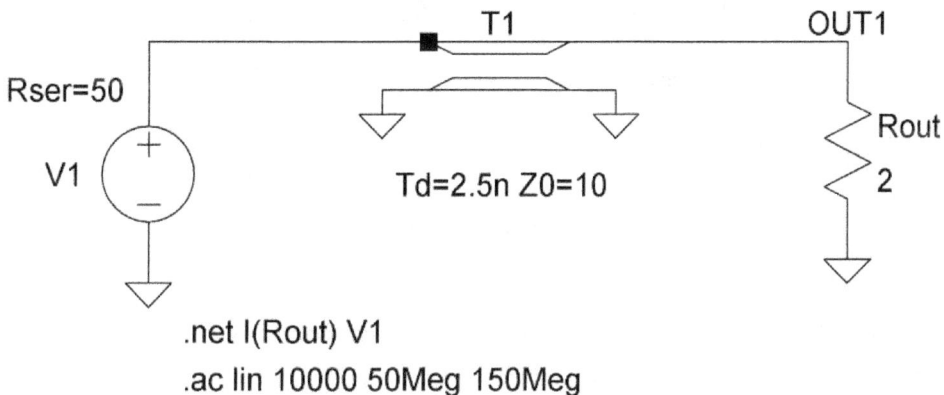

Figure 6-3 Schematic of the quarter-wave matching network

Simulate the schematic and display the input reflection coefficient, S11, and forward transmission, S21 in a rectangular plot. Also measure the matching bandwidths manually at 3 and 20 dB return loss.

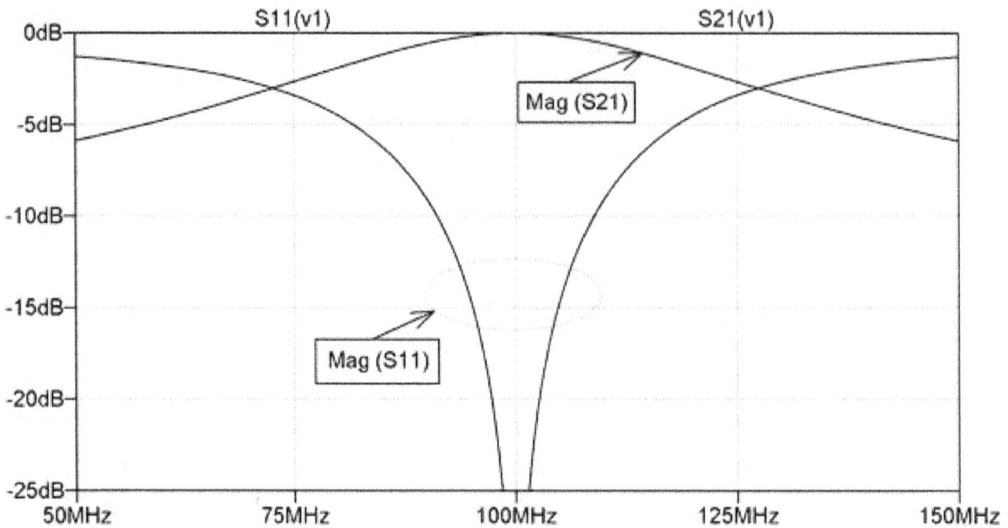

Figure 6-4 Bandwidth measurements

Place cursors at -3 and -20 dB points to measure the matching bandwidth at 3 and 20 dB return loss. (For the measurement see the Appendix Figures 6-4A and 6-4B).

$$BW_{-3dB} = 127.4 - 72.5 = 52.9 \qquad MHz$$

$$BW_{-20dB} = 102.6 - 97.3 = 5.4 \qquad MHz$$

The Q factor of the matching network is calculated by using Equation (6-9),

$$Q = (96.1/52.9) = 1.75$$

Note that the measured Q factor indicates a broadband matching bandwidth at 3 dB return loss.

Example 6.2 Design a quarter-wave network intended to match a 50 Ω source to a 150 Ω load. The design frequency is 100 MHz. Compare the calculated Q factor and the fractional bandwidths, at 3 and 20 dB return loss, with the measurements.

Solution: The schematic of the quarter-wave matching network is shown in Figure 6-5.

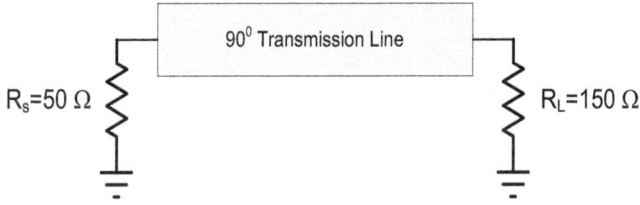

Figure 6-5 Quarter-wave matching network ($R_L > R_S$)

Using Equation (6-2) the characteristic impedance of the quarter-wave matching network is:

$$Z_1 = \sqrt{(50)(150)} = 86.6 \ \Omega$$

The quarter-wave matching network is shown in Figure 6-6.

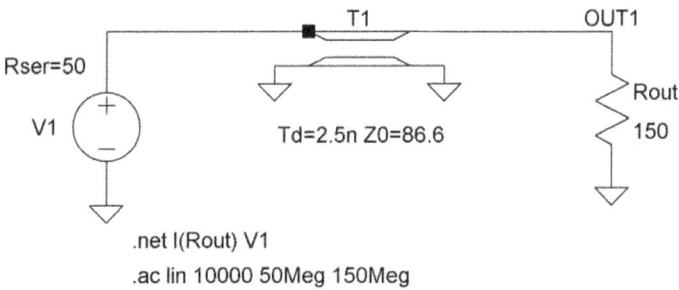

.net I(Rout) V1
.ac lin 10000 50Meg 150Meg

Figure 6-6 Schematic of the matching quarter-wave network

Using Equations developed in section 6.2, calculate the characteristic impedance, the Q factor, and the bandwidths of the matching network at 3 and 20 dB return loss. The procedure follows.

1. Enter design parameters and normalized load impedance

RS = 50; RL = 150; f = 100e6
r = RL/RS

2. Calculate Z1, FBW, BW and Q factor

Z1 = RS*sqrt(r)
FBW20dB = 2-(4/PI)*acos((0.2*sqrt(r))/(sqrt(0.99)*(abs(1 - r))))

BW20dB = (f*FBW20dB)

FBW3dB = 2-(4/PI)*acos((2*sqrt(r))/(abs(1 - r)))

BW3dB = (f*FBW3dB)

Qe = 1/FBW3dB

The MATLAB solutions are given in Table 6-2.

Name	Value
BW20dB	22.28e+6
BW3dB	NaN
FBW20dB	0.223
FBW3dB	NaN
Qe	NaN
RL	150
RS	50
Z1	66.603
f	100e+6
r	3

Table 6-2: MATLAB Solutions

The calculation results show that the bandwidth at 20 dB return loss is 22.2 MHz. This is over four times the bandwidth that was achieved when the load impedance was 2 Ω.

Next simulate the matching network in Figure 6-6 from 50 MHz to 150 MHz and display the input reflection coefficient, S11, and the forward transmission, S21, in a rectangular plot. Place cursors at -20 dB points to measure the matching bandwidths at -20 dB return loss.

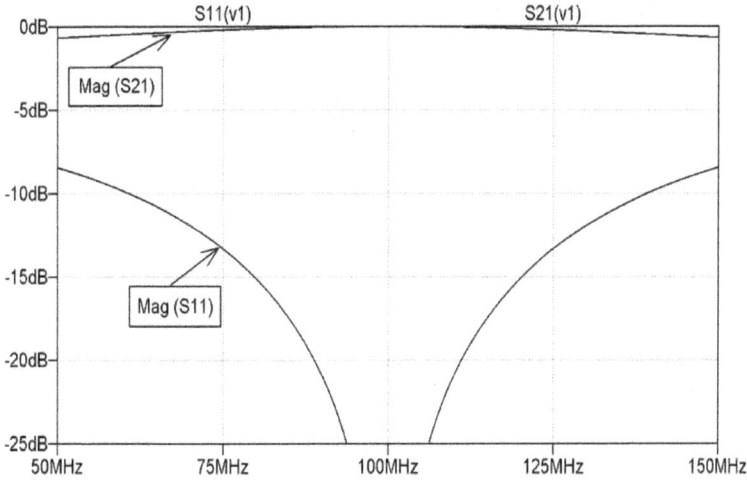

Figure 6-7 Magnitude of S11 and S21 (Courtesy of Keysight ADS)

The measured bandwidth at 20 dB return loss is:

$$111.1 - 88.89 = 22.21 \text{ MHz}$$

This agrees well with the calculated bandwidth.

6.3 Quarter-Wave Impedance Matching Bandwidth

Equation (6-6) shows that the achievable bandwidth in a quarter-wave matching network is related to the ratio of the load to the source impedance (mismatch ratio) as well as the value of the input reflection coefficient. It is insightful to examine the relationship between these quantities when one of the parameters is swept in value.

Quarter-Wave Network Matching Bandwidth and Power Loss

The fractional bandwidth and power loss of a quarter-wave matching network can be calculated in an Equation Editor as a function of the input reflection coefficient, Γ_{IN}, and the normalized load resistor.

The procedure is listed below.

1. Enter equations for input reflection coefficient, power loss, and the conversion of reflection coefficient to return loss in dB.

2. Enter the desired values for the source and load resistors

3. Normalize the load resistor with respect to source resistor

4. Use Equation (6-6) to calculate the fractional bandwidth.

Example 6.3: For a 50 Ω source and 2 Ω load resistors, calculate the fractional bandwidth and power loss from $\Gamma = 0.1$ to $\Gamma = 0.707$.

Solution: Follow the above procedure to calculate the fractional bandwidth and power loss when $R_S = 50 \Omega$ and $R_L = 2 \Omega$.

1. Enter reflection coefficient, power loss, and return loss Equations

vswr=Sweep1_Data.VSWR1

ReflCoef=(vswr-1)/(vswr+1)

Powerloss=(1-(1-(abs(ReflCoef)^2)))*100

RLdB=-20*log(abs(ReflCoef))

2. Enter the source and load resistor values
RS=50; RL=2
r=RL/RS

3. Calculate the Fractional Bandwidth

FBW= (2-(4/PI)*acos((2*(ReflCoef)*sqrt(r))/(sqrt((1-(ReflCoef)^2))*abs(1-r))))

Sweep the parameters to display the fractional bandwidth and power loss for reflection coefficients from 0.1 to 0.707, as shown in Table 6-3.

	ReflCoef	FBW ...		RLdB	Powerloss
1	0.1	0.053		20	1
2	0.11	0.059		19.172	1.21
3	0.12	0.064		18.417	1.44
4	0.13	0.07		17.721	1.69
5	0.14	0.075		17.077	1.96
6	0.15	0.081		16.478	2.25
7	0.16	0.086		15.917	2.56
8	0.17	0.092		15.391	2.89
9	0.18	0.097		14.895	3.24
10	0.19	0.103		14.425	3.61
11	0.2	0.108		13.979	4
12	0.3	0.167		10.458	9
13	0.4	0.233		7.959	16
14	0.5	0.309		6.021	25
15	0.6	0.405		4.437	36
16	0.707	0.547		3.012	49.985

Table 6-3: Fractional bandwidth and power loss for $R_L = 2\ \Omega$

Table 6-3 shows that the fractional bandwidths at 0.1 reflection coefficient, corresponding to 20 dB return loss, is 5.3 % with 1% power loss while at 0.707 reflection coefficient, corresponding to 3 dB return loss, the fractional bandwidth is 54.7 % with 49.985 % power loss. Notice that the fractional bandwidths, at 3 and 20 dB return loss, are the same as calculated in Example 6.2.

Example 6.4: For a 50 Ω source and 150 Ω load resistors, calculate the fractional bandwidth and power loss from $\Gamma = 0.1$ to $\Gamma = 0.707$.

Solution: Following the procedure, the calculations are shown in Figure 6-15. Sweep the parameters and display the result, as shown in Table 6-3.

1. Enter reflection coefficient, power loss, and return loss Equations

vswr=Sweep1_Data.VSWR1

ReflCoef=(vswr-1)/(vswr+1)

Powerloss=(1-(1-(abs(ReflCoef)^2)))*100

RLdB=-20*log(abs(ReflCoef))

2. Enter the source and load resistor values and normalized load

RS=50; RL=150;

r=RL/RS

3. Calculate the Fractional Bandwidth

FBW = (2-(4/PI)*acos((2*(ReflCoef)*sqrt(r))/(sqrt((1-(ReflCoef)^2))*abs(1-r)))))

The reflection coefficient, power loss, and return loss are given in the following Table.

	ReflCoef	FBW (rad)	RLdB	Powerloss
1	0.1	0.223	20	1
2	0.11	0.246	19.172	1.21
3	0.12	0.269	18.417	1.44
4	0.13	0.292	17.721	1.69
5	0.14	0.315	17.077	1.96
6	0.15	0.339	16.478	2.25
7	0.16	0.362	15.917	2.56
8	0.17	0.386	15.391	2.89
9	0.18	0.411	14.895	3.24
10	0.19	0.435	14.425	3.61
11	0.2	0.46	13.979	4
12	0.3	0.733	10.458	9
13	0.4	1.091	7.959	16
14	0.5	2	6.021	25
15	0.6	1.#QO	4.437	36
16	0.707	1.#QO	3.012	49.985

Table 6-4: Fractional bandwidth and power loss measurements for R_s=50 Ω and R_L=150 Ω

Table 6-2 shows that the fractional bandwidth at $\Gamma = 0.1$, corresponding to 20 dB return loss, is 22.3 % with a 1% power loss. The fractional bandwidth is the same as calculated in the previous example. The higher fractional bandwidth in this example, compared to its value in the previous example, is due to the lower ratio of the load to source resistor.

6.4 Single-Stub Impedance Matching Networks

The single-stub, also known as line and stub, matching network is a popular narrowband transmission line technique used to match real or complex load impedance to real source impedance. This technique is frequently used in distributed matching circuit designs using microstrip or stripline. The single-stub matching network, shown in Figure 6-8, consists of a series transmission line connected directly to the load impedance and a shunt stub (short-circuited or open-circuited transmission line) attached to the source impedance. The characteristic impedance of both matching elements has the same value as the source impedance, R_s. This section demonstrates both analytical and graphical techniques to design fixed frequency single-stub matching networks.

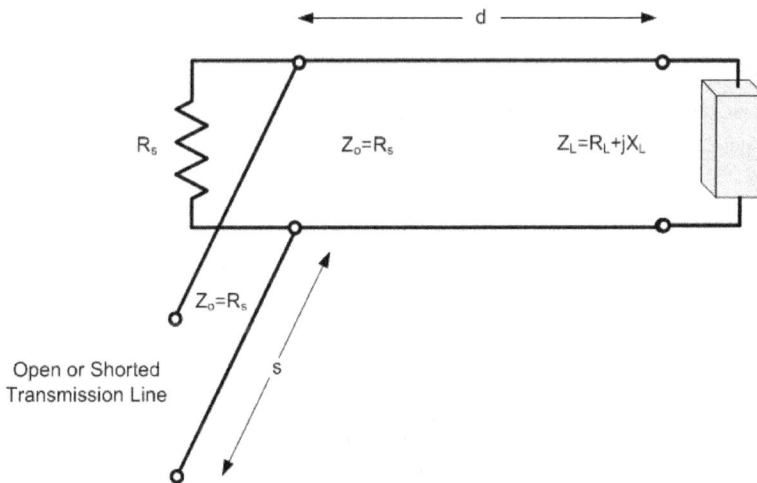

Figure 6-8 Single-stub matching network

To start the design of the single-stub matching network, attach a transmission line section, with characteristic impedance equal to R_S, to the load and determine its electrical length in such a way that the normalized admittance at a distance d from the load falls on the unit conductance circle of the Smith Chart. Calculation of the electrical lengths for the series line and shunt stub are given separately.

Analytical Design of the Series Transmission Line

The input impedance of a lossless transmission line of length d and characteristic impedance Z_0 terminated in an arbitrary load Z_L, was given in chapter 2 in Equation (2-46) and repeated here for convenience.

$$Z_{IN} = Z_o \frac{Z_L + jZ_o \tan \beta d}{Z_o + jZ_L \tan \beta d}$$

Setting $Z_0 = R_S$ and $\tan \beta d = t$, the input admittance of the network can be written as:

$$Y_{IN} = \frac{1}{Z_{IN}} = \frac{R_S + jZ_L t}{R_S(Z_L + jR_S t)} = G_{IN} + jB_{IN} \tag{6-13}$$

Substituting the normalized load impedance, $Z_L/R_s = r + jx$ into Equation (6-13) and separating its real and imaginary parts we get:

$$G_{IN} = \frac{r(1+t^2)}{R_S(r^2 + x^2 + t^2 + 2xt)} \tag{6-14}$$

And,

$$B_{IN} = \frac{xt^2 + (r^2 + x^2 - 1)t + x}{R_S(r^2 + x^2 + t^2 + 2xt)} \tag{6-15}$$

The value of d, which implies t, can be obtained by setting the input conductance, G_{IN}, equal to source conductance:

$$\frac{r(1+t^2)}{R_S(r^2 + x^2 + t^2 + 2xt)} = \frac{1}{R_S} \tag{6-16}$$

Equation (6-16) can be rearranged as:

$$(r-1) \cdot t^2 - 2xt - \left(r^2 + x^2 - r\right) = 0 \tag{6-17}$$

Notice that the quadratic Equation (6-17) has two solutions for t. For a wider bandwidth and lower loss usually the smaller value of t is selected. The two solutions for t are:

$$t_1 = \frac{x + \sqrt{r(r^2 + x^2 - 2r + 1)}}{r - 1} \qquad (6\text{-}18)$$

And,

$$t_2 = \frac{x - \sqrt{r(r^2 + x^2 - 2r + 1)}}{r - 1} \qquad (6\text{-}19)$$

With $\tan \beta d = t$, and $\beta \lambda = 2\pi$, we have $d = \dfrac{\lambda}{2\pi} \tan^{-1} t$ and the two solutions

for d are:

$$d_1 = \frac{\lambda}{2\pi} \tan^{-1} t_1 \qquad t_1 \geq 0 \qquad (6\text{-}20)$$

$$d_2 = \frac{\lambda}{2\pi} \tan^{-1} t_2 \qquad t_2 \geq 0 \qquad (6\text{-}21)$$

To specify the lengths of d_1 and d_2 in electrical degrees, we get:

$$d_1 = \frac{360}{2\pi} \tan^{-1} t_1 \qquad t_1 \geq 0 \qquad (6\text{-}22)$$

$$d_2 = \frac{360}{2\pi} \tan^{-1} t_2 \qquad t_2 \geq 0 \qquad (6\text{-}23)$$

Because at every half wavelength the input impedance of a transmission line repeats, there are an infinite number of transmission line lengths that matches the load to source impedance. Usually the shorter length is selected to improve the matching bandwidth. If t_1 or t_2 is negative, we add half a wavelength to each line to get positive d_1 and d_2.

$$d_1 = \frac{360(\pi + \tan^{-1} t_1)}{2\pi} \qquad t_1 < 0 \qquad (6\text{-}24)$$

$$d_2 = \frac{360(\pi + \tan^{-1} t_2)}{2\pi} \qquad t_2 < 0 \qquad (6\text{-}25)$$

Analytical Design of the Open-Circuited Shunt Stub

To calculate the electrical length of the shunt stub, substitute t_1 and t_2 in Equation (6-15) to determine B_1 and B_2.

$$B_1 = \frac{xt_1^2 + (r^2 + x^2 - 1)t_1 + x}{R_S(r^2 + x^2 + t_1^2 + 2xt_1)} \qquad (6\text{-}26)$$

$$B_2 = \frac{xt_2^2 + (r^2 + x^2 - 1)t_2 + x}{R_S(r^2 + x^2 + t_2^2 + 2xt_2)} \qquad (6\text{-}27)$$

The electrical lengths of the open circuited stubs are found by setting the susceptance of the stubs equal to the negative of the input susceptance.

$$so_1 = \frac{-\lambda\left(\tan^{-1}(R_S B_1)\right)}{2\pi} \qquad (6\text{-}28)$$

$$so_2 = \frac{-\lambda\left(\tan^{-1}(R_S B_2)\right)}{2\pi} \qquad (6\text{-}29)$$

If either stub length in Equation (6-28) or (6-29) is negative, add one half wavelength to obtain a positive stub length. For short-circuited stubs, the two solutions are:

$$SS_1 = \frac{\lambda \left(tan^{-1} \left(\frac{1}{R_S B_1} \right) \right)}{2\pi} \tag{6-30}$$

$$SS_2 = \frac{\lambda \left(tan^{-1} \left(\frac{1}{R_S B_2} \right) \right)}{2\pi} \tag{6-31}$$

Single-Stub Impedance Matching Network Design

Example 6.5: Design an open-circuited single-stub network to match a load resistance $Z_L = 2 - j5$ Ω to a resistive source $R_S = 50$ Ω at 100 MHz. Calculate the fractional bandwidths of the matching network at 3 and 20 dB return loss. Display the response and compare the calculations with measurements.

Solution: The design example is shown in Figure 6-9.

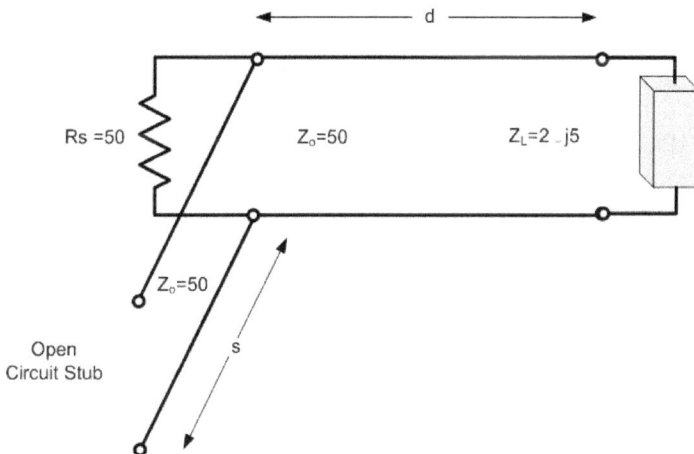

Figure 6-9 Complex load to resistive source matching network

Using the equations developed in section 6.4, calculate the electrical lengths of the line and stub matching transmission lines.

The calculation in MATLAB script follows.

1. Enter design parameters and normalized load impedance

RS=50; RL=2; XL=-5; f=100e6

r = RL/RS

x = XL/RS

2. Calculate t1, t2, d1 and d2, B1, B2, so1 and so2.

t1 = (x + sqrt (r*(r^2+x^2-2*r+1)))/(r-1)

t2 = (x – sqrt (r*(r^2+x^2-2*r+1)))/(r-1)

d1 = 360*(atan (t1))/(2*pi)

d2 = 360*(atan (t2))/(2*pi)

B1 = (x*t1^2 + (r^2 + x^2-1)*t1 - x)/(RS*(r^2 + x^2 + t1^2 + 2*x*t1))

B2 = (x*t2^2 + (r^2 + x^2-1)*t2 - x)/(RS*(r^2 + x^2 + t2^2 + 2*x*t2))

so1 = 360*(pi – atan (RS*B1))/(2*pi)

so2 = -360*atan (RS*B2)/(2*pi)

The MATLAB solutions for the calculation of line and stub lengths are given in Table 6-5.

Name	Value
B1	0.097
B2	-0.097
RL	2
RS	50
XL	-5
d1	-5.536
d2	16.975
f	100e+6
r	0.04
so1	101.707
so2	78.293t1
t1	-0.097
t2	0.305
x	-0.1

Table 6-5: MATLAB Solutions

The calculation results in Table 6-5 show that the problem has two solutions for line lengths and two solutions for the stub lengths.

Either solution can be used in the design of the single-stub matching network. However, for both line and stub, usually the shorter line lengths are chosen. For this example we select the shorter electrical lengths in the second solution; d2 = 16.875 for the line and so2 = 78.293 for the open-circuited stub. By converting the electrical lengths to time delays the schematic of the matching network is shown in Figure 6-10.

Rser=50

V1

Td=2.17481n Z0=50

T2

T1   OUT1

Td=0.47152778n Z0=50

Rout
2

Cout
318.31pF

.net I(Rout) V1

.ac lin 10000 90Meg 110Meg

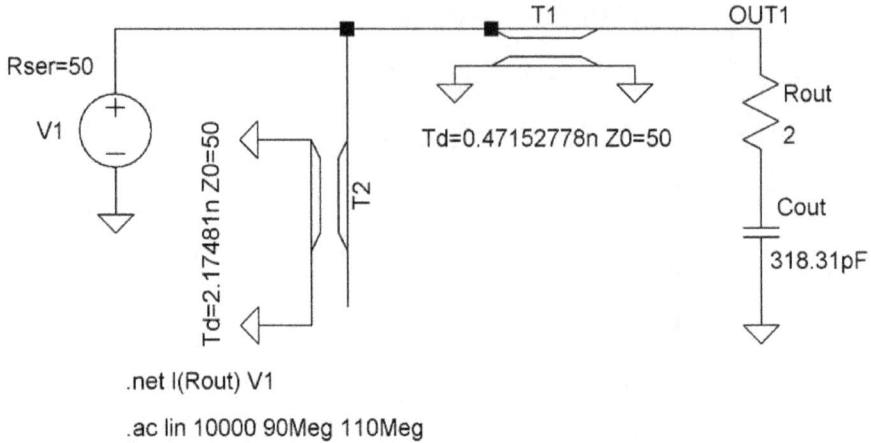

Figure 6-10 Schematic of the single-stub matching network

Simulate the schematic and display S11 in a rectangular plot.

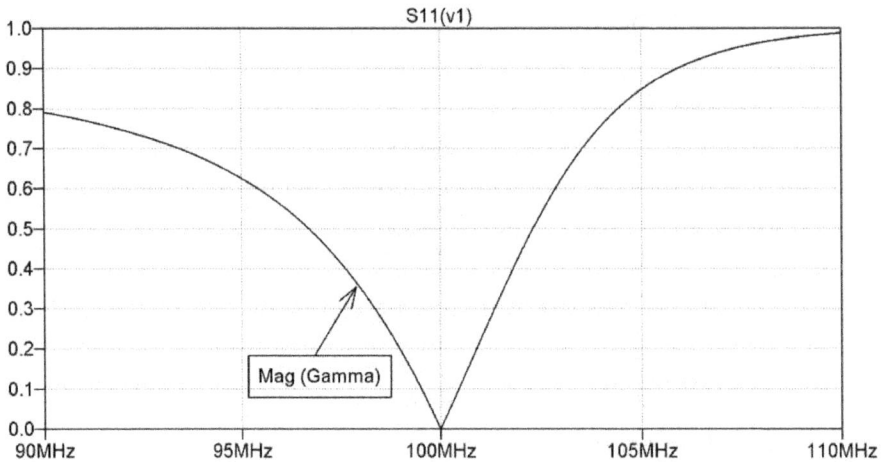

S11(v1)

Mag (Gamma)

Figure 6-11 Response of the single-stub matching network

Figure 6-11 shows that the single-stub matching network is perfectly matching the source to the load at 100 MHz.

Automated Calculation of Line and Stub Lengths

The conditional IF-THEN statements can be used in the Equation Editor to choose the proper solution. For the series transmission line, the shorter line length is chosen. For the shunt transmission line, the non-negative solution is chosen. The IF-THEN statement is used to redefine the line and stub calculations as shown in Figure 6-12.

```
1     % Enter Design Parameters
2     RS=50; RL=2; XL=-5; r=RL/RS; x=XL/RS
3     %Calculate t1, t2, d1, d2, so1, so2, ss1, ss2
4     t1 = (x + sqrt(r*(r^2+x^2-2*r+1))) / (r-1)
5     t2 = (x - sqrt(r*(r^2+x^2-2*r+1))) / (r-1)
6     If t > 0 then
7     d1= (360/(2*PI))*atan(t1)
8     else
9     d1= 360*(PI + atan(t1)) / (2*PI)
10    endif
11    If t2 >0 then
12    d2= (360/(2*PI))*atan(t2)
13    else
14    d2= 360*(PI + atan(t2)) / (2*PI)
15    endif
16    If d1 <d2 then
17    SeriesLineLength = d1
18    else
19    SeriesLineLength = d2
20    endif
21    % Calculate B1 and B2
22    B1 = (x*t1^2 + (r^2 + x^2 -1)*t1 – x) / (RS*(r^2 + x^2 +2*x*t1))
23    B2 = (x*t2^2 + (r^2 + x^2 -1)*t2 – x) / (RS*(r^2 + x^2 +2*x*t1))
24    so1 = -360*(atan(RS*B1))
25    so2 = -360*(atan(RS*B1))
26    If so1 < 0 then
27    OpenShuntStub = so2
28    else
29    OpenShuntStub = so1
30    else
31    ss1 = (360*atan(1 / (RS + B1))) / (2*PI)
32    ss2 = (360*atan(1 / (RS + B2))) / (2*PI)
33    If ss1 < 0 then
34    Shorted Shunt Stub = ss2
35    else
36    Shorted Shunt Stub = ss1
37    endif
```

Figure 6-12 Automatic calculation **of** line and stub lengths

6.5 Graphical Design of Single-Stub Matching Networks

Building on the techniques covered in chapter 3.7 this section demonstrates the ease in which the single-stub matching network can be graphically designed using the Smith Chart. The Smith Chart combined with tunable elements is a very powerful tool in which matching networks can be graphically designed without the need to solve the exact Equations.

Graphical Design of the Open-Circuited Stub

In this section a 50 Ω source will be matched to a 5 - 25j Ω load at a frequency of 1000 MHz. Both open-circuited and short-circuited shunt stubs will be considered.

Example 6.6 Consider matching the 5 - j25 Ω load impedance to 50 Ω source resistor.

Solution: Create a new schematic in LTspice and place a transmission line between the source and the load terminations. Wire up the schematic as shown in Figure 6-13.

Since we intend to place a shunt element after the series transmission line, enable the admittance circles on the Smith Chart and make the length of the line tunable to move the load impedance (point A) to point B on the unit conductance circle on the top half of the Smith chart. As the schematic shows, an electrical length of 42.5° places the impedance on the unit conductance circle at point B. The equivalent time delay is 0.118 nanosecond.

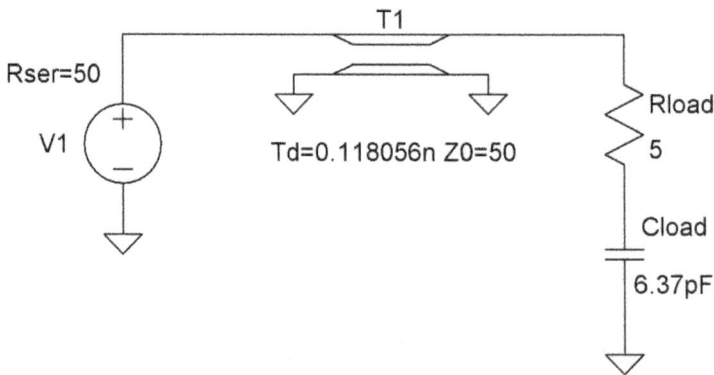

Figure 6-13 Adding series transmission line (electrical length=42.5°)

The movement of the impedance from point A to point B is displayed in Figure 6-14.

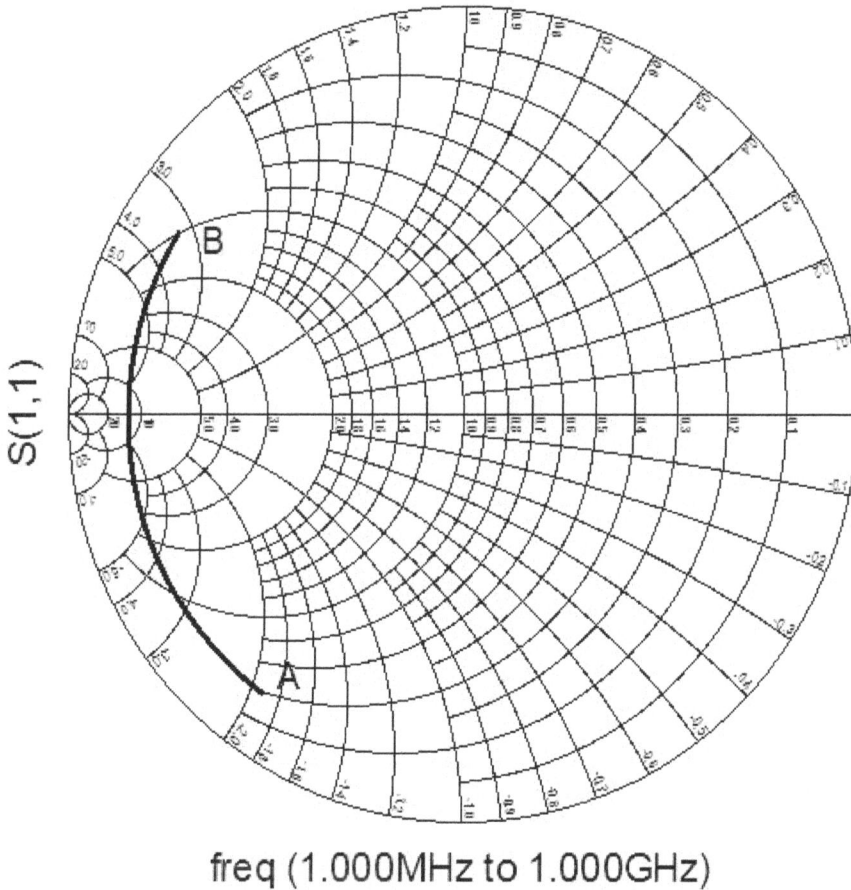

freq (1.000MHz to 1.000GHz)

Figure 6-14 Point B on the unit conductance circle (electrical length=42.5°)

Note that at point B the normalized impedance is z = 0.088 + 0.28j Ω.

To move point B to the center of the Smith chart, add an open-circuited shunt transmission line and tune the length of the transmission line until the impedance moves to the center of the chart (Z = 50 Ω). As the schematic of Figure 6-15 shows, a 73° length of transmission line, equivalent to 0.203 nanosecond time delay, would be required to move point A to point B.

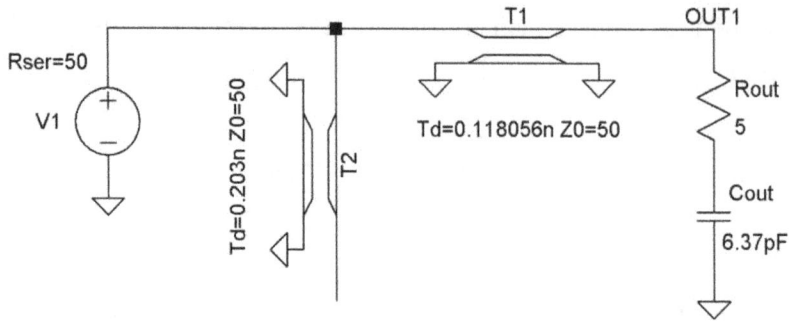

Figure 6-15 Open-circuited shunt transmission line (73°) is added

The movement of point B to the center of the Smith chart is shown in Figure 6-16.

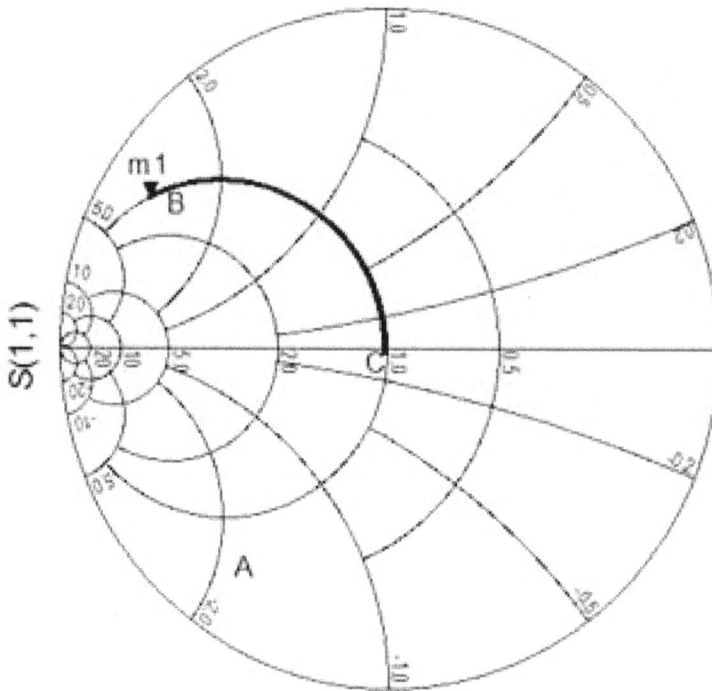

Figure 6-16 Moving point B to the center of Smith chart (73°)

Graphical Design of the Short-Circuited Stub

In this Section a short-circuited shunt stub will be used.

Example 6.7 Consider matching the 5 - j25 Ω load impedance to 50 Ω source resistor using short-circuited shunt stubs.

Solution: A short-circuited shunt transmission line can be used to perform the function of the shunt stub. Using a short-circuited shunt transmission line, the series transmission line should intersect the unit conductance circle on the bottom half of the Smith Chart.

Add a series transmission line of 11° electrical length to move the impedance at point A to intersect the unit conductance circle at point B as shown in Figure 6-20. Then add the short-circuited shunt transmission line as shown in Figure 6-18. Tune the length to 17° to move the impedance to the center of the Smith Chart.

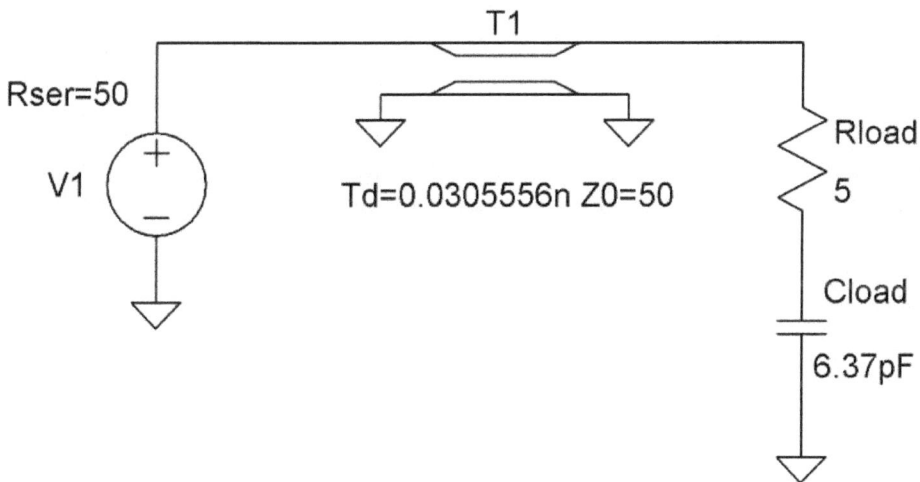

Figure 6-17 A series transmission line (11°) added to Z= 5 - j25 Ω

Simulate the schematic and notice the movement of the impedance from point A to point B.

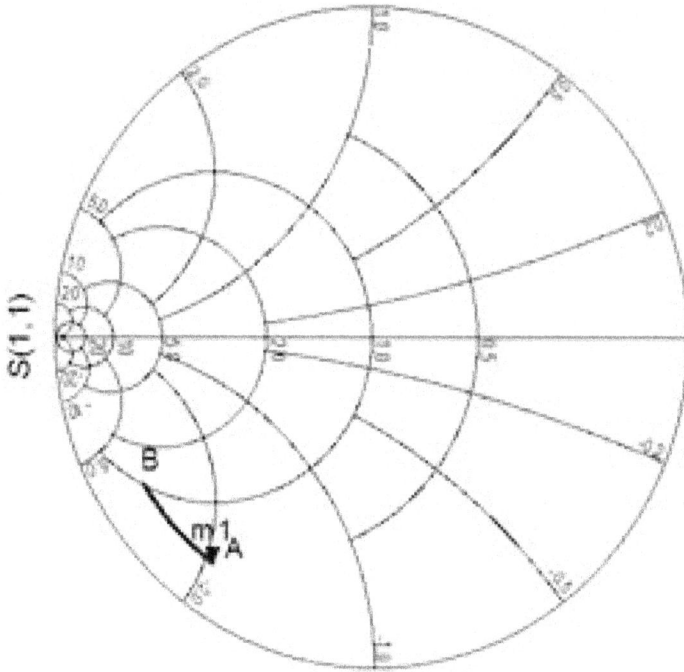

Figure 6-18 Adding series transmission line (11°) to 5-j25 Ω

To move point B to the center of Smith chart, add a short-circuited shunt transmission line and tune the length of the line until the impedance moves to the center of the chart (50Ω). As the schematic of Figure 6-19 shows, a 17° length of transmission line would be required.

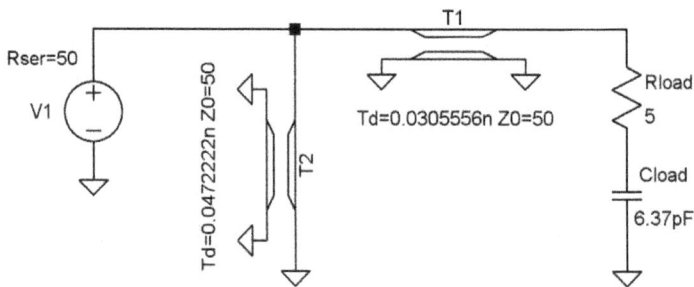

Figure 6-19 Adding short-circuited shunt transmission line (17°)

Simulate the schematic and notice the movement of the impedance from point B to the center of Smith chart.

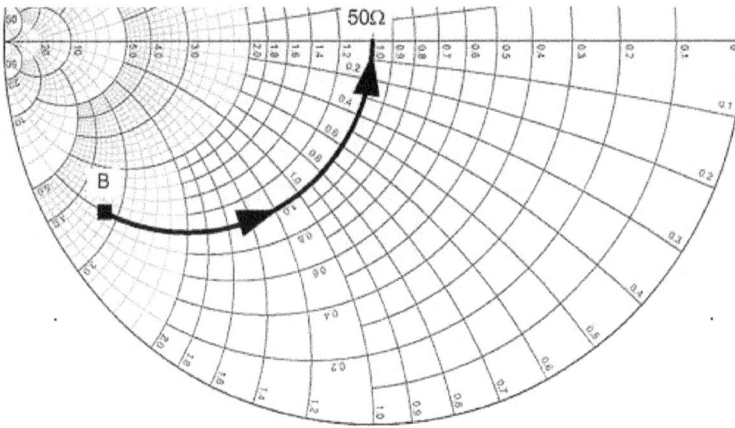

Figure 6-20 Adding short-circuited shunt transmission line (17°)

6.6 Design of Cascaded Single-Stub Matching Networks

When the impedance of the load and the source are both complex functions, we can define a virtual resistor and design single-stub networks to match the complex impedances to the virtual resistor. The final matching network is obtained by cascading the single-stub matching networks. The procedure is demonstrated in the following example.

Example 6.8: Design single-stub networks to match a complex load Z_L = 10 - j5 Ω to a complex source Z_S = 50 - j15 Ω at 100 MHz. Calculate the electrical lengths of the lines and the fractional bandwidths of the matching network at 3 dB and 20 dB return loss. Display the simulated response and verify the calculations with simulation.

First Solution: The following procedure is used to design the impedance matching networks.

1. Calculate the intermediate resistor R1.

$$R1 = \sqrt{R_L R_S} = \sqrt{(10)\,(50)} = 22.36\ \Omega$$

2. Design a single-stub matching network between *R1* and the source impedance

3. Design a second single-stub matching network between *R1* and the load impedance

4. Cascade the two matching networks

a. Enter design parameters and normalized load impedance

RS = 22.36; RL = 50; XL = -15; f = 100e6

r = RL/RS

x = XL/RS

b. Use MATLAB Script to calculate t1, t2, d1, d2, B1, B2, so1, and so2

t1 = (x + sqrt (r*(r^2 + x^2 - 2*r + 1)))/(- 1)

t2 = (x – sqrt (r*(r^2 + x^2-2*r + 1)))/(r - 1)

d1 = 360*(atan (t1))/(2*pi)

d2 = 360*(pi + atan (t2))/(2*pi)

B1 = (x*t1^2 + r^2+x^2 - 1)*t1 - x)/(RS*(r^2 + x^2 + t1^2 + 2*x*t1))

B2 = (x*t2^2 + (r^2 + x^2 - 1)*t2 - x)/(RS*(r^2 + x^2 + t2^2 + 2*x*t2))

so1 = 360*(pi – atan (RS*B1))/(2*pi)

so2 = -360*atan (RS*B2)/(2*pi)

The MATLAB solutions are given in Table 6-6.

Name	Value
B1	0.042
B2	-0.042
RL	50
RS	22.36
XL	-15
d1	49.204
d2	114.019
f	100e+6
r	2.236
so1	136.755
so2	43.245
t1	1.159
t2	-2.244
x	-0.671

Table 6-6: MATLAB Solutions

From MATLAB solutions in Table 6-6 we select d2=114.019 degree and so2= 43.43.245 degree transmission lines. Figure 6-21 shows the schematic of line and stub matching network.

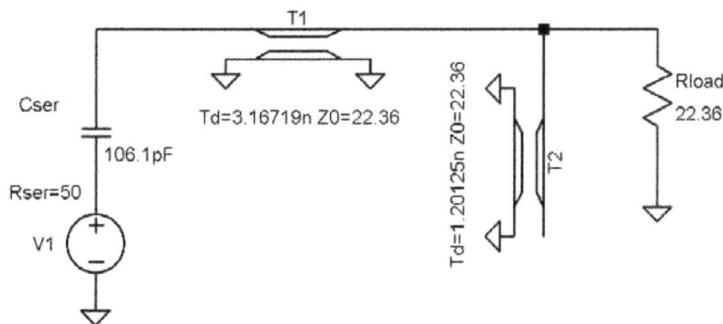

Figure 6-21 Schematic of the first matching network

Second Solution: Design the line and stub matching network between R1 and the load impedance

1. Enter design parameters and normalize the load impedance

RS = 22.36; RL = 10; XL = -5; f = 100e6

r = RL/RS

x = XL/RS

2. Calculate t1 and t2, d1 and d2, B1 and B2, so1 and so2,

t 1 = (x + sqrt (r*(r^2 + x^2 - 2*r + 1)))/(r - 1)

t2 = (x-sqrt (r*(r^2 + x^2 - 2*r + 1)))/(r - 1)

d1 = 360*(pi + atan (t1))/(2*pi)

d2 = 360*(atan (t2))/(2*pi)

B1 = (x*t1^2 + (r^2+x^2-1)*t1-x)/(RS*(r^2+x^2+t1^2+2*x*t1))

B2 = (x*t2^2+(r^2+x^2-1)*t2-x)/(RS*(r^2+x^2+t2^2+2*x*t2))

so1 = 360*(pi-atan (RS*B1))/(2*pi)

so2 = -360*atan (RS*B2)/(2*pi)

The MATLAB solutions are given in Table 6-7.

Name	Value
B1	0.04
B2	-0.04
RL	10
RS	22.36
XL	-5
d1	162.418
d2	48.39
f	100e+6
r	0.447
so1	138.278
so2	41.722
t1	-0.317
t2	1.126
x	-0.224

Table 6-7: MATLAB Solutions

From MATLAB solutions in Table 6-7 we select d2=48.39 degree and so2= 41.722 degree transmission lines. Figure 6-22 shows the schematic of line and stub matching network.

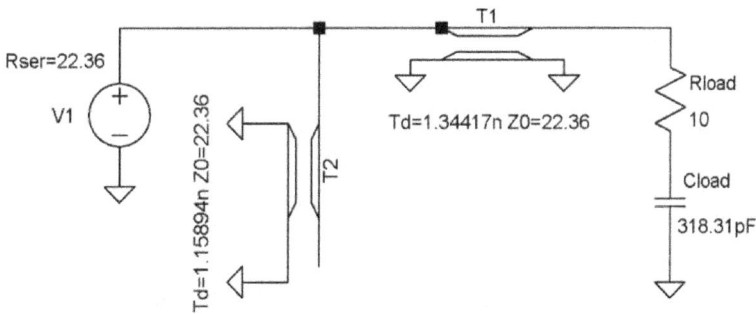

Figure 6-22 Schematic of the second matching network

Connect both matching networks to obtain the complete matching network between the source and the load impedances, as shown in Figure 6-23.

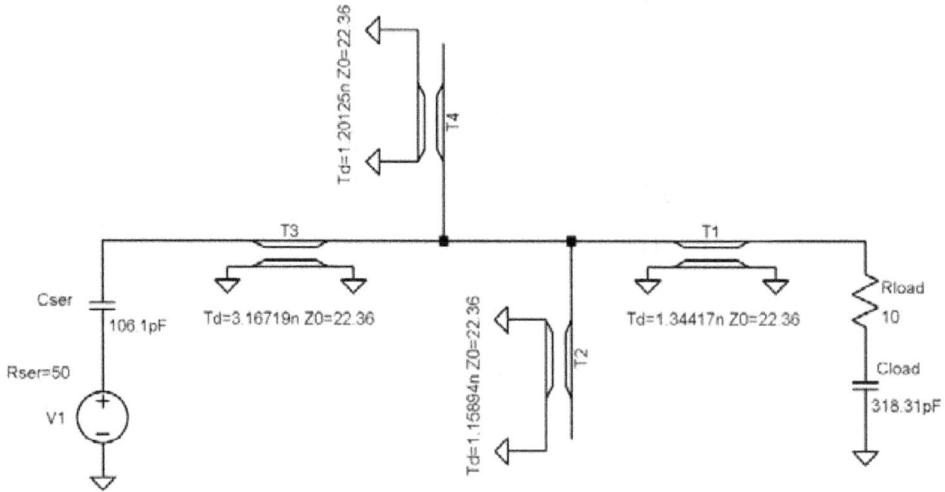

Figure 6-23 Cascading single-stub matching networks

Simulate the schematic and measure the -3 dB and -20 dB bandwidth[7].

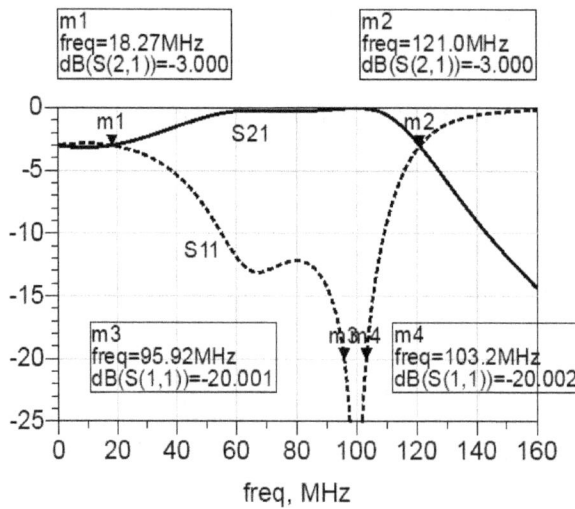

Figure 6-24 Bandwidth measurement [7]

Notice that the -3dB and -20dB matching bandwidths are:

BW_{-3dB} = 121 - 18.27 = 102.73   MHz

BW_{-20dB} = 103.2 - 95.92 = 7.28   MHz.

6.7 Broadband Quarter-Wave Matching Network Design

In Examples 6.1 and 6.2, the bandwidth of a single quarter-wave transformer matching network is less than 10 % which is considered to be a narrowband matching network. We can increase the bandwidth by cascading two or more quarter-wave transformers to achieve a broadband matching network. To design a broadband matching network with N quarter-wave transformers first use Equation (6-32) to calculate the characteristic impedance of each quarter-wave transformer then cascade all the sections into one matching network. Let $r = R_L/R_S$ and N the number of quarter-wave networks. The characteristic impedance of each section can be calculated from Equation (6-32).

$$Z_n = R_S(r)^{(2n-1)/2N} \quad n = 1, 2, \dots, N \qquad (6\text{-}32)$$

Example 6.9: Design a three-section quarter-wave transformer network to match a load resistance $R_L = 2\ \Omega$ to a resistive source $R_S = 50\ \Omega$ at 100 MHz. Calculate the characteristic impedance, the Q factor, and the fractional bandwidths of the three-section quarter-wave transformer matching network..

Solution: Use the following equations to calculate the characteristic impedances, the Q factor, and the bandwidths.

Z1=RS*r^(1/(2*N))

Z2=RS*r^(3/(2*N))

Z3=RS*r^(5/(2*N))

R1=RS*r^(1/N)

R2=RS*r^(2/N)

Q=1/FBW3dB

For this example RS=50; RL=2; f=100e6; r=RL/RS; and N=3, therefore, the characteristic impedances are:

$$Z1 = R_S(r)^{1/2N} = 50(0.04)^{1/6} = 29.24 \quad \text{Ohm}$$

$$Z2 = R_S(r)^{3/2N} = 50(0.04)^{3/6} = 10.00 \quad \text{Ohm}$$

$$Z3 = R_S(r)^{5/2N} = 50(0.04)^{5/6} = 3.42 \quad \text{Ohm}$$

And,

$$R_1 = 50\,(0.04)^{\frac{1}{3}} = 17.1 \ \Omega \qquad\qquad \text{Ohm}$$

$$R_2 = 50\,(0.04)^{\frac{2}{3}} = 5.848 \ \Omega \qquad\qquad \text{Ohm}$$

The quarter-wave matching network is shown in Figure 6-25.

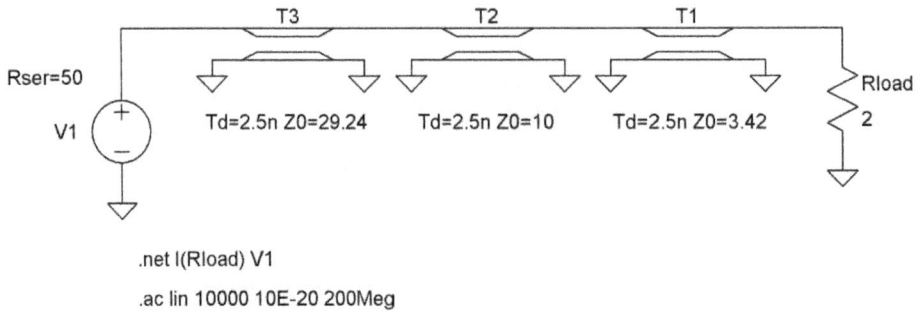

Figure 6-25 Three-section quarter-wave matching network

Simulate the schematic and display the response in Figure 6-26.

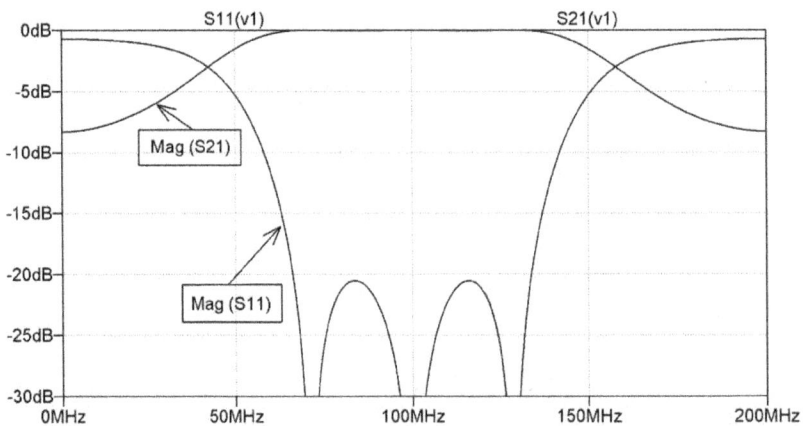

Figure 6-26 Bandwidth measurements (Courtesy of Keysight ADS)

By placing cursors at 3 dB and 20 dB return loss, we can measure the matching bandwidth and the fractional bandwidth of the matching network at 3 and 20 dB return loss. (For the measurement see the Appendix Figures 6-26A and 6-26B).

$$BW_{3dB} = 157.7 - 42.31 = 114.89 \text{ MHz}$$

$$FBW_{3dB} = \frac{114.89x100}{\sqrt{(157.7).(42.31)}} = 140.6\,\%$$

Similarly the bandwidth at the 20 dB return loss is:

$$BW_{20dB} = 133.8 - 66.2 = 67.6 \quad \text{MHz}$$

$$FBW_{20dB} = \frac{67.6x100}{\sqrt{(133.8).(66.2)}} = 71.8\,\%$$

The measured Q factor is,

$$Q = \frac{1}{1.406} = 0.71$$

The wide bandwidth and lower Q factor is an indication that by adding two more quarter-wave sections the Q factor has reduced to less than one half and the 3-dB bandwidth has more than doubled compared to a single quarter-wave matching network.

Example 6.10: Design a broadband quarter-wave network to match a complex load, $Z_L = 10 - j5\ \Omega$ to 50 Ω source impedance at 100 MHz. Calculate the bandwidth at 20 dB return loss and compare with the bandwidth of a single-stub matching network.

Solution: First we design a 5 section quarter-wave matching network between the source resistor and the complex load impedance. Then we replace the quarter-wave transformer adjacent to the load with a single-stub matching network that matches the complex load to the real resistor.

Design of the 5 section quarter-wave matching network between R_L and R_S is shown in the MATLAB script as follows.

1. Enter design parameters and normalize the load impedance

RS = 50; RL = 10; f = 100e6

r= RL/RS

N=5

2. Calculate characteristic impedances and intermediate resistors

Z1=RS*r^(1/(2*N))

Z2=RS*r^(3/(2*N))

Z3=RS*r^(5/(2*N))

Z4=RS*r^(7/(2*N))

Z5=RS*r^(9/(2*N))

R1=RS*r^(1/N)

R2=RS*r^(2/N)

R3=RS*r^(3/N)

R4=RS*r^(4/N)

The MATLAB solutions are given in Table 6-8.

Name	Value
N	5
R1	36.239
R2	26.265
R3	19.037
R4	13.797
RL	10
RS	50
XL	-5
Z1	42.567
Z2	30.852
Z3	22.361
Z4	16.207
Z5	11.746

Table 6-8: MATLAB Solutions

The MATLAB solutions in Table 6-8 shows that the characteristic impedances of five quarter-wave transformers are Z1= 42.567 Ohm, Z2= 30.852 Ohm, Z3= 22.361 Ohm, Z4= 16.207 Ohm, and Z5= 11.746 Ohm. Figure 6-27 shows the schematic of the quarter-wave matching network.

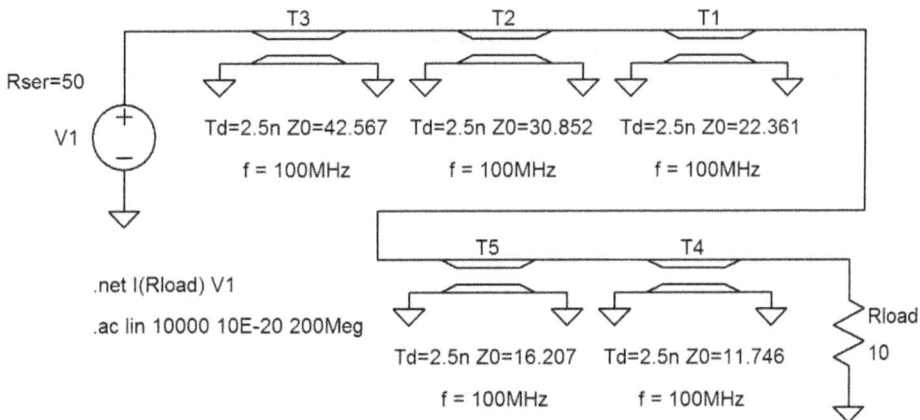

Figure 6-27 Five-section quarter-wave transformer matching network

Next replace the quarter-wave section adjacent to the load with a single-stub matching network. The design of the single-stub matching network is achieved by the solution of the following equations.

1. Enter design parameters and normalize the load impedance

RS = 13.797; RL = 10; XL = -5; f = 100e6

r = RL/RS

x = XL/RS

2. Calculate t1 and t2, d1 and d2, B1 and B2, so1 and so2,

t1 = (x + sqrt(r*(r^2 + x^2 - 2* r+ 1)))/(r - 1)

t2 = (x - sqrt(r*(r^2 + x^2-2*r + 1)))/(r - 1)

d1 = 360*(pi + atan (t1))/(2*pi)

d2 = 360*(atan (t2))/(2*pi)

B1 = (x*t1^2 + (r^2 + x^2 - 1)*t1 - x)/(RS*(r^2 + x^2 + t1^2 + 2*x*t1))

B2 = (x*t2^2 + (r^2 + x^2 - 1)*t2 - x)/(RS*(r^2 + x^2 + t2^2 +2 *x*t2))

so1 = 360*(pi - atan(RS*B1))/(2*pi)

so2 = -360*atan(RS*B2)/(2*pi)

Note that the intermediate resistor, R4=13.797 Ohm, is used as the source resistor.

The MATLAB solutions are given in Table 6-9.

Name	Value
d1	174.808
d2	69.845
f	100e+6
r	0.725
so1	151.875
so2	28.125
t1	-0.091
t2	2.725
x	-0.362

Table 6-9: MATLAB Solutions

Select d2= 69.845 degrees for the line length and so2=28.125 degrees for the stub length and convert them to time delays at f =100 MHz.

L1=69.845   Line length in degrees

L2=28.125   Stub length in degrees

Td1= d2/ (360*f) = 1.94014   Line delay in nanosecond

Td2= so2/ (360*f) = 0.78125   Stub delay in nanosecond

The line and stub matching network is shown in Figure 6-28.

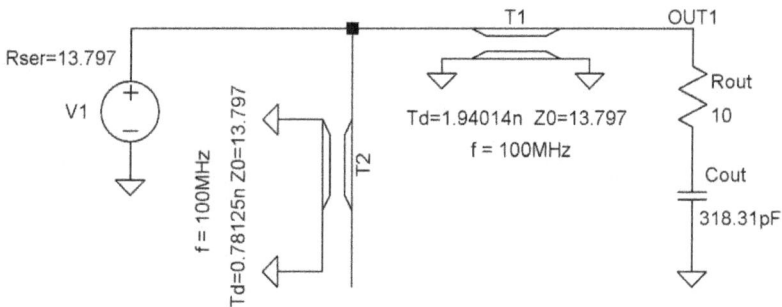

Figure 6-28 Schematic of the single-stub matching network

Now cascade the line and stub with four quarter-wave networks to form the final design of the matching network in Figure 6-29.

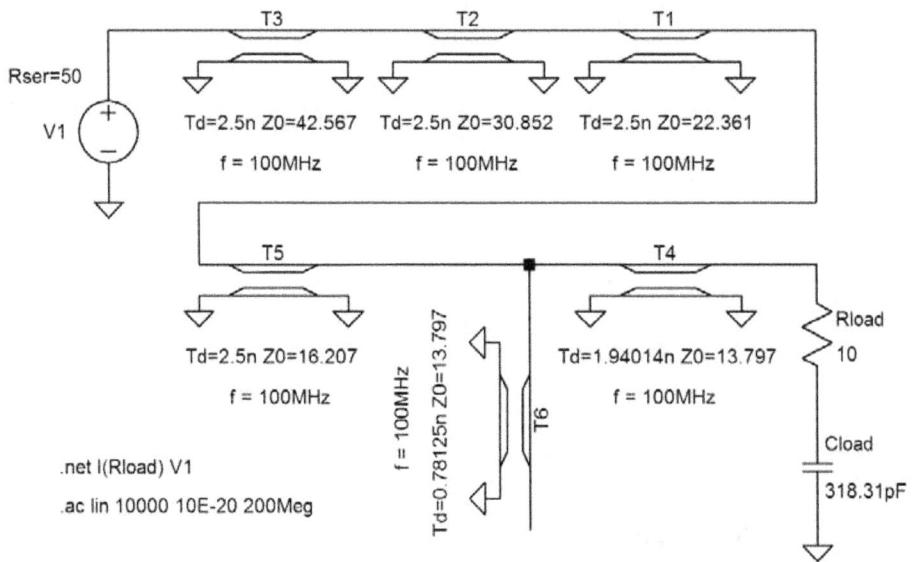

Figure 6-29 Schematic of the broadband matching network

Simulated the schematic and display S11 in a rectangular plot.

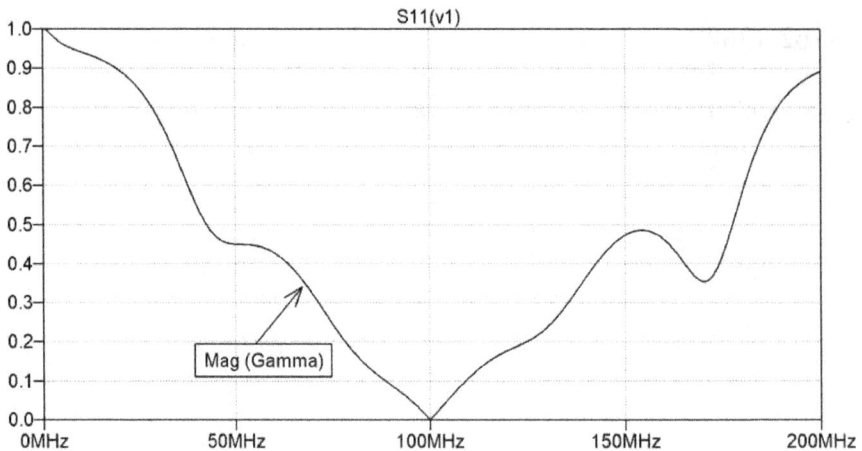

Figure 6-30 Response of the broadband matching network

Note that the source impedance is perfectly matched to the load impedance at 100 MHz.

Example 6.11: Design a broadband network to match a complex load, $Z_L = 150 - j30$ Ohm to 50 Ω source impedance at 100 MHz. Display the simulated response and measure the bandwidth at 20 dB return loss. Compare the results with the singe-stub matching network.

Solution: Use the same method as in Example 6.6.1 to design the broadband matching network. The design of the broadband quarter-wave transformer matching network with five quarter-wave sections is shown in Figure 6-33.

1. Enter design parameters and normalize the load impedance

RS=50; RL=150; f=100e6

r=RL/RS

N=5

2. Calculate characteristic impedances and intermediate resistors
Z1 = RS*r^(1/(2*N))

Z2 = RS*r^(3/(2*N))

Z3 = RS*r^(5/(2*N))

Z4 = RS*r^(7/(2*N))

Z5 = RS*r^(9/(2*N))

R1 = RS*r^(1/N)

R2 = RS*r^(2/N)

R3 = RS*r^(3/N)

R4 = RS*r^(4/N)

The MATLAB solutions are given in Table 6-10.

Name	Value
B1	2.604e-3
B2	-2.604e-3
N	5
R1	62.287
R2	77.592
R3	96.659
R4	120.411
R5	150
RS	50
XL	-30
Z1	55.806
Z2	69.519
Z3	86.603
Z4	107.883
Z5	134.394

Table 6-10: MATLAB Solutions

The MATLAB solutions in Table 6-10 show that the characteristic impedances of five quarter-wave transformers are Z1= 55.806 Ohm, Z2= 69.515 Ohm, Z3= 86.603 Ohm, Z4= 107.883 Ohm, and Z5= 134.394 Ohm.

Figure 6-31 shows the schematic of the quarter-wave matching network.

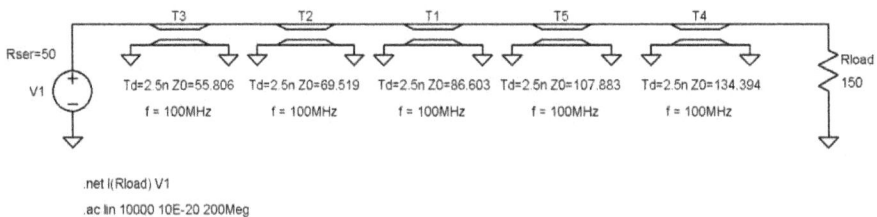

Figure 6-31 Five-section quarter-wave matching network

The simulated response of the quarter-wave matching network is shown in Figure 6-32.

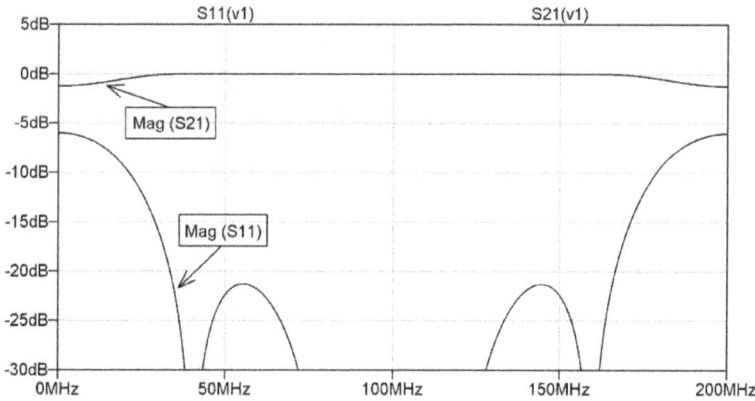

Figure 6-32 Response of the matching network

Next replace the quarter-wave transformer adjacent to the load with a single-stub matching network. The design of the single-stub matching network is shown in the following equations. Note that the intermediate resistor, R4 = 120.411 Ω, represents the source resistor in the design of the line and stub matching network.

1. Enter design parameters and normalize the load impedance

RS=120.411; RL=150; XL=-30; f=100e6

r=RL/RS

x=XL/RS

4. Calculate t1 and t2, d1 and d2, B1 and B2, so1 and so2,

t1 = (x + sqrt (r*(r^2 + x^2 - 2*r + 1)))/(r - 1)

t2 = (x − sqrt (r*(r^2 + x^2 - 2*r + 1)))/(r - 1)

d1 = 360*(atan (t1))/(2*pi)

d2 = 360*(pi + atan (t2))/(2*pi)

B1 = (x*t1^2 + (r^2 + x^2 - 1)*t1 - x)/(RS*(r^2 + x^2 + t1^2 + 2*x*t1))

B2 = (x*t2^2 + (r^2 + x^2 - 1)*t2 - x)/(RS*(r^2 + x^2 + t2^2 + 2*x*t2)) so1d = 360*(pi-atan (RS*B1))/(2*pi)

so2d = -360*atan (RS*B2)/(2*pi)

The MATLAB solutions are given in Table 6-11.

Name	Value
d1	22.922
d2	111.013
f	100e+6
r	3
so1	139.676
so1d	162.592
so2	50.324
so2d	17.408
t1	0.576
t2	-2.603
x	-0.249

Table 6-11: MATLAB Solutions

Select d2= 111.013 degrees for the line length and so2d=17.408 degrees for the stub length and convert them to time delays at f =100 MHz.

Td1= d2 / (360*f) = 3.083   Line delay in nanosecond

Td2= so2d / (360*f) = 0.483   Stub delay in nanosecond

The line and stub matching network is shown in Figure 6-33.

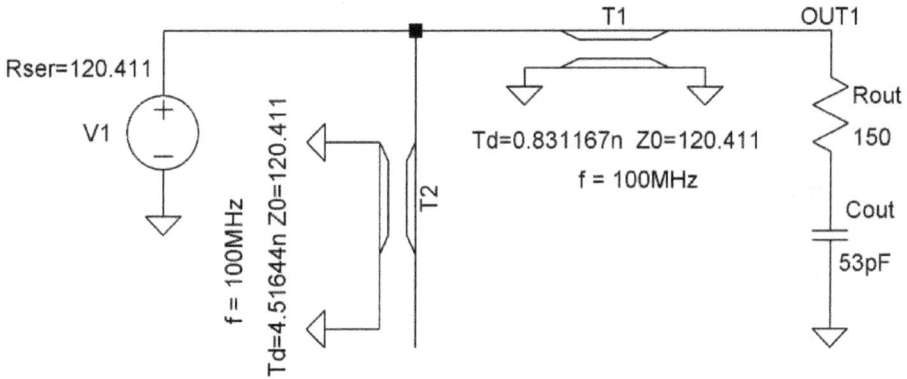

Figure 6-33 Schematic of the single-stub matching network

Cascade the two matching networks to form the final design of broadband matching network, as shown in Figure 6-34.

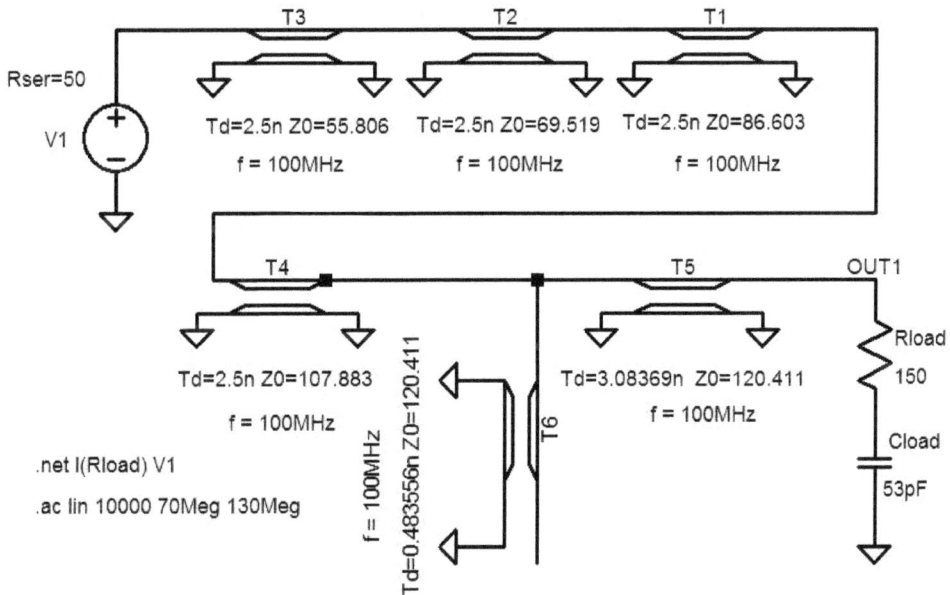

Figure 6-34 Schematic of the broadband matching network

The simulated response of the quarter-wave matching network[7] is shown in Figure 6-35

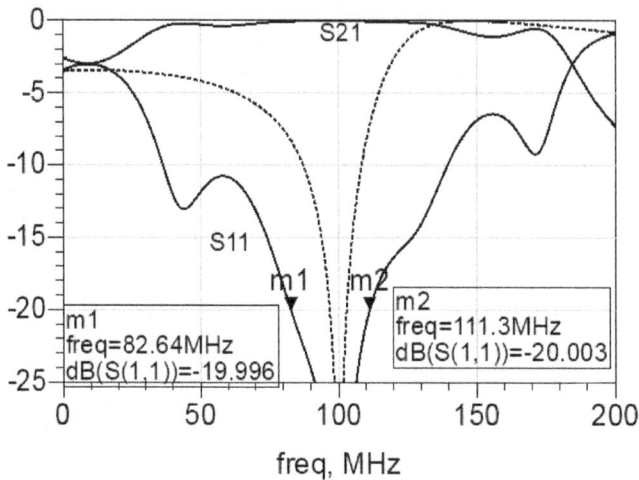

Figure 6-35 Response of the broadband matching network

Figure 6-38 shows that the bandwidths of the broadband matching network at 20 dB return loss are:

$$BW_{20dB} = 111.3 - 82.6 = 28.7 \text{ MHz}$$

$$FBW_{20dB} = \frac{28.7 x 100}{\sqrt{(111.3).(82.6)}} = 29.9 \%$$

The same numbers for the narrowband single-stub matching network are:

$$BW_{20dB} = 102.8 - 97.1 = 5.7 \text{ MHz}$$

$$FBW_{20dB} = \frac{5.7 x 100}{\sqrt{(102.8).(97.1)}} = 5.7 \%$$

Notice that the fractional bandwidth at 20 dB return loss is about 29.9 % as opposed to only 5.7 % for the narrowband matching network. This is an indication that, by adding four quarter-wave sections to the single-stub matching network, we have increased the matching bandwidth at 20 dB return loss more than five times over a single-stub matching network.

References and Further Readings

[1]   Ali Behagi and Manou Ghanevati, *Fundamentals of RF and Microwave Circuit Design,* Practical Analysis and Design Tools, Techno Search, Ladera Ranch, California 2017

[2]   Guillermo Gonzales, *Microwave Transistor Amplifiers – Analysis and Design*, Second Edition, Prentice Hall Inc., Upper Saddle River, NJ.

[3]   Randy Rhea, *The Yin-Yang of Matching: Part 1 – Basic Matching Concepts*, High Frequency Electronics, March 2006

[4]   Steve C. Cripps, *RF Power Amplifiers for Wireless Communications*, Artech House Publishers, Norwood, MA. 1999.

[5]   David M. Pozar, *Microwave Engineering*, Third Edition, John Wiley & Sons, New York, 2005

[6]   R. Ludwig, P. Bretchko, *RF Circuit Design*, Theory and Applications, Prentice Hall, Upper Saddle River, NJ, 2000

[7]   Ali Behagi, *RF and Microwave Circuit Design*, A Design Approach Using (**ADS**), Techno Search, Ladera Ranch, CA 2015.

Problems

6-1.   Design a quarter-wave transmission line to match a load resistance $R_L = 5 \ \Omega$ to a resistive source $R_S = 75 \ \Omega$ at 500 MHz. Calculate the characteristic impedance of the quarter-wave line, the Q factor and the fractional bandwidths of the matching network at 3 and 20 dB return loss. Display the simulated response and compare the calculations with measurements. For the quarter-wave matching network, calculate the fractional bandwidth and power loss from $\Gamma = 0.1$ to $\Gamma = 0.707$.

6-2.   Design a quarter-wave transmission line to match a load resistance $R_L = 100 \ \Omega$ to a resistive source $R_S = 25 \ \Omega$ at 600 MHz. Calculate the characteristic impedance of the quarter-wave line, the Q factor and the fractional bandwidths of the matching network at 3 dB and 20 dB return loss. Display the simulated response and compare the calculations with measurements. For the quarter-wave matching network, calculate the fractional bandwidth and power loss from $\Gamma = 0.1$ to $\Gamma = 0.707$.

6-3.   Design a single-stub network to match a load resistance $Z_L = 5 - j5$ Ω to a resistive source $R_S = 75$ Ω at 700 MHz. Calculate the electrical lengths of the matching line and stub and the fractional bandwidths of the matching network at 3 and 20 dB return loss. Display the simulated response and compare the calculations with measurements.

6-4.   Design a single-stub network to match a load resistance $Z_L = 100 + j20$ Ω to a resistive source $R_S = 40$ Ω at 800 MHz. Calculate the electrical lengths of the matching line and stub and the fractional bandwidths of the matching network at 3 dB and 20 dB return loss. Display the simulated response and compare the calculations with measurements.

6-5.   Design a single-stub network to match a complex load $Z_L = 20 + j5$ Ω to a complex source $Z_S = 50 + j20$ Ω at 1000 MHz. Calculate the electrical lengths of the matching line and stub and the fractional bandwidths of the matching network at 3 dB and 20 dB return loss. Display the simulated response and verify the calculations with measurements.

6-6.   Design a three-section quarter-wave network to match a load resistance $R_L = 5$ Ω to a resistive source $R_S = 50$ Ω at 100 MHz. Calculate the characteristic impedance of the quarter-wave line, the Q factor and the fractional bandwidths of the matching network at 3 and 20 dB return loss. Display the simulated response and compare the measurements with the singe quarter-wave matching network.

6-7.   Design a three-section quarter-wave network to match a load resistance $R_L = 100$ Ω to a resistive source $R_S = 25$ Ω at 900 MHz. Calculate the fractional bandwidths of the matching network at 3 and 20 dB return loss. Display the simulated response and compare the measurements with the singe quarter-wave matching network.

6-8.   Design a broadband network to match a complex load, $Z_L = 25 + j10\,\Omega$ to 75 Ω source impedance at 300 MHz. Measure the bandwidth at 20 dB return loss and compare it with the results of singe-stub matching network.

6-9.   Design a broadband network to match a complex load, $Z_L = 100 + j20$ to 40 Ω source impedance at 1200 MHz. Display the simulated response and measure the bandwidth at 20 dB return loss. Compare the results with the singe-stub matching network.

Chapter 7

Single Stage Amplifier Design

7.1 Introduction

Modern amplifier design involves the use of high frequency transistors to use DC power to control and amplify RF energy. These transistors are fundamentally of the bipolar and field effect transistor (FET) variety. There are many subtypes of bipolar and FET devices that the reader is encouraged to explore. For the many types of transistors the impedance matching techniques, covered in chapters 5 and 6, are very useful for the design and simulation of linear transistor amplifiers. There are many different ways in which transistor amplifiers are impedance matched depending on the function or purpose of the amplifier circuit. In this chapter we design two single stage amplifiers that require two different impedance matching techniques. Namely,

Maximum Gain Amplifier Design

Low Noise Figure Amplifier Design

Most amplifier designs actually involve selective mismatch of the transistor to its source and load impedance to accomplish its intended purpose. Only the maximum gain amplifier requires a conjugate matching network design. The specific matching techniques for each category of amplifiers are listed below:

1. Maximum Gain Amplifier

In a maximum gain amplifier the input and the output are simultaneously conjugate matched to achieve maximum gain.

2. Low Noise Figure Amplifier

In a low noise amplifier the transistor's input is mismatched to achieve a specific Noise Figure.

7.2 Maximum Gain Amplifier Design

Sections 7.2 through 7.12 cover the design of the maximum gain amplifier at 1 GHz using the Avago's AT-41486 transistor. The AT-41486 is a general purpose NPN bipolar transistor that offers excellent high frequency performance. The data Sheet of the AT-41486 transistor is given in the Appendix. This transistor is housed in a low cost surface mount .085" diameter plastic package using Avago's 10 GHz Self-Aligned-Transistor (SAT) process. The pin connection of the transistor is shown in Figure 7-1.

Pin Connections

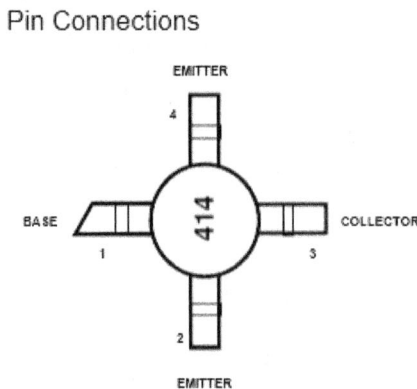

Figure 7-1: AT-41486 pin connections. (Courtesy of Data Sheet Library)

The maximum gain amplifier is conjugately matched at the input and output to achieve a good return loss over a narrow bandwidth. In general a device can only be conjugately matched if it is unconditionally stable. If the device is potentially unstable in the band of interest, a conjugate impedance match cannot be realized unless we add additional circuitry to stabilize the device. In sections 7-3 through 7-12 we use the AT-41486 transistor in LTspice to design a maximum gain amplifier at 1 GHz. To achieve this goal it is required that we take the following steps.

- Create a symbol for the AT-41486 transistor
- Bias the transistor at Vce = 8VDC and Ic = 25 mA

- Add stability components to make the transistor unconditionally stable

- Calculate GTmax and S parameters at 1 GHz

- Design the input and output impedance matching networks

- Assemble the maximum gain amplifier and measure GTmax

Creating a Symbol for the AT41486 Transistor

Example 7-1: Create a symbol in LTspice for the AT41486 transistor Spice model and save it with .asy extension.

Solution: The following steps are used to generate the transistor symbol.

1) Start LTspive IV and select File > New Symbol.
2) Select Edit > Add Pin/Port > OK to add ports. Repeat the step for a total of three ports, as shown in Figure 7-2.
3) Select Draw > Line to draw lines connecting the Ports similar to lines in Figure 7-2.
4) Select Draw > Text to open the Edit Text on the Symbol window. Type NPN and press OK.
5) Repeat the above step to add Unnn, as shown in Figure 7-2.

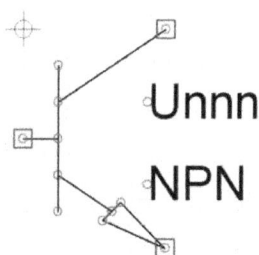

Figure 7-2: Transistor symbol

6) Press Ctrl-A to open the Symbol Attribute Editor window, as show in Figure 7-3.

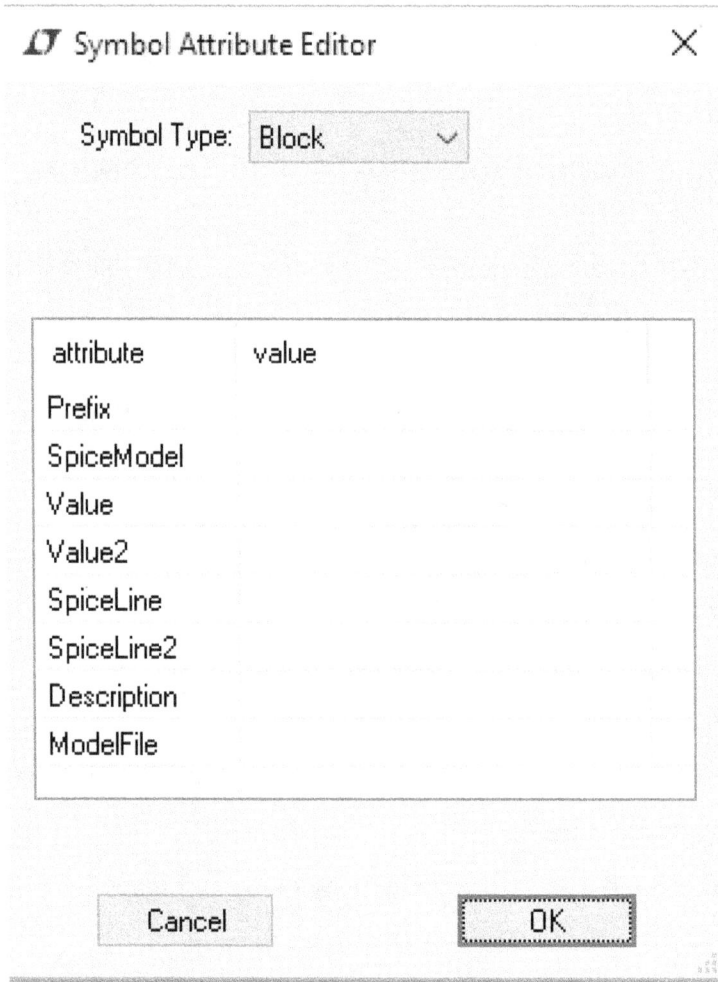

Figure 7-3: Symbol Attribute Editor

7) In the Symbol Type box select Cell. Type X for Prefix and NPN for Value. Type Bipolar NPN transistor for Description, as shown in Figure 7-4. Press OK.

8) Select File > Save as AT-41486.asy in your LTspice directory.

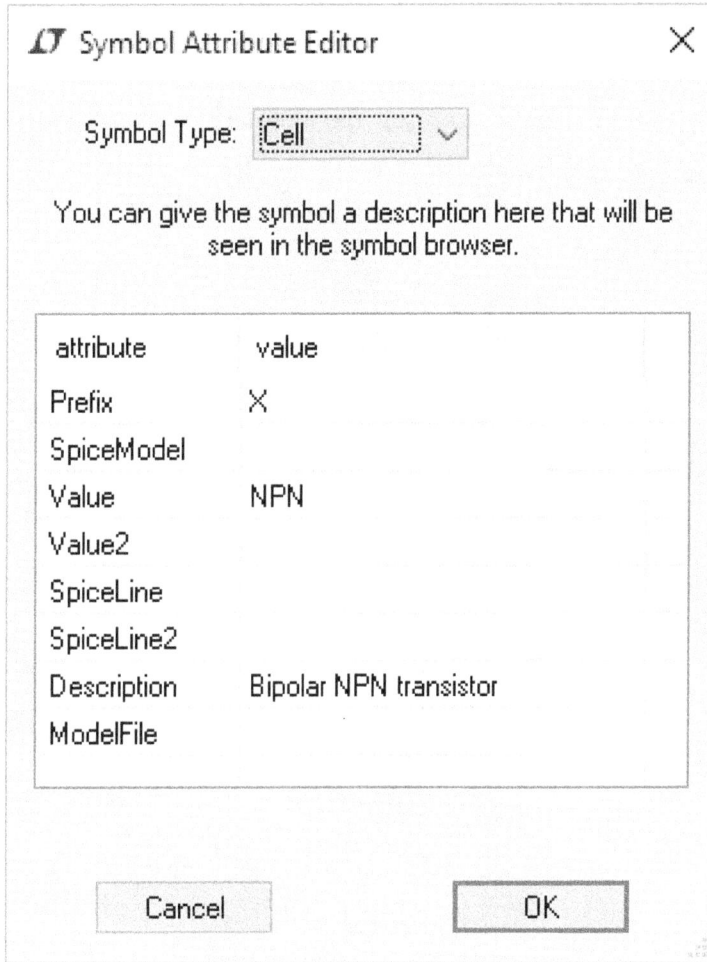

Figure 7-4: Transistor Symbol Attributes

Creating Schematic and Plotting the DC IV Curves

Example 7-2: Create a new schematic for the AT14846 transistor and plot the DC IV curves. Determine the base to emitter voltage at Ic = 25 mA and Vce = 8 VDC.

Solution: To create a new schematic and plotting the DC IV curves in LTspice do the followings:

1) Open a new schematic in LTspice by selecting File > New Schematic.

2) Select Edit > Component to open the Select Component Symbol window.

3) Select npn and press OK to insert the Bipolar NPN transistor, as shown in Figure 7-5.

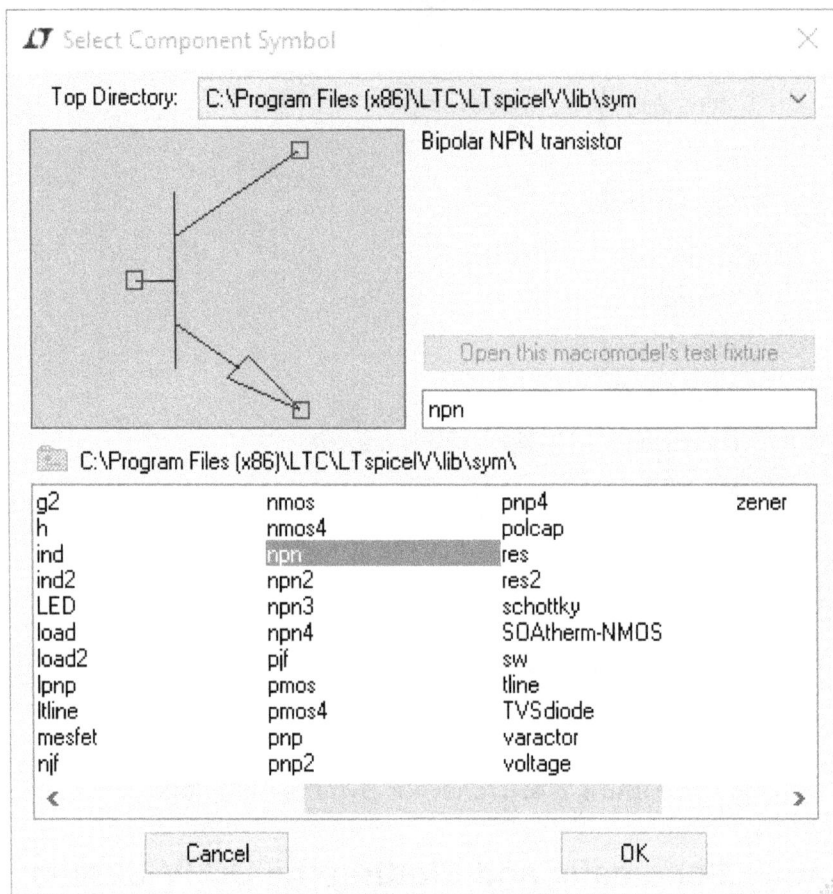

Figure 7-5: Selecting Bipolar NPN Transistor

4) Right click on Q1 and change to U1, then click OK.

5) Right click on NPN and change to AT41486, then click OK.

6) Select Edit > SPICE Directive to open the Edit Text on the Schematic window, as shown in Figure 7-6.

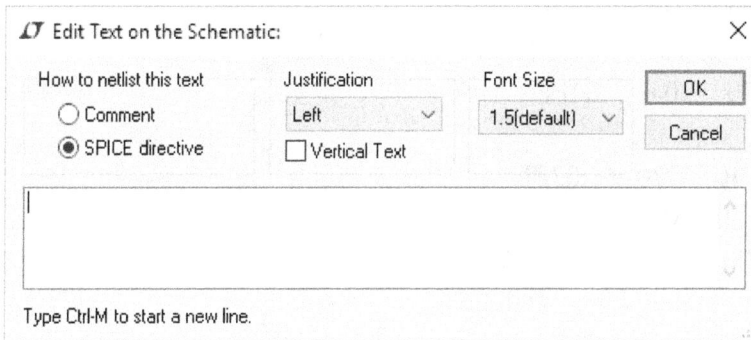

Figure 7-6: Edit Text on the Schematic window

7) Type in the following spice model for the transistor in the window provided. If you already have the transistor Spice model saved somewhere in your computer, you can copy and paste the spice model inside the window.

```
*SPICE model for AT-41486
*
.SUBCKT AT41486 60    20    40
LL1    20    25    .55NH
* PI NETWORK TO SIMULATE TRANSMISSION LINE T1
C1T1   25    0     .06PF
LT1    25    30    .48NH
C2T1   30    0     .06PF
*
LLB    30    35    .55NH
CCEB   30    50    .02PF
LL2    40    45    .06NH
* PI NETWORK TO SIMULATE TRANSMISSION LINE T2
C1T2   45    0     .04PF
LT2    45    50    .08NH
C2T2   50    0     .04PF
*
LLE    50    55    .1NH
CCEC   50    70    .03PF
LL3    60    65    .25NH
* PI NETWORK TO SIMULATE TRANSMISSION LINE T3
```

```
C1T3   65    0     .04PF
LT3    65    70    .30NH
C2T3   70    0     .04PF
*
CCBC   30    70    .03PF
* CALL DIE MODEL
XDIE   70    35    55    AT414
.ENDS

* DIE MODEL (excludes bond wires)
.SUBCKT AT414   75    20    85
CCB    20    60    .032PF
DCD1   20    60    DMOD   572
RRB1   20    25    1.07
DCD2   25    60    DMOD   680
RRB2   25    30    3.2
DCD3   30    60    DMOD   340
RRB3   30    35    2.7
RRC    60    75    5
RRE    80    85    .24
CCE    60    85    .032PF
QINT   60    35    80    QDIS   420
.ENDS
*
.MODEL  DMOD   D(IS=1E-25, CJO=2.45E-16, VJ=.76, M=.53, BV=45,
IBV=1E-9)
.MODEL  QDIS   NPN (BF=100, BR=2.5, IS=1.65E-18, VA=20,
TF=12PS,
+         CJE=2.4E-15, VJE=1.01, MJE=0.6, PTF=25, XTB=1.818,
+         VTF=6, ITF=3E-4, IKF=1.3E-4, XTF=4, NF=1.03, ISE=5E-15,
+         NE=2.5)
```

8) Add DC voltage sources V1 and V2 to the transistor schematic.

9) Select Edit > Component > voltage > OK. The first voltage source is V1.

Repeat to insert the second voltage source V2.

10) Wire up the schematic and add grounds, as shown in Figure 7-7.

```
*SPICE model for AT-41486
*
.SUBCKT AT41486 60    20    40
LL1   20    25    .55NH
* PI NETWORK TO SIMULATE TRANSMISSION LINE T1
C1T1  25    0     .06PF
LT1   25    30    .48NH
C2T1  30    0     .06PF
*
LLB   30    35    .55NH
CCEB  30    50    .02PF
LL2   40    45    .06NH
* PI NETWORK TO SIMULATE TRANSMISSION LINE T2
C1T2  45    0     .04PF
LT2   45    50    .08NH
C2T2  50    0     .04PF
*
LLE   50    55    .1NH
CCEC  50    70    .03PF
LL3   60    65    .25NH
* PI NETWORK TO SIMULATE TRANSMISSION LINE T3
C1T3  65    0     .04PF
LT3   65    70    .30NH
C2T3  70    0     .04PF
*
CCBC  30    70    .03PF
* CALL DIE MODEL
XDIE  70    35    55    AT414
.ENDS

* DIE MODEL (excludes bond wires)
.SUBCKT AT414  75    20    85
CCB   20    60    .032PF
DCD1  20    60    DMOD   572
RRB1  20    25    1.07
DCD2  25    60    DMOD   680
RRB2  25    30    3.2
DCD3  30    60    DMOD   340
RRB3  30    35    2.7
RRC   60    75    5
RRE   80    85    .24
CCE   60    85    .032PF
QINT  60    35    80    QDIS   420
.ENDS
*
.MODEL  DMOD   D(IS=1E-25, CJO=2.45E-16, VJ=.76, M=.53, BV=45, IBV=1E-9)
.MODEL  QDIS   NPN (BF=100, BR=2.5, IS=1.65E-18, VA=20, TF=12PS,
+        CJE=2.4E-15, VJE=1.01, MJE=0.6, PTF=25, XTB=1.818,
+        VTF=6, ITF=3E-4, IKF=1.3E-4, XTF=4, NF=1.03, ISE=5E-15,
+        NE=2.5)
```

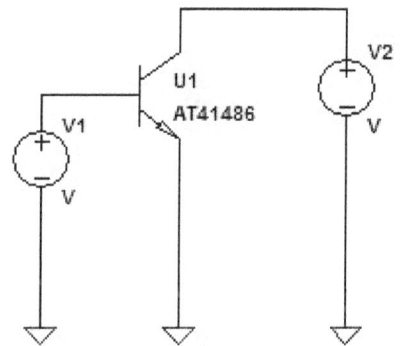

Figure 7-7: AT-41486 transistor Spice model with schematic

11) Select Simulate > Edit Simulation Cmd to open the Edit Simulation Command window.

12) Select DC sweep and then select V1 for the 1st Source and Linear for Type of Sweep, if already is not selected.

13) Type in 0.78 for the Start Value, 0.85 for the Stop Value, and 0.012 for the Increment, as shown in Figure 7-8.

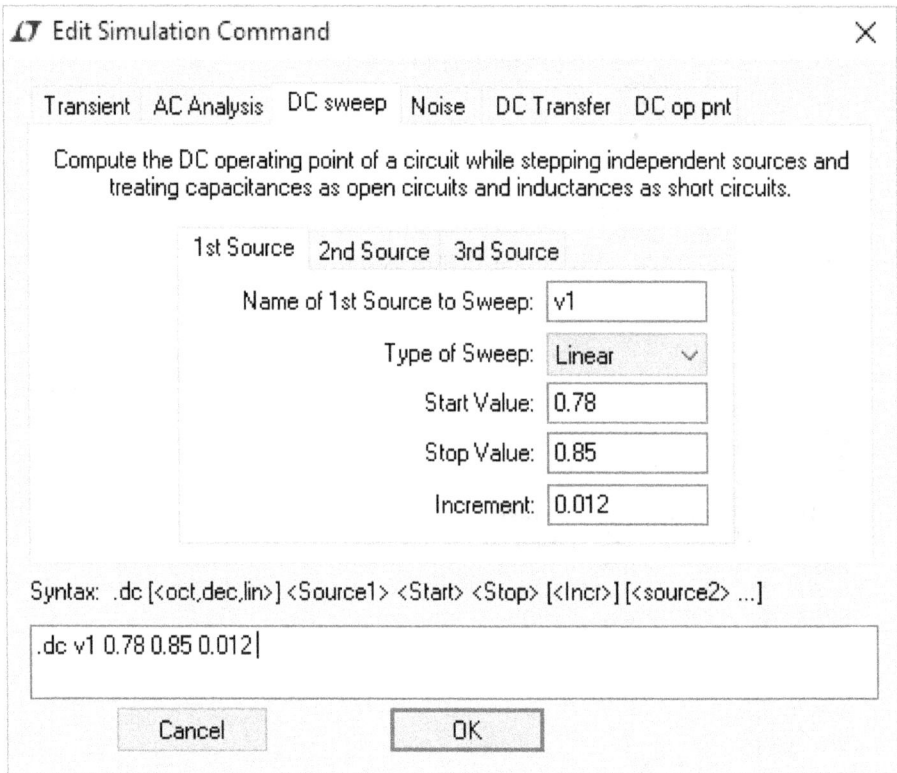

Figure 7-8: Selecting v1 values

14) Select V2 for the 2nd Source and Type in 0 for the Start Vale, 10 for the Stop Value, and 0.5 for the Increment, as shown in Figure 7-9.

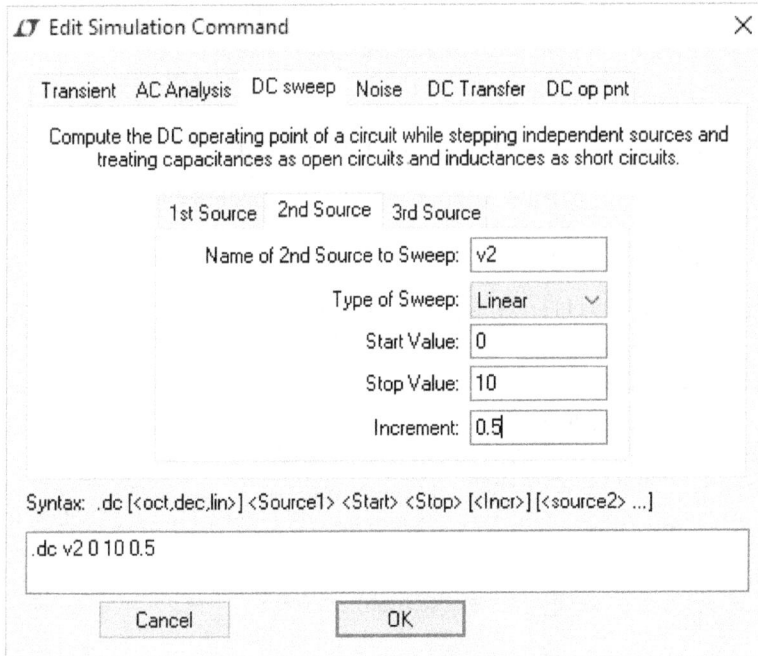

Figure 7-9: Selecting v2 values

The schematic for DC simulation is shown in Figure 7-10.

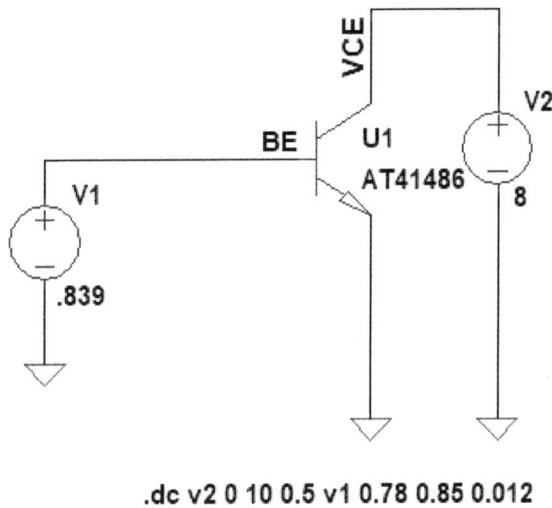

.dc v2 0 10 0.5 v1 0.78 0.85 0.012

Figure 7-10: Transistor schematic for DC simulation

The AT41486 schematic in Figure 7-10 is now ready for simulation and plotting the DC IV curves. The procedure follows.

15) Select Simulate > Run to open the Plot window.

16) Right click on the Plot window and select Visible Traces to open the Select Visible Waveforms window.

17) Select Vbe and ix[U1:C], as shown in Figure 7-11.

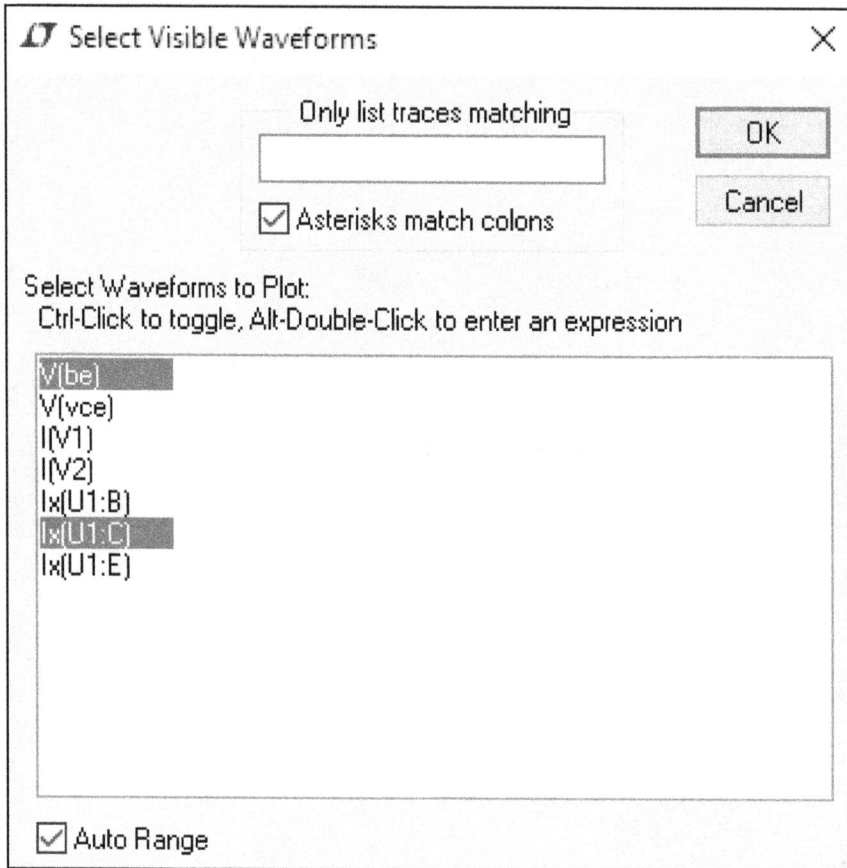

Figure 7-11: Selection of V(be) and collector current

18) Press OK to display V(be) and the collector current in DC IV curves, as shown in Figure 7-12.

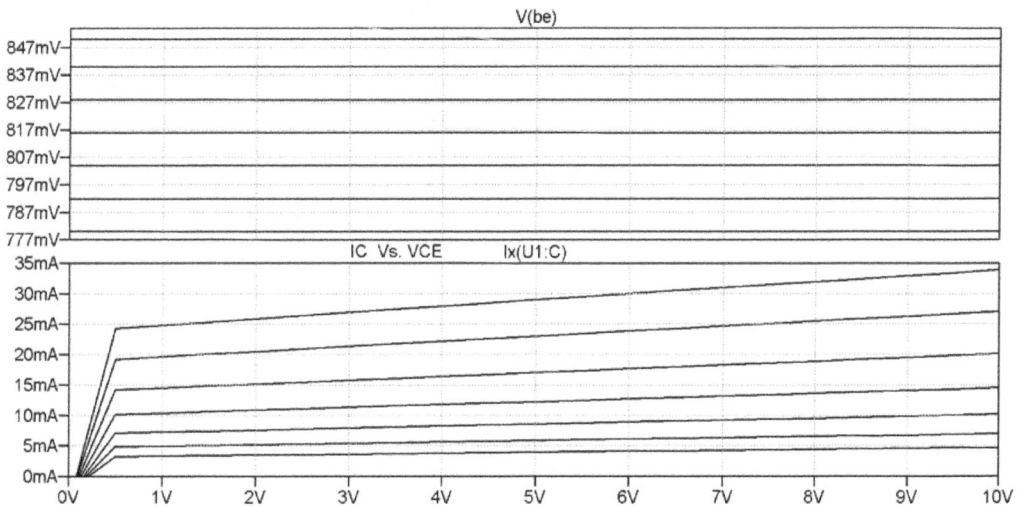

Figure 7-12: DC-IV plots of AT-41486 transistor

Figure 7-12 shows that the collector current of 25 mA is achieved at Vce = 8 VDC when the base to emitter voltage, Vbe, is approximately 0.837 VDC.

Biasing the Transistor at Vce = 8 VDC and Ic = 25 mA

Example 7-3: Obtain a more accurate value for Vbe by adjusting the Vbe increments until the desired collector current Ic = 25 mA is achieved at Vce = 8 VDC.

Solution: To achieve this goal the following steps are used.

1) From the schematic diagram select Simulate > Edit Simulation Cmd to open the Edit Simulation Command. Select DC op pnt and then press OK to place .op syntax on the schematic. Right click on V1 and enter 0.8393 as DC value for V1 (Vbe).

2) Right click on v2 and enter 8 as DC value for v2 (Vce).

3) Select Simulate > Run to simulate the schematic.

4) A table of all Operating Point currents and voltages will appear. The value of Ic (i.e., I(v2)) is about 24.87 mA.

5) Continue to incrementally increase Vbe value until Ic is 25 mA is reached.

The operating point data is shown in Table 7-1.

```
        --- Operating Point ---

V(be):              0.8393           voltage
V(vce):             8                voltage
I(V2):              -0.0250326       device_current
I(V1):              -0.000248777     device_current
Ix(u1:60):          0.0250326        subckt_current
Ix(u1:20):          0.000248777      subckt_current
Ix(u1:40):          -0.0252814       subckt_current
```

Table 7-1 DC operating point for AT41486

From Table 7-1 we see that Ic =25 mA is achieved with Vbe = 0.8393V.

Transistor Stability Considerations

Most Microwave transistors are potentially unstable at some frequency. This does not necessarily make them undesirable for use as an amplifier. A potentially unstable transistor does not mean that it will definitely oscillate in a circuit. Referring to the device in Figure 7-5 there may exist some combination of input or output reflection coefficient, Γ_S or Γ_L, that, when presented to the transistor, may indeed make the device oscillate. An oscillation condition is indicated by $|\Gamma_{IN}| >$ 1 or $|\Gamma_{OUT}| > 1$. This can also be viewed as a positive value for the Return Loss, S_{11} or S_{22}, in dB. This is referred to as negative resistance which is actually a design goal in the design of microwave oscillators. The conditions for unconditional stability for an RF transistor are given by equations (7-1) through (7-4) [2].

$$|\Gamma_S| < 1 \qquad\qquad (7\text{-}1)$$

$$|\Gamma_L| < 1 \tag{7-2}$$

$$|\Gamma_{IN}| = \left| S_{11} + \frac{S_{12}S_{21}\Gamma_L}{1 - S_{22}\Gamma_L} \right| < 1 \tag{7-3}$$

$$|\Gamma_{OUT}| = \left| S_{22} + \frac{S_{12}S_{21}\Gamma_S}{1 - S_{11}\Gamma_S} \right| < 1 \tag{7-4}$$

Typically there exists a select set of reflection coefficient values for Γ_S and Γ_L that will give the device this negative resistance characteristic. One method of dealing with this problem is to select values for Γ_S and Γ_L that avoid these unstable regions of impedance. This technique is demonstrated in Section 7.3.

Another technique is to introduce a network that forces the device to be unconditionally stable for any values of Γ_S and Γ_L. This technique is often preferable as it will be shown that a transistor can be made unconditionally stable over a very wide frequency range. This is important because even though a device may be stable in its desired frequency band, it may become unstable at some frequency outside of the band of operation.

A simultaneous numerical solution of Equations (7-1) through (7-4) for all values of Γ_S and Γ_L can prove that there are two conditions for unconditional stability. The stability factor, K, and stability measure, B1, are given by the following Equations.

$$K = \frac{(1 - |S_{11}|^2 - |S_{22}|^2 + |(S_{11}) \cdot (S_{22}) - (S_{12}) \cdot (S_{21})|^2)}{(2 \cdot |(S_{12}) \cdot (S_{21})|)} > 1 \tag{7-5}$$

$$B_1 = 1 + |S_{11}|^2 - |S_{22}|^2 - |(S_{11}) \cdot (S_{22}) - (S_{12}) \cdot (S_{21})|^2 > 0 \tag{7-6}$$

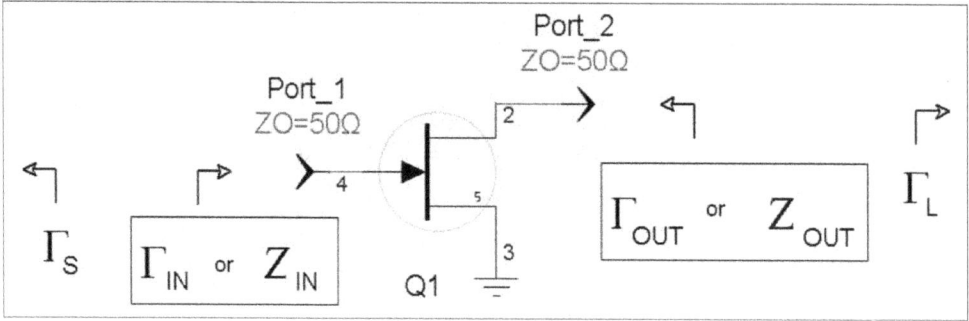

Figure 7-13: Transistor input and output reflection coefficients

When the transistor is unconditionally stable a simultaneous conjugate match can be defined. The conjugate match exists when $\Gamma_{IN} = \Gamma_S{}^*$ and $\Gamma_{OUT} = \Gamma_L{}^*$.

The simultaneous conjugate match reflection coefficients are often referred to as:Γ_{MS} andΓ_{ML}. The maximum gain that can be obtained with a

simultaneous conjugate matched amplifier is determined from the transducer gain equation, G_{Tmax}.

$$G_{T\,max} = \frac{\left(1-|\Gamma_{MS}|^2\right)|S_{21}|^2\left(1-|\Gamma_{ML}|^2\right)}{\left|\left(1-S_{11}\Gamma_{MS}\right)\left(1-S_{22}\Gamma_{ML}\right)-S_{12}S_{21}\Gamma_{ML}\Gamma_{MS}\right|^2} = \frac{|S_{21}|}{|S_{12}|}\left(K-\sqrt{K^2-1}\right) \quad (7\text{-}7)$$

If K is exactly equal to one, then G_{Tmax} becomes the Maximum Stable Gain designated as G_{MSG}. G_{MSG} is the maximum value that G_{Tmax} can achieve.

$$G_{MSG} = \frac{|S_{21}|}{|S_{12}|} \quad (7\text{-}8)$$

Stabilizing the Transistor

As the Equations (7-5) through (7-7) show, the stability factor K and the stability measure B_1 are functions of the transistor S parameters.

In the next Example we prepare the transistor circuit for S parameter simulation and plotting the stability parameters.

Example 7-4: Prepare the schematic for stability analysis. Simulate the schematic and plot the stability parameters K and B1 from 10 MHz to 10 GHz.

Solution: To prepare the transistor Spice model for stability and S parameter analysis do the followings:

1) Create a new schematic in LTspice and place the AT- 41486 transistor model on the schematic.

2) Select Edit > Component to open the Select Component Symbol window.

3) Select npn > OK to insert the Bipolar NPN transistor.

4) Attach an 8 VDC voltage source to the collector and 0.8393 VDC voltage source to the base.

5) Insert a new voltage source V3 from LTspice library in the schematic. Right click on the voltage source and select Advanced to open the Independent Voltage Source window.

6) Type in 1 for the AC Amplitude and 50 for the Series Resistance.

7) Attach a 50-Ohms load resistor. Right click on the load resistor designator and change R1 to Rout.

8) Insert 1 μF DC blocking capacitors and 1 μH AC blocking inductors as shown in the schematic.

9) Wire up the schematic as shown in Figure 7-14.

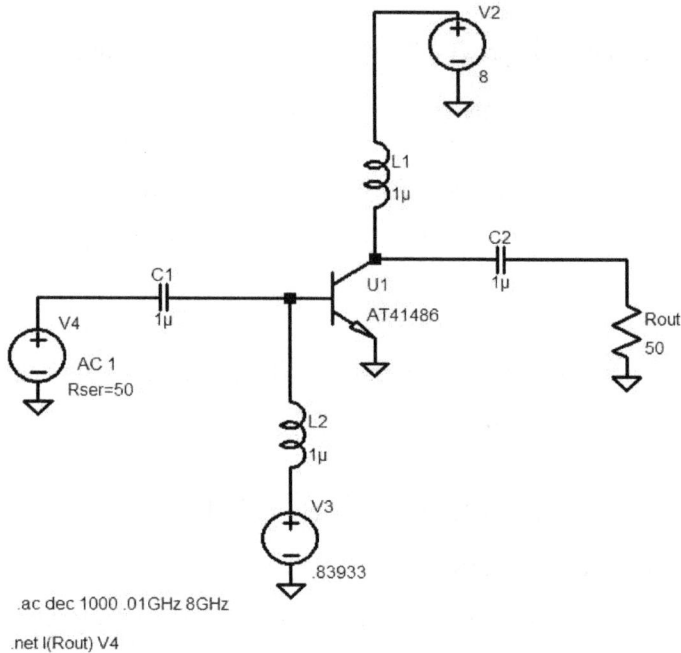

Figure 7-14: Transistor schematic for AC analysis

10) Select Edit > SPICE Directive to open the Edit Text on the Schematic window shown in Figure 7-6.

11) Copy and paste the transistor spice parameters.

12) Wire up the schematic as shown in Figure 5-13.

13) Select Simulate > Edit Simulation Cmd to open the Edit Simulation Command window. Click on AC Analysis tab. Select Decade for the Type of Sweep. Type in 1000 for the Number of points, 10MEG for Start Frequency, and 10G for Stop Frequency. Then press OK.

14) Select Edit > SPICE Directive (or press S on the keyboard) to open the SPICE Directive window. Type in .net I(Rout) V3 and press OK.

15) Select Simulate > Run. A blank waveform window opens.

16) Right click on the blank waveform window (or simply Ctrl + A on the computer keyboard) and select Add Trace to open the Add Traces to Plot window, as shown below.

17) In the Expression(s) to add box type in the functions for K and B_1 by using Equations (7-5) and (7-6) in MATLAB scripting format together with the S-parameter at 1 GHz. Press OK to plot K and B_1 values, as shown in Figure 7-15.

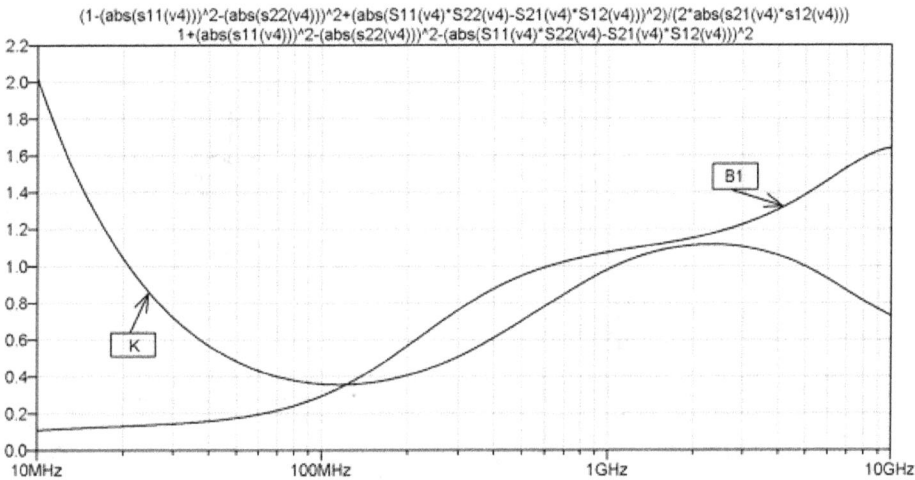

$(1-(abs(s11(v4)))\wedge2-(abs(s22(v4)))\wedge2+(abs(S11(v4)*S22(v4)-S21(v4)*S12(v4)))\wedge2)/(2*abs(s21(v4)*s12(v4)))$
$1+(abs(s11(v4)))\wedge2-(abs(s22(v4)))\wedge2-(abs(S11(v4)*S22(v4)-S21(v4)*S12(v4)))\wedge2$

Figure 7-15: Stability parameters plot for the AT41486 device

Note that stability factor K is less than 1 for most of the frequency band from 10 MHz up to 10 GHz indicating that the transistor is potentially unstable over most of its frequency range. In the next Example we add stabilizing elements to the transistor schematic to make it unconditionally stable.

7.3 Making the Transistor Unconditionally Stable

Example 7-5: Design stabilizing elements to make the transistor unconditionally stable.

Solution: The effective stabilization network for medium and high power transistors employs a parallel RC circuit at the input of the device, as shown in the Figure 7-16. Keep the resistor value at 50 Ohm to match with the source impedance and make the capacitor value C variable {C}. Sweep the capacitor value from 0.1 pF to 0.5 pF in 0.1 pF steps. Add .the STEP command to the schematic and plot the stability parameters while varying the variable capacitors.

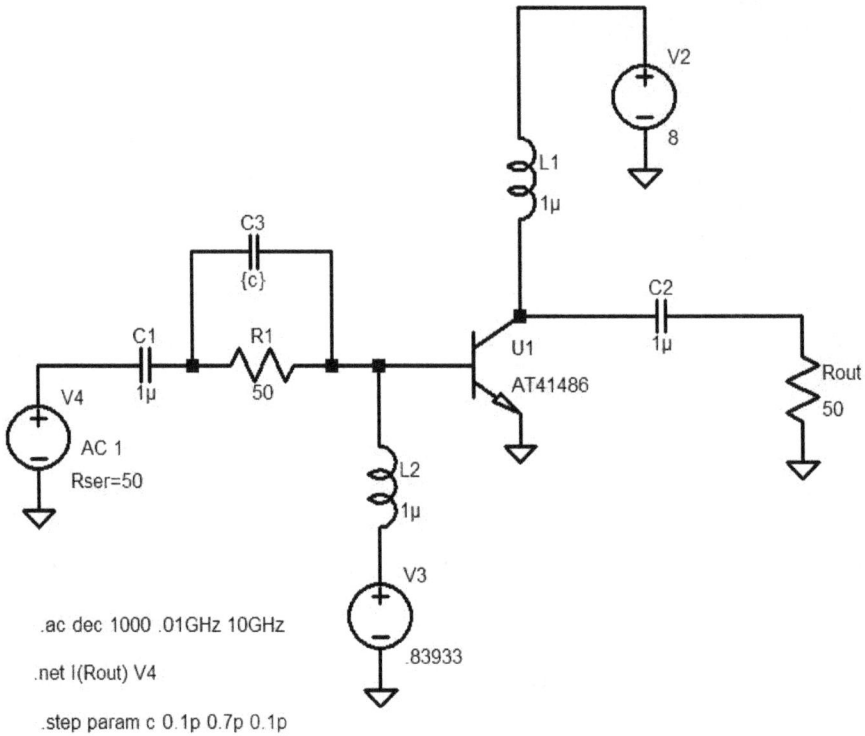

Figure 7-16: Transistor stability analysis

The following plot in Figure 7-17 shows the swept K and B_1 parameters from 10 MHz to 10 GHz.

$(1-(abs(s11(v4)))^2-(abs(s22(v4)))^2+(abs(S11(v4)*S22(v4)-S21(v4)*S12(v4)))^2)/(2*abs(s21(v4)*s12(v4)))$

$1+(abs(s11(v4)))^2-(abs(s22(v4)))^2-(abs(S11(v4)*S22(v4)-S21(v4)*S12(v4)))^2$

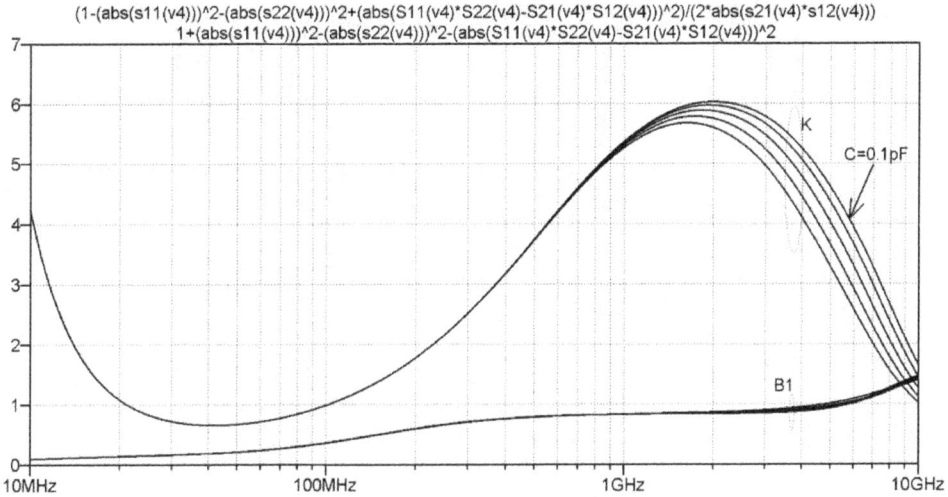

Figure 7-17: Stability parameters with parallel RC network variation

Figure 5-17 shows that the effect of the variable capacitor is negligible at low frequencies bat at frequencies above 1 GHz the stability factor K varies as the capacitor values vary between 0.1 and 0.5 pF. Therefore, the capacitor sweep cannot make the transistor unconditionally stable at frequencies below 100 MHz.

To make the device unconditionally stable, we keep the base shunt capacitor at 0.1 pF but add shunt resistors to the base and collector inductors. To see the effect of shunt resistors on the device stability, we make the resistors variable and sweep them from 200 to 400 ohms in 100 ohm steps, as shown in Figure 7-18.

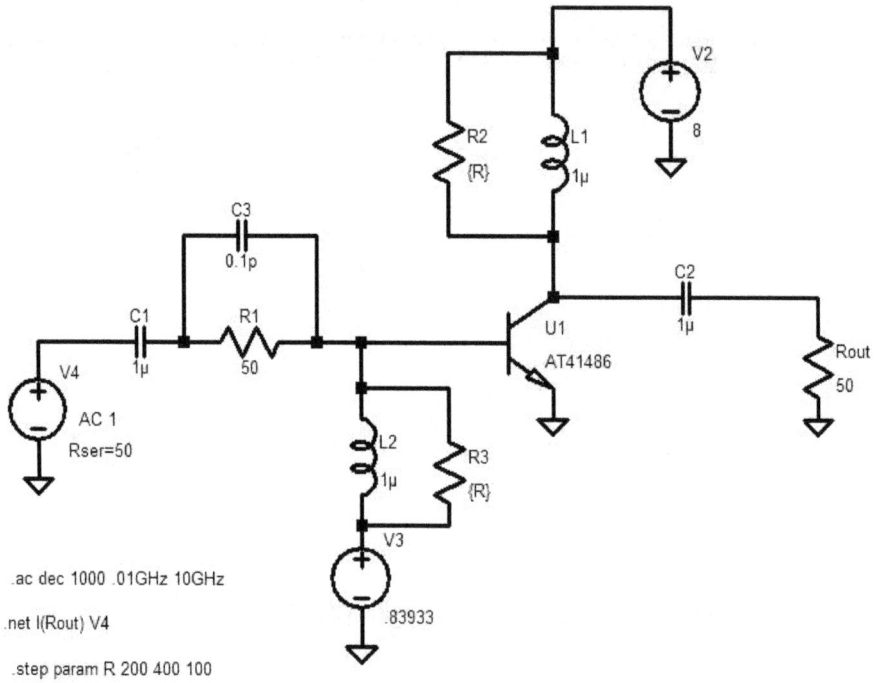

Figure 7-18: Tuning R2 for transistor unconditional stability

After sweeping the resistor values we can see the effect of shunt resistors on stability factors in Figure 7-19.

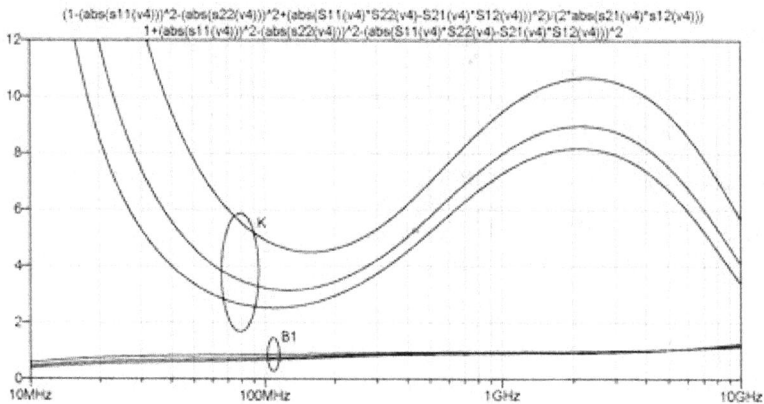

Figure 7-19: Stability parameters of unconditionally stability

Examination of Figure 7-19 shows that the device is unconditionally stable across the entire frequency range for the resistor values of 100, 200, and

300 Ohms. It is also noticed that stability increases as the resistor values are decreased.

Obviously each resistor value gives a different GTmax value. We decided to make the GTmax value equal to 13.34 dB at 1 GHz. This is exactly the GTmax value that we get using the published linear S parameters at 1 GHz when the transistor is biased at Vce = 8 VDC and Ic = 25 mA. By lowering the sweep increments we achieved this goal at R = 278 Ohm, as shown in Figure 7-21.

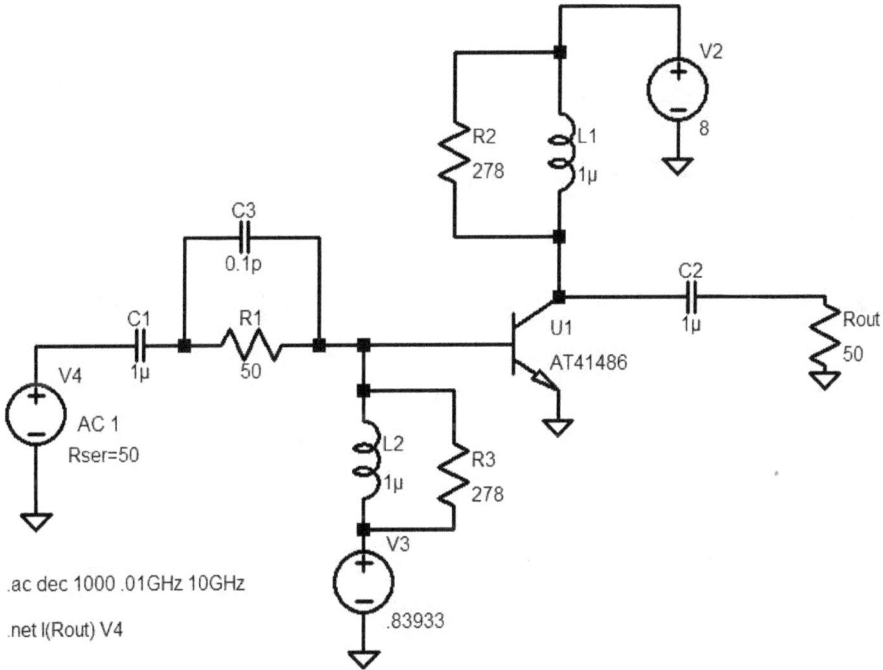

Figure 7-20: Stabilized Transistor schematic

Simulate the schematic and plot the stability parameters from 10 MHz to 10 GHz.

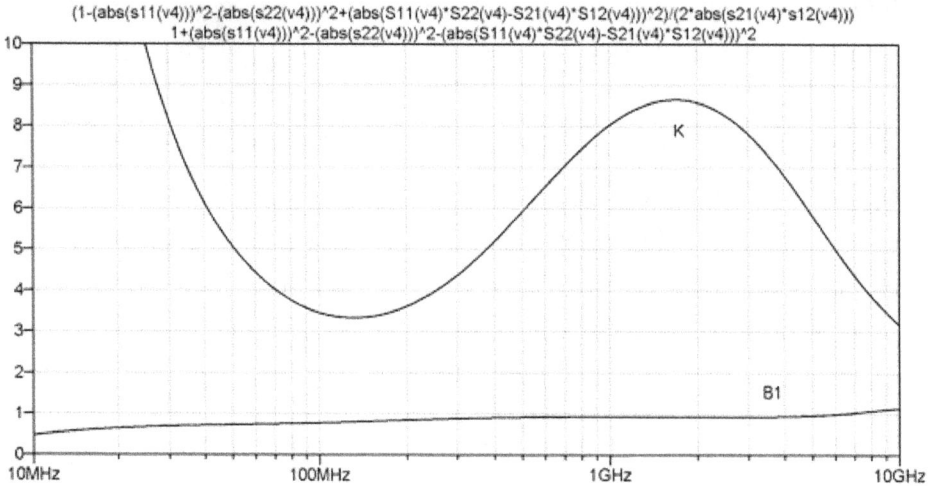

Figure 7-21: Stability parameters of unconditionally stable device

Note that in Figure 7-21 the stability parameters $B_1 > 0$ and $K > 1$ from 10 MHz to 10 GHz. This means that the transistor is unconditionally stable.

7.4 Calculating the GTmax and S Parameters at 1 GHz

In this section, the maximum transducer power gain, GTmax, of the device is calculated. Example 7-6 shows the details of how this can be accomplished in LTspice.

Example 7-6: Simulate the schematic and plot the, GTmax, from 0.5 GHz to 1.5 GHz. Show that the GTmax at 1 GHz is 13.34 dB.

Solution: The following steps shows how to simulate the schematic and plot the GTmax from 0.5 GHz to 1.5 GHz,

1. Go to Simulate > Edit Simulation Cmd.

2. Change the start and stop frequencies to 0.5 and 1.5 GHz

3. Select Run > Simulation.

4. Right click on the new waveform window and select Add trace.

5. Enter the expression for GTmax in the window under *Expression(s)*.

The simulation result is shown in Figure 7-22.

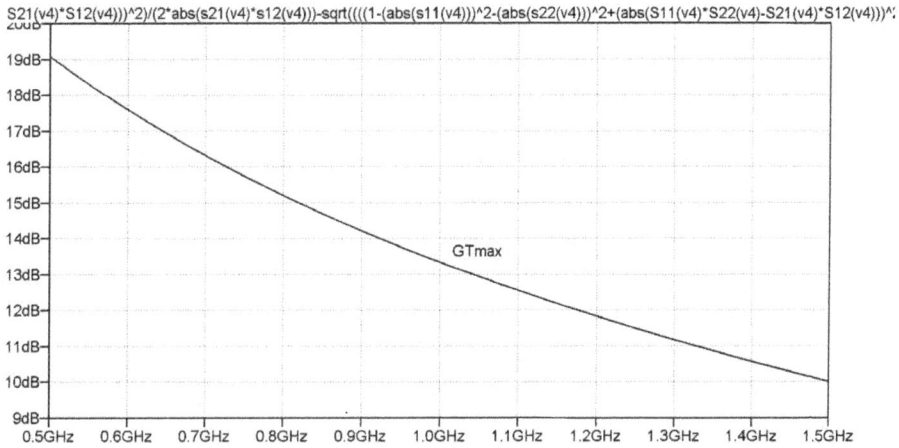

S21(v4)*S12(v4)))^2)/(2*abs(s21(v4)*s12(v4)))-sqrt((((1-(abs(s11(v4)))^2-(abs(s22(v4)))^2+(abs(S11(v4)*S22(v4)-S21(v4)*S12(v4)))^:

Figure 7.22: GTmax plot for the stabilized device

The data point in Figure 7-23 shows that GTmax is 13.3447 dB at 1 GHz.

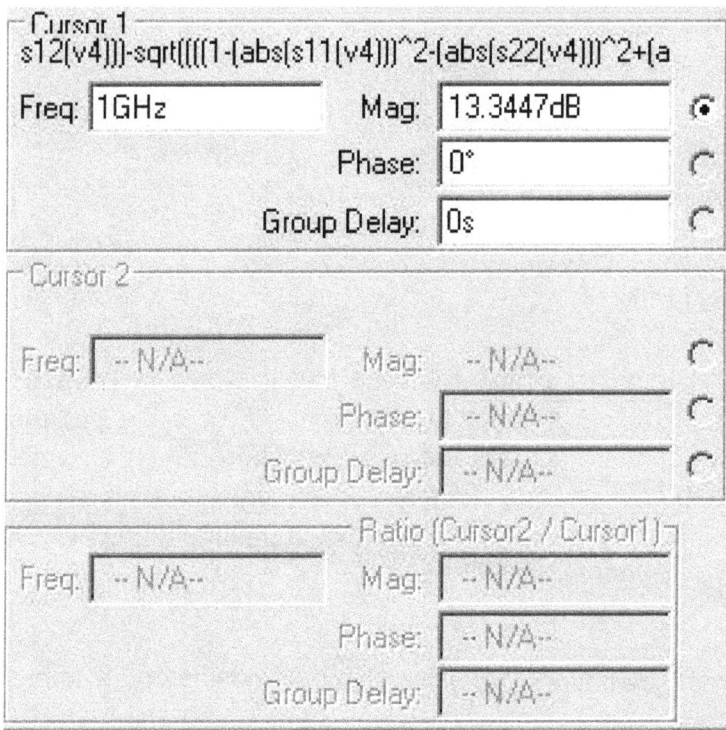

Figure 7.23: GTmax value of 13.3447 dB at 1 GHz.

Example 7-7: Measure the transistor S-parameters at 1 GHz.

Solution: Use the following steps to measure the S-parameters at 1 GHz.

1. Change the start and stop frequencies to 1 GHz and 3 GHz respectively

2. Change the number of points to 3, select Type of Sweep to linear, and then press Run simulation

3. Click File > Exports to open the "Select Traces to Export" window

4. Hold the Ctrl key on your keyboard

5. In the Select Traces to Export window, select S11(v4), S12(v4), S21(v4), and S22(v4)

6. In the above window, select the proper format for the tabulated data

7. Use the top-down arrow to select Real and Imaginary > OK.

8. Go to the directory where the generated .txt file is stored. Open the file and organize the S parameters at 1 GHz.

$S11 = 0.121 - j*0.056$
$S12 = 0.0062 + j*0.011$
$S21 = -0.051 + j*4.435$
$S22 = 0.253 - j*0.111$

Once the S parameters of the stabilized transistor at 1 GHz are determined we can use MATLAB script in Equations (7-5), and (7-6) to calculate the stability parameters K and B1, Equation (7-7) to calculate the maximum transducer gain GTmx, and the equations given in Guillermo Gonzalez[5] to calculate simultaneous match input and output impedances at 1 GHz.

Example 7-8: Calculate (a) The stability parameters K and B1, (b) The maximum transducer gain GTmax, (c) The simultaneous match input and output impedances at 1 GHz.

Solution: Substitute S parameters in Equations (7-5), (7-6) and (7-7), to calculate stability parameters K, B1, and maximum transducer gain GTmx. Also use Gonzalez[5] equations to calculate the simultaneous match input and output impedances,

(a) Using MATLAB script to calculate stability parameters K and B1

N=(1-(abs(S11))^2-(abs(S22))^2+(abs((S11)*(S22)-(S21)*(S12)))^2)

D = (2*(abs((S12)*(S21))))

K=N/D

B1=1+(abs(S11))^2-(abs(S22))^2-(abs((S11)*(S22)-(S12)*(S21)))^2

(b) Calculation of GTmax

GTmx=(abs(S21)*(K-sqrt(K^2-1)))/(abs(S12))

(c) Calculation of simultaneous match input-output impedances

Delta=(S11)*(S22)-(S12)*(S21)

Delta2=abs(Delta)^2

C1=S11-(Delta)*conj(S22)

Gama1=(B1-sqrt((B1)^2-4*(abs(C1)^2)))/(2*C1)

B2=1+(abs(S22))^2-(abs(S11))^2-(abs((S11)*(S22)-(S12)*(S21)))^2

C2=S22-(Delta)*conj(S11)

Gama2=(B2-sqrt((B2)^2-4*(abs(C2)^2)))/(2*C2)

Z1sm=50*(1+Gama1)/(1-Gama1)

Z2sm=50*(1+Gama2)/(1-Gama2)

The MATLAB solutions for (a), (b), and (c) are given in Table 7-2.

Name	Value
B1	0.933
B2	1.05
C1	0.096 - 0.05j
C2	0.241 - 0.109j
D	0.114
Delta	0.074 - 0.055j
Delta2	8.491e-3
GTmx	21.601
Gama1	0.104 + 0.055j
Gama2	0.247 + 0.111j
K	8.051
N	0.914
R1	39.764
R2	31.623
R3	25.149
RS	50
S11	0.121 − 0.056j
S12	6.2e-3 + 0.011j
S21	-0.051 + 4.435j
S22	0.253 - 0.111j
Z1sm	61.184 + 6.789j
Z2sm	79.89 + 19.21j

Table 7-2: MATLAB Solutions

The MATLAB solutions in Table 7-2 shows that: (a) the stability factor K= 8.051>1, (b) the maximum transducer gain GTmx = 21.601 at 1 GHz. Use the following Equation to convert the GTmx to dB.

$$(GTmx)_{dB} = 10log (GTmx) \qquad (7-9)$$

Therefore, GTmx = 10 log (21.601) = 13.34 dB.

(c) The simultaneous match input and output impedances are.

$Z_{MS} = 61.184 + j6.789$ Ohms

$Z_{ML} = 79.887 + j19.21$ Ohms.

7.5 Designing the Input Impedance Matching Network

The design of the amplifier can now be started with the design of simultaneous match source and load impedance matching networks.

Example 7-9: Design the input impedance matching network.

Solution: Use the Chapter 5 Equations (5-16), (5-18), (5-36) and (5-37) to design the amplifier input impedance matching circuit.

Let Z0=50, RL=61.184, XL=-6.789, f=1000e6, r=RL/Z0, and x=XL/Z0

B3=(x + sqrt(r*(r^2 + x^2-r)))/ (Z0*(r^2 + x^2))

X3=Z0*sqrt ((r^2 + x^2 - r)/r)

CP1=B3/ (2*pi*f)

LS1=X3/ (2*pi*f)

The MATLAB solutions are given in Table 7-3.

Name	Value
B3	6.097e-3
C1	581.3e-15
CP1	970.4e-15
L2	1.976e-9
LS1	3.888e-9
RL	61.184
X3	24.431
XL	-6.789
Z0	50
f	1e+9
r	1.224
x	-0.136

Table 7-3: MATLAB Solutions

The MATLAB solutions in Table 7-1 show that the parallel capacitor CP1=0.9704 pF and the series inductor LS1 = 3.888 nH. To verify the design of the input matching network at 1 GHz; create a new schematic in LTspice and assemble the matching elements as shown in Figure 7-24.

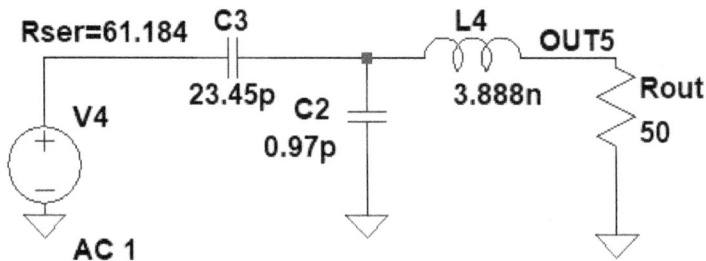

.ac lin 10000 0.5GHz 1.5GHz
.net I(Rout) V4

Figure 7-24: Amplifier input matching network

Simulate the schematic and display S11 from 0.5 to 1.5 GHz.

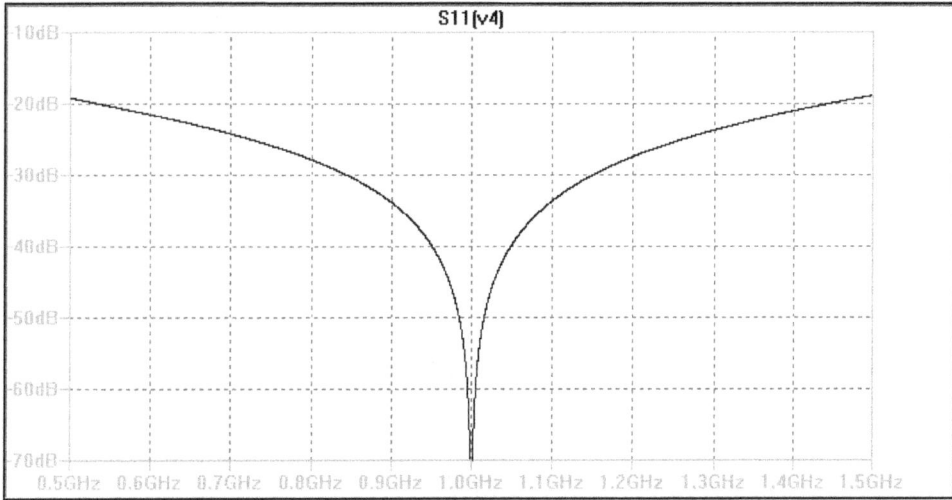

Figure 7-25: Amplifier input reflection coefficient S11

The simulated response in Figure 7-25 shows that the input impedance of the transistor is perfectly matched to the source impedance at 1 GHz.

7.6 Designing the Output Impedance Matching Network

Example 7-10: Design and verify the simultaneous match output impedance matching circuit.

Solution: Use the Chapter 5 Equations (5-16), (5-18), (5-36) and (5-37) to design the amplifier output impedance matching circuit.

Let Z0=50, RL=79.88, XL=-19.21, f=1000e6, r=RL/Z0, and x=XL/Z0

B3=(x + sqrt(r*(r^2 + x^2 - r)))/ (Z0*(r^2+x^2))

X3=Z0*sqrt ((r^2 + x^2 - r)/r)

CP1=B3/ (2*pi*f)

LS1=X3/ (2*pi*f)

The MATLAB solutions are given in Table 7-4.

Name	Value
B3	6.985e-3
C1	581.3e-15
CP1	1.112e-12
L2	1.976e-9
LS1	6.611e-9
RL	79.885
X3	41.536
XL	-19.21
Z0	50
f	1e+9
r	1.598
x	-0.384

Table 7-4: MATLAB Solutions

The MATLAB solutions in Table 7-2 show that the parallel capacitor CP1 = 1.112 pF and the series inductor LS1 = 6.611 nH.

To verify the design of the output matching network at 1 GHz; create a new schematic in LTspice and assemble the matching elements as shown in Figure 7-26.

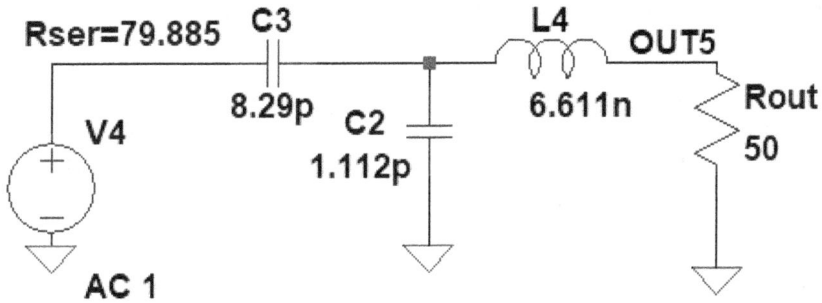

Figure 7-26: Amplifier input matching network

Simulate the schematic and display S11 from 0.5 to 1.5 GHz.

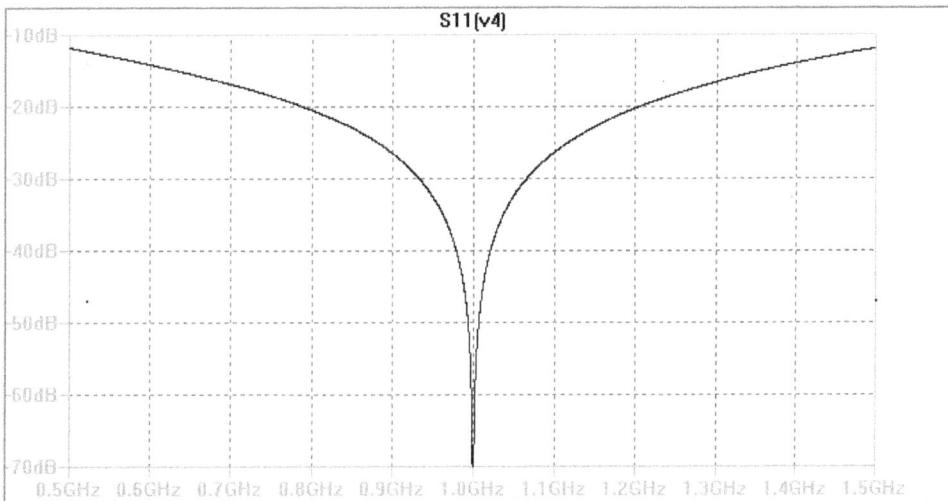

Figure 7-27: Amplifier input reflection coefficient S11

The simulated response in Figure 7-27 shows that the output impedance of the transistor is perfectly matched to the load impedance at 1 GHz.

7.7 Assemble the Amplifier and Measure GTmax

Example 7-11: Assemble and simulate the maximum gain amplifier from 0.5 to 1.5 GHz. Measure the amplifier gain at 1 GHz and verify that it is 13.34 dB.

Solution: The maximum gain amplifier with the input and output matching networks is shown in Figure 7-28. Note that inductors L3 and L4 each include a shunt resistor value of 278 ohms (not shown in the schematic).

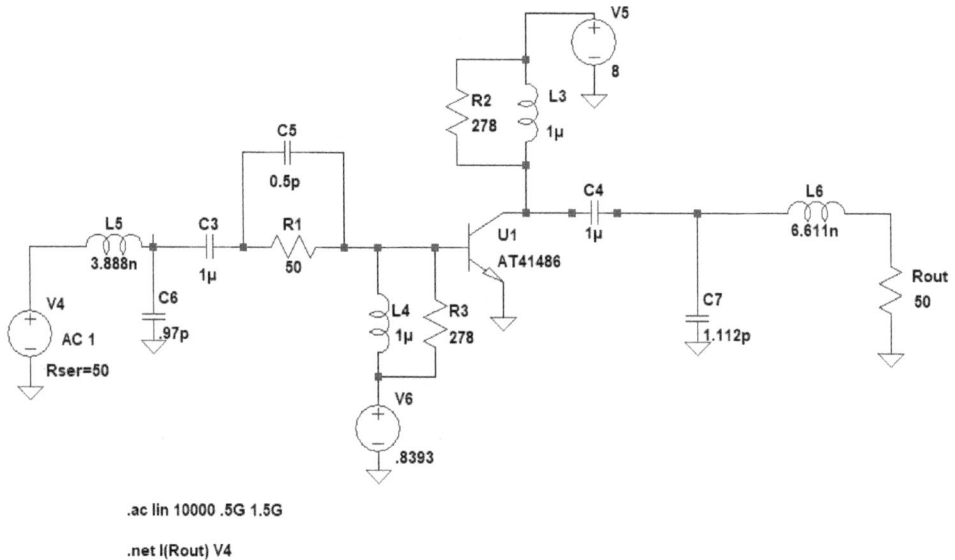

Figure 7-28: Maximum gain amplifier

The simulated S21 and GTmax is shown in Figure 7-29. Examination of the plot reveals that S21 and GTmax curves cross each other at 1 GHz.

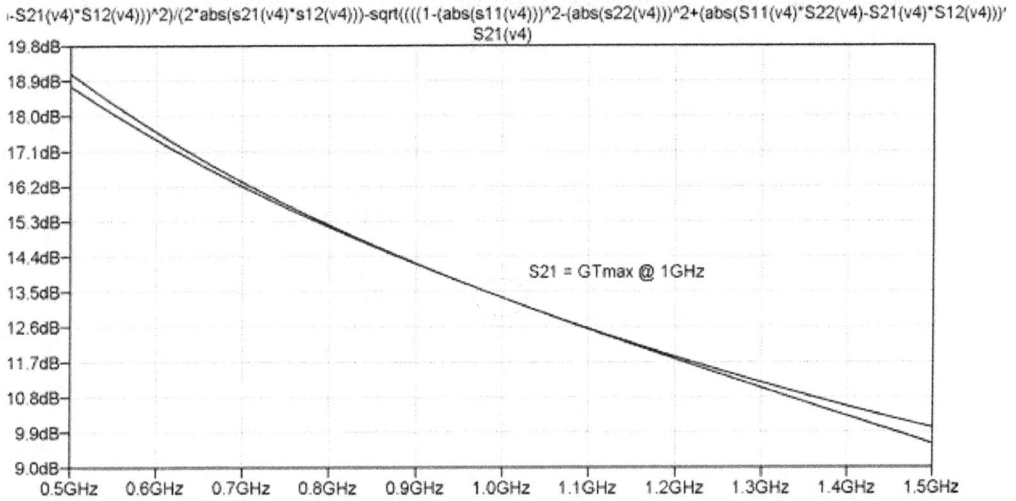

Figure 7-29: Gain plots for S21 and GTmax

The corresponding values for S21 and GTmax are shown in Figure 7-30.

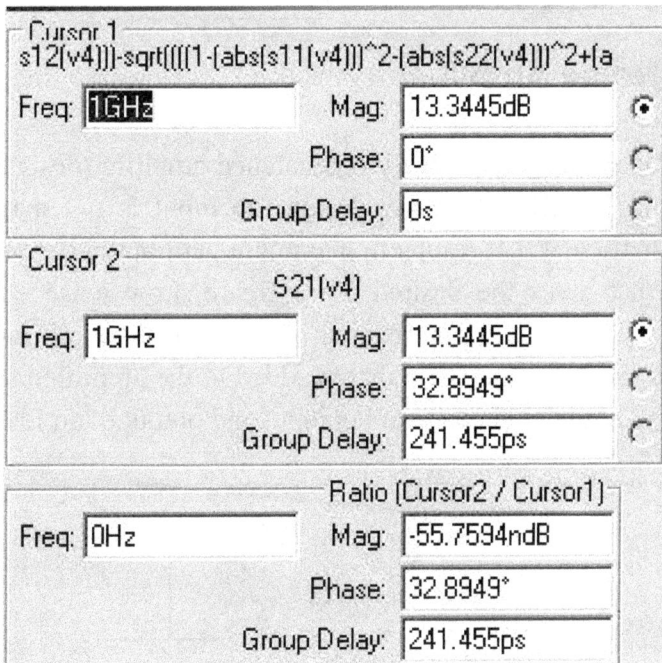

Figure 7-30: S21 and GTmax values at 1 GHz

Data point in Figure 7-30 shows that both S21 and GTmax are equal to 13.34 dB at 1 GHz

The simulated response of the amplifier in Figure 7-31 shows that both S21 and S22 are perfectly matched at 1 GHz.

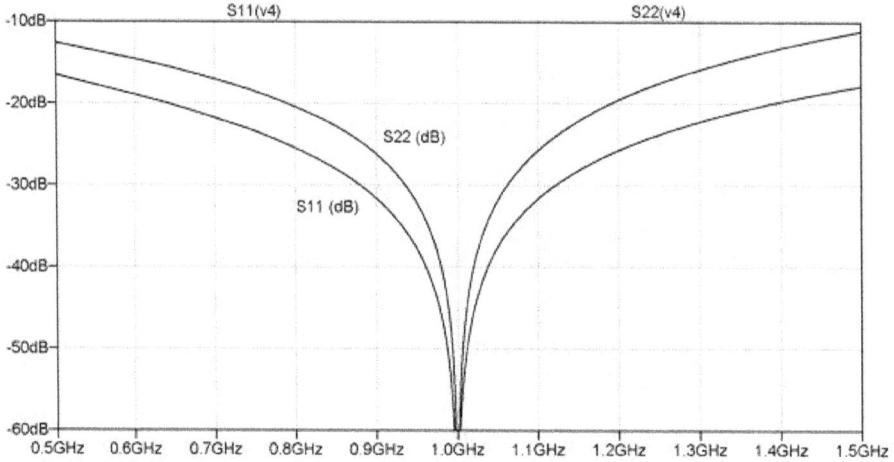

Figure 7-31: Plots of input and output return loss.

7.8 Low Noise Amplifier

One important case of selectively mismatched amplifier design is the Low Noise Amplifier, LNA. In LNA design the input is not matched to the reflection coefficient that results in maximum gain but rather to a reflection coefficient that gives the desired noise figure. Low noise amplifiers are frequently used as the input stage in a radio receiver or satellite down converter to minimize the noise that is added to the amplified signal. Figure 7-32 depicts a signal with noise at the input and output of an LNA.

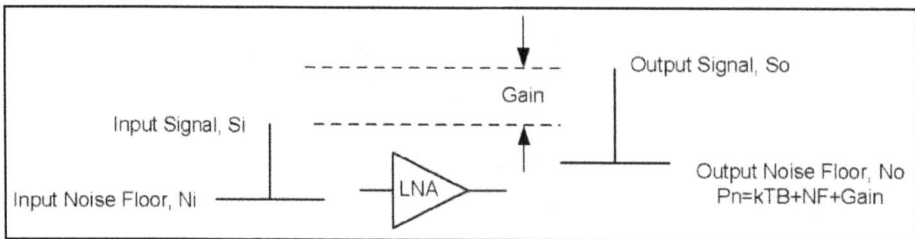

Figure 7-32: Signal and noise through the LNA

All transistors will add some amount of noise to the input signal. Low Noise transistors are optimized so that they will add a minimum amount of noise to the signal. The signal at the output of a linear LNA is simply the signal level at the input plus the gain of the LNA in dB. However the noise at the output will be increased by the gain and the noise figure of the transistor.

The noise figure of the LNA is the degradation of the signal to noise ratio of the input of the LNA to the signal to noise ratio of the output as given by Equation (7-14).

$$NF_{dB} = 10 \cdot \log \left| \frac{\left(\frac{S_i}{N_i} \right)}{\left(\frac{S_o}{N_o} \right)} \right| \qquad (7\text{-}14)$$

The ratio $\left(\frac{S_i}{S_o} \right)$ is inversely proportional to the gain of the LNA, which is a function of frequency. Therefore any measurement of noise must be referred to a specific bandwidth. The thermal noise power is also a function of temperature and is defined by Equation (7-15).

$$P_n = kTB \qquad (7\text{-}15)$$

Where,

k= Boltzmann's constant, 1.374×10^{-23} J/°K
T = Temperature of the input noise source (290 °K = room temperature)
B = Bandwidth of Measurement in Hz

If we normalize the measurement bandwidth to 1 Hz i.e., B=1 the thermal noise can be calculated using Equation (7-16):

$$P_n = \left(1.374 \cdot 10^{-23} \right) (290)(1) = 3.984 \cdot 10^{-21} \; Watts \, / \, Hz$$

Or: $3.984 \cdot 10^{-18} \; mW \, / \, Hz$

Converting mW to dBm, we can express the thermal noise floor as:

$$P_{no} = 10 \cdot \log \left(3.984 \cdot 10^{-18} \right) = -173.9 \; dBm / Hz$$

Working with the thermal noise floor normalized to a 1 Hz bandwidth it is straightforward to calculate the output noise from the amplifier using the amplifier's noise figure and gain as given by Equation (7-16).

$$P_n(1Hz) = kTB_{1Hz} + Gain_{dB} + NoiseFigure_{dB} \qquad \text{dB/Hz} \quad (7\text{-}16)$$

By using Equation (7-17) we can normalize the noise power in a one Hertz bandwidth to any measurement bandwidth.

$$P_n(B) = P_n(1Hz) + 10 \cdot \log(B) \qquad (7\text{-}17)$$

Where B is the measurement bandwidth in Hz

Now, suppose that the noisy LNA with a voltage gain of A_V is connected to an antenna, as shown in Figure 7-33 [10]. How can we determine the noise figure?

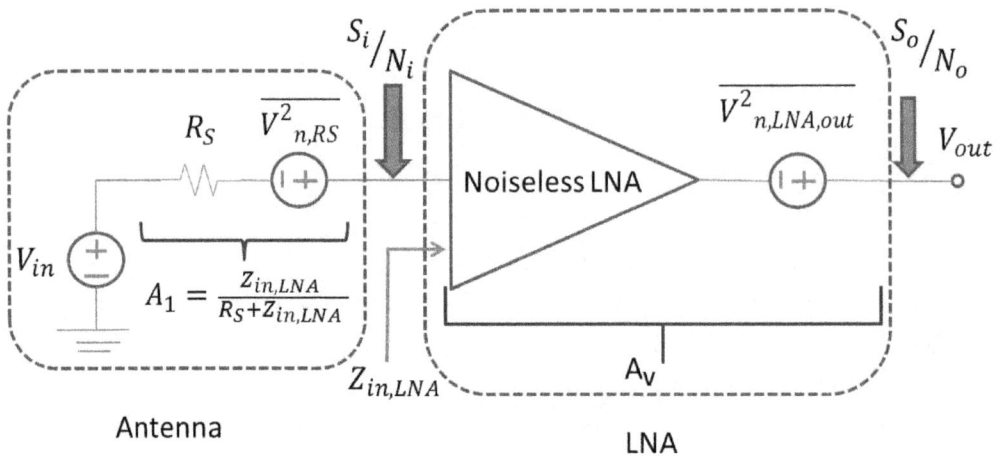

Figure 7-33: Front-end receiver including an antenna and LNA

In Figure 7-33 V_{in} is the rms input voltage source representing the received signal to the antenna. R_S is the source impedance equivalent to antenna radiation resistance. The $\overline{V^2_{n,RS}}$ represents the antenna's thermal noise and

$\overline{V^2_{n,LNA,out}}$ represents the output noise of the LNA. A_V is the LNA's voltage gain and A_1 is the voltage gain between source and antenna input. In the case of infinite LNA input impedance, A_1 reduces to unity. Otherwise, it is less than one and it acts as an attenuator. Therefore, the total voltage gain from input to output is the product of A_1 and A_V. Note that the unit of noise voltage is $[V^2/Hz]$. For the above configuration, the following can be observed for signal and noise at the LNA input:

$$S_i = |A_1|^2 \, V^2_{in}$$

(7-18)

$$N_i = |A_1|^2 \, \overline{V^2_{n,RS}}$$

(7-19)

Similarly, the signal and noise at LNA output can be written as:

$$S_o = |A_1|^2 \, A^2_V \, V^2_{in}$$

(7-20)

$$N_o = \overline{V^2_{n,RS}} \, |A_1|^2 \, A^2_V + \overline{V^2_{n,LNA,out}} = \overline{V^2_{n,out}}$$

(7-21)

Substituting these equations in Equation (7-15), the noise figure can be written as:

$$NF_{dB} = 10 \log \left[\frac{\overline{V^2_{n,out}} \, |A_1|^2 A_V{}^2 + \overline{V^2_{n,LNA,out}}}{\overline{V^2_{n,RS}} \, |A_1|^2 A_V{}^2} \right]$$

(7-22)

Or:

$$NF_{dB} = 10 \log \left[\frac{\overline{V^2_{n,out}}}{4KTR_S \, |A_1|^2 A^2_V} \right],$$

(7-23)

Where in the above equation,

$\overline{V^2_{n,out}}$ is the total noise at LNA output and:

$4KTR_S = \overline{V^2_{n,RS}}$ is the Thermal noise due to source resistance.

From Equation (7-24) it can be seen that noise figure is the ratio of total noise at the LNA output divided by the product of the source impedance thermal noise and the square of the total voltage gain from input to out. As a result, the noise figure can also be expressed as the total noise at the input divided by the source impedance thermal noise.

$$NF_{dB} = 10log\left[\frac{\overline{V^2_{n,out}}}{4KTR_S |A_1|^2 A^2_V}\right] = 10log\left[\frac{\overline{V^2_{n,in}}}{4KTR_S}\right], \qquad (7\text{-}24)$$

Where

$$\frac{\overline{V^2_{n,out}}}{|A_1|^2 A^2_V} = \overline{V^2_{n,in}} = \text{the total equivalent noise at LNA input.} \qquad (7\text{-}25)$$

Equations (7-25) and (7-26) are used in LTspice to calculate LNA's noise figure.

Noise Figure Analysis

In order to perform noise figure simulation the device S parameter file must include noise parameters. Table 7-3 shows an S parameter file with noise parameters appended to the end of the file. In the S parameter file the noise parameters include the optimum noise figure, NF$_{opt}$, as a function of frequency. The NF$_{opt}$ is the lowest possible noise figure that can be achieved when the transistor input is matched to optimum reflection coefficient Γ_{opt}. The third and the fourth column give the magnitude and angle of the Γ_{opt}. The final column gives the normalized noise resistance for the device. These noise parameters are usually provided by the device manufacturer on the S parameter data sheet, as shown in Table 7-3 for the AT-41486 device.

Matching the device to Γ_{opt} can potentially lead to very poor input return loss and instability. Just as we have seen in the example of section 7.4.2 the designer must be aware of the location of Γ_{opt} with respect to the input stability circle. In the LNA design we do not add resistive loading to the input of the device to improve the stability because this would increase the thermal noise power and will increase the device noise figure. In certain critical designs the poor input return loss is accepted and an isolator is added to the input of the LNA to provide a good input return loss. This also involves tradeoffs as the losses in the isolator add directly to the noise figure.

As a tradeoff a reflection coefficient can be chosen that corresponds to a noise figure that is slightly greater than minimum noise figure. In this case the resulting noise figure can be calculated for any source reflection coefficient, Γs, by using the following equation [10].

$$ NF = NF_{min} + \frac{4R_n}{Z_0 \left| 1 + \Gamma_{opt} \right|^2} \cdot \frac{\left| \Gamma_s - \Gamma_{opt} \right|^2}{\left(1 - \left| \Gamma_s \right|^2 \right)} \qquad (7\text{-}26) $$

```
|!AT-41486
!S AND NOISE PARAMETERS at Vce=8V   Ic=10mA.   LAST UPDATED
07-21-92

# ghz s ma r 50

0.1    .74    -38    25.46    157    .011    68    .94    -12
0.5    .59   -127    12.63    107    .031    47    .60    -29
1.0    .56   -168     6.92     84    .041    46    .49    -29
1.5    .57    169     4.72     69    .049    49    .45    -32
2.0    .62    152     3.61     56    .058    43    .42    -39
2.5    .63    142     2.91     47    .068    52    .40    -42
3.0    .64    130     2.41     37    .078    52    .39    -50
3.5    .68    122     2.06     26    .093    51    .37    -60
4.0    .71    113     1.80     16    .106    48    .35    -70
4.5    .74    105     1.59      7    .125    48    .35    -84
5.0    .77     99     1.42     -4    .139    43    .35    -98
5.5    .79     93     1.27    -13    .153    38    .35   -114
6.0    .81     87     1.13    -22    .170    34    .35   -131

!FREQ   NFopt      GAMMAopt         Rn/Zo
!GHZ     dB     MAG     ANG          -

0.1     1.3    .12        3        0.17
0.5     1.3    .10       16        0.17
1.0     1.4    .04       43        0.16
2.0     1.7    .12     -145        0.16
4.0     3.0    .44      -99        0.40
```

Table 7-5 AT41486 S parameter file with noise parameters included

Low Noise Amplifier Design

Design of a low noise amplifier starts with establishing its DC operating point conditions. To ensure stable operation over all frequencies, where there is a potential for oscillation, stability analysis is then performed and the device is made unconditionally stable. Design for specific parameters such as gain, noise figure, and input-output return loss at a given biasing conditions is then followed. In the previous section we have used the AT-41486 transistor in LTspice to design a maximum gain amplifier at 1 GHz.

To design the LNA network it is required that we take the following steps.

- Use the transistor symbol designed in Example 7-1 and bias the transistor at Vce = 8VDC and Ic = 10 mA
- Add stability components to make the transistor unconditionally stable
- Calculate the transistor Noise Figure and S parameters at 1 GHz
- Design the input and output impedance matching networks
- Assemble the LNA and measure the Noise Figure at 1 GHz

The LNA design now follows.

Biasing the Transistor at Vce = 8 V and Ic =10 mA

Example 7-12: Find the Vbe value for biasing the AT41486 transistor at Vce = 8 VDC and Ic =10 mA.

Solution: From the DC IV curves in Figure 7-12 it is clear that Vbe is approximately 0.807 VDC at Vce = 8 VDC and Ic = 10 mA.

Next follow the steps in Example 7-3 to get a more accurate value for Vbe, at Vce = 8 VDC and Ic = 10 mA. Sweeping the base voltage we get the operating data point shown in Table 7-6.

```
          --- Operating Point ---

V(n005):        0.8055          voltage
V(n002):        7.99999         voltage
V(n004):        4.0275e-017     voltage
V(n001):        8               voltage
V(n003):        4e-016          voltage
V(n006):        0.8055          voltage
I(C2):          -7.99999e-018   device_current
I(C1):          8.055e-019      device_current
I(L2):          -8.41361e-005   device_current
I(L1):          0.0100001       device_current
I(Rout):        7.99999e-018    device_current
I(V1):          -8.41361e-005   device_current
I(V2):          -0.0100001      device_current
I(V4):          8.055e-019      device_current
Ix(u1:60):      0.0100001       subckt_current
Ix(u1:20):      8.41361e-005    subckt_current
Ix(u1:40):      -0.0100842      subckt_current
```

Table 7-6: Operating point data (Ic = I_{L1} = 10 mA)

From the operating point data in Table 7-6 we see that Vbe = 0.8055 VDC at Vce=10 VDC and Ic=10 mA.

Unconditionally Stabilize the Transistor

Example 7.13: Perform stability analysis and make the transistor unconditionally stable.

Solution: To determine the device stability, create a new schematic in LTspice and select the Vbe = 0.8055 VDC, as shown in Figure 7-34.

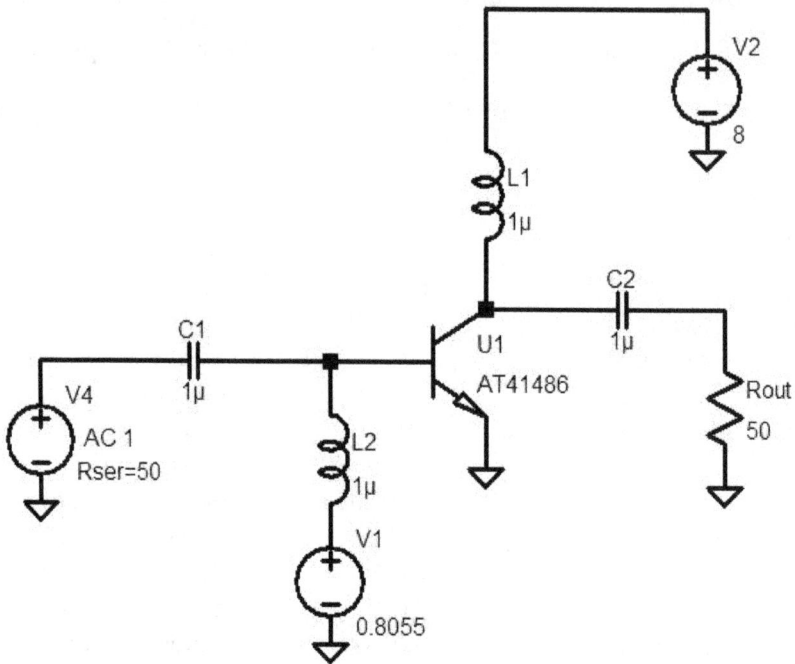

Figure 7-34: Transistor AC analysis biased at Vc = 8V, and Ic = 10mA

Simulate the schematic and plot K and B1 as shown in Figure 7-35.

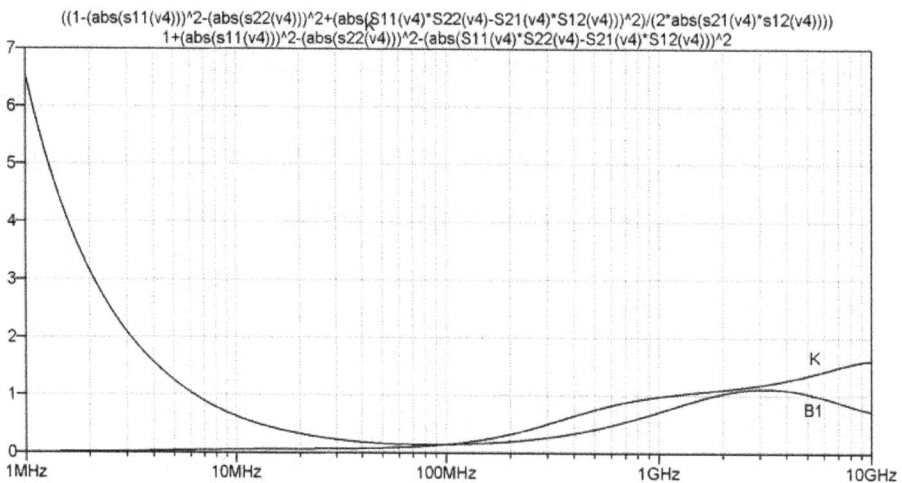

Figure 7-35: Stability parameters K and B1 at 8V, 10mA

Figure 7-35 shows that K is less than 1 for most of the frequency range indicating that the transistor is not stable at these frequencies.

To make the transistor unconditionally stable add a variable shunt resistor at the device output and sweep the resistor from 100 to 170 Ohms with 35 Ohm steps, as shown in Figure 7-36.

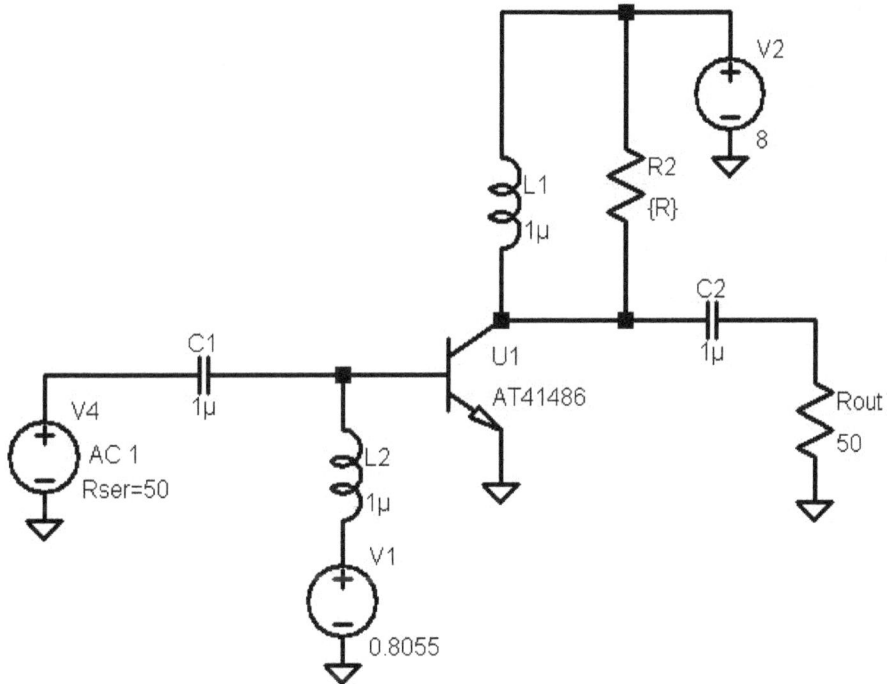

.ac dec 1000 1Meg 10G

.net I(Rout) V4 ; Rin & Rout determined from V4 and Rout

.step param R 100 170 35

Figure 7-36: Schematic diagram of the LNA for tuning R2

Plot the stability parameters K and B1as shown in Figure 7-37.

$$\frac{((1-(abs(s11(v4)))^2-(abs(s22(v4)))^2+(abs(S11(v4)*S22(v4)-S21(v4)*S12(v4)))^2)/(2*abs(s21(v4)*s12(v4))))}{1+(abs(s11(v4)))^2-(abs(s22(v4)))^2-(abs(S11(v4)*S22(v4)-S21(v4)*S12(v4)))^2}$$

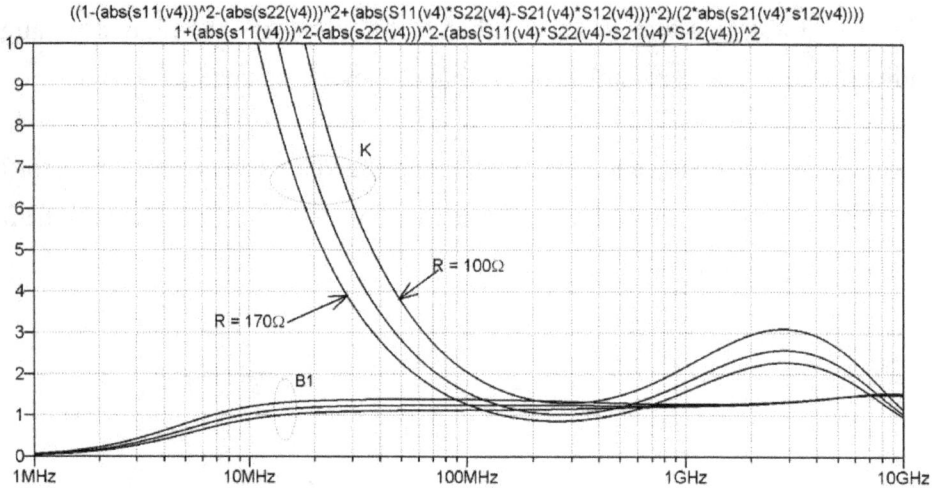

Figure 7-37: K and B_1 plots as a function of R2 tuning.

Figure 7-37 shows that R2 = 135 is the smallest resistor value that makes the transistor unconditionally stable. Therefore, we select R2=135 Ohm.

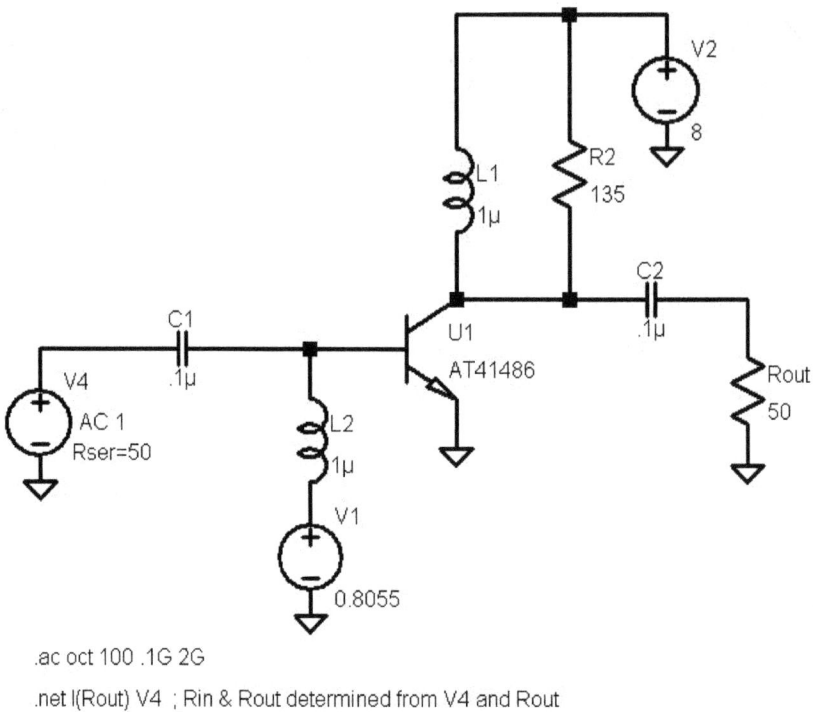

.ac oct 100 .1G 2G

.net I(Rout) V4 ; Rin & Rout determined from V4 and Rout

Figure 7-38: Schematic diagram of the unconditionally stable LNA.

Simulate the schematic and plot K and B1 in Figure 7-39.

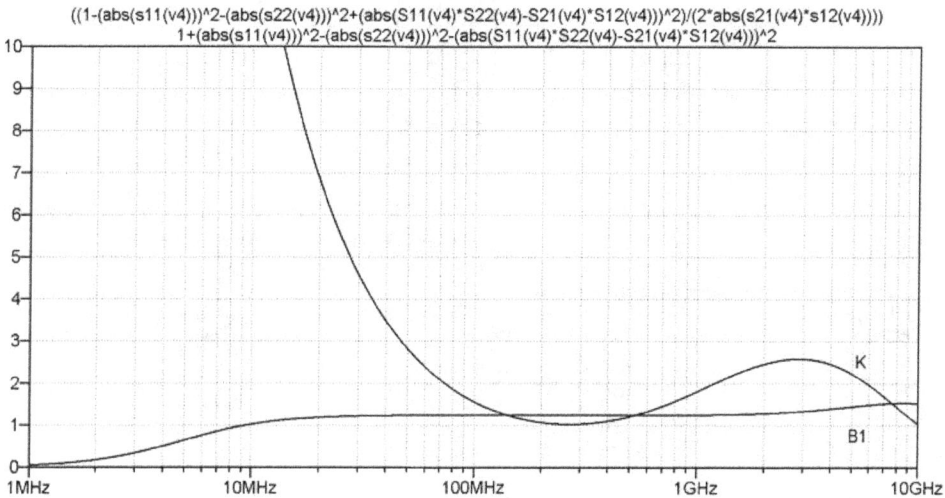

Figure 7-39: K and B1 Plots when R2=135 Ohm.

Figure 7-39 shows that K>1 and B1>0 for all the frequencies from 1 MHz to 10 GHz which means the transistor is unconditionally stable.

Now that the transistor being properly biased and unconditionally stabilized we want to see the gain of the transistor at 1 GHz.

Plot and Measure the Transistor Gain at 1 GHz

Example 7-14: Plot the transistor gain and measure the approximate gain of the transistor at 1 GHz.

Solution: Add the output port, OUT, to be ready for a two port analysis, as shown in Figure 7-40.

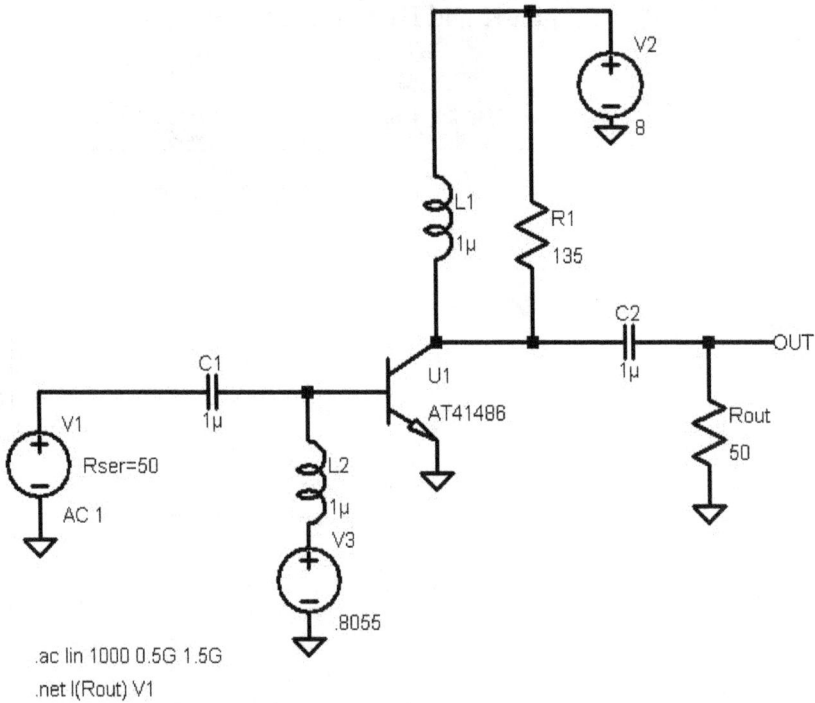

Figure 7-40: LNA simulation from 0.5 to 1.5 GHz

Simulate the schematic and plot the forward gain from 0.5 to 1.5 GHz.

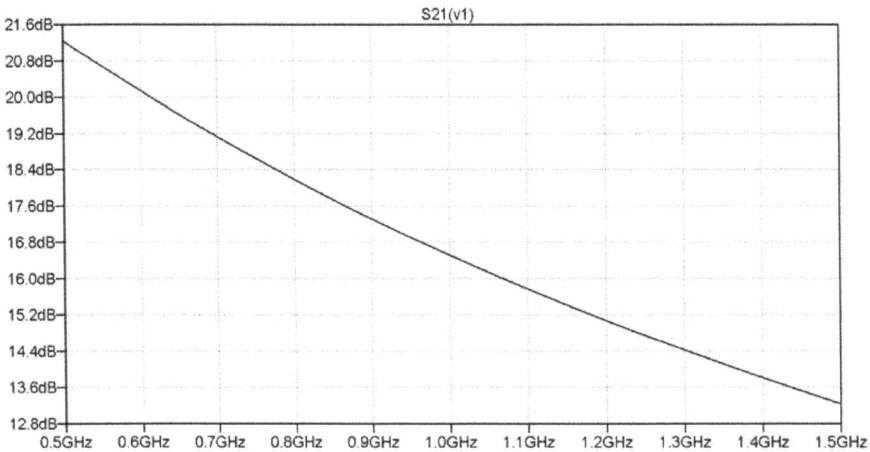

Figure 7-41: Transistor gain from 0.5 to 1.5 GHz.

Figure 7-41 shows that the transistor gain at 1 GHz is about 16.6 dB.

Plot and Measure the LNA Noise Figure at 1 GHz

Next we want to plot the transistor Noise Figure and measure it at 1 GHz.

Example 7-15: Plot the stabilized transistor Noise Figure and measure its value at 1GHz.

Solution: To simulate the schematic and plot the noise figure use the following steps:

1) Construct the schematic of the stabilized device.

2) Apply an input AC voltage (V1) with 50 ohm source impedance.

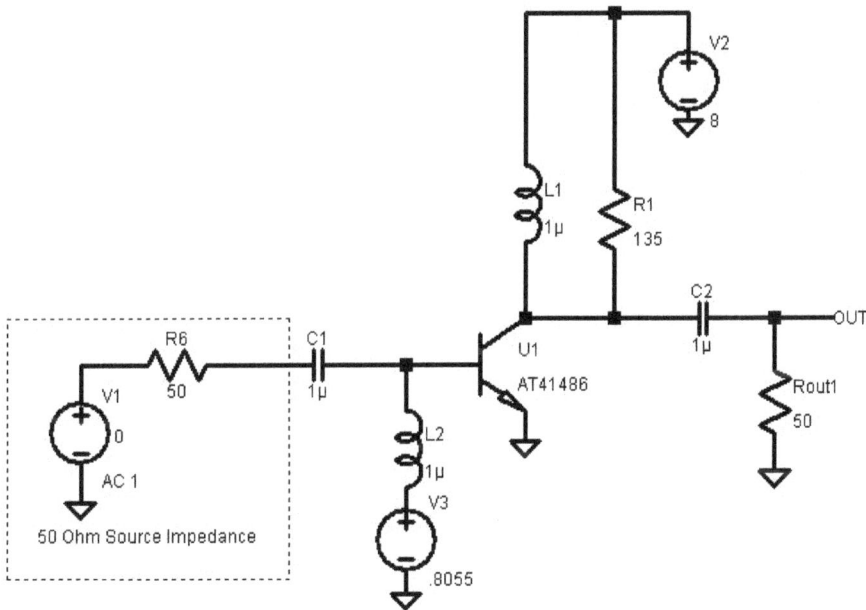

.noise V(OUT) V1 lin 100 500Meg 1.5G

Figure 7-42. LNA Schematic for noise figure calculation

3) Select Simulate > Edit Simulation Cmd to open the Edit Simulation Command window

4) Select the Noise tab and complete the boxes as shown in Figure 7-43

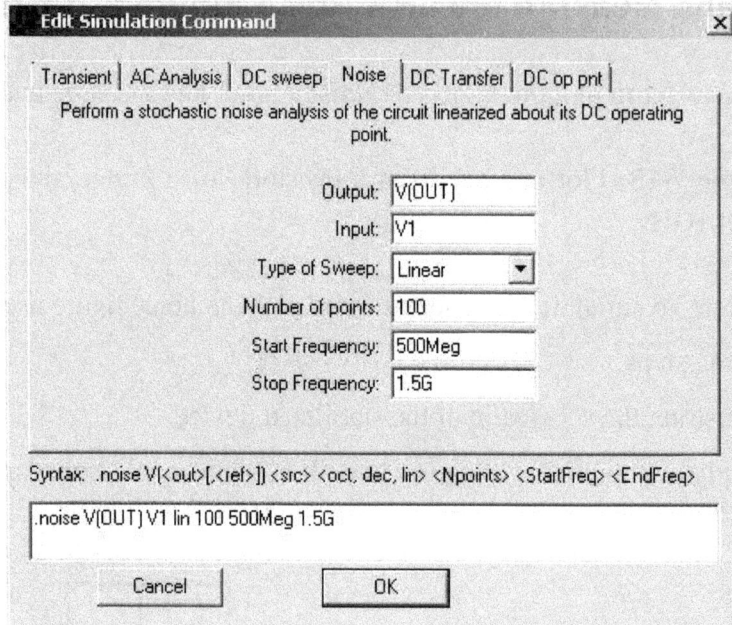

Figure 7-43. Noise measurement in Edit Simulation Command.

5) Select Simulate > Run.

6) In Add Traces to Plot window, type in

    10*log10(V(inoise)*V(inoise)/(4*k*300.15*50))

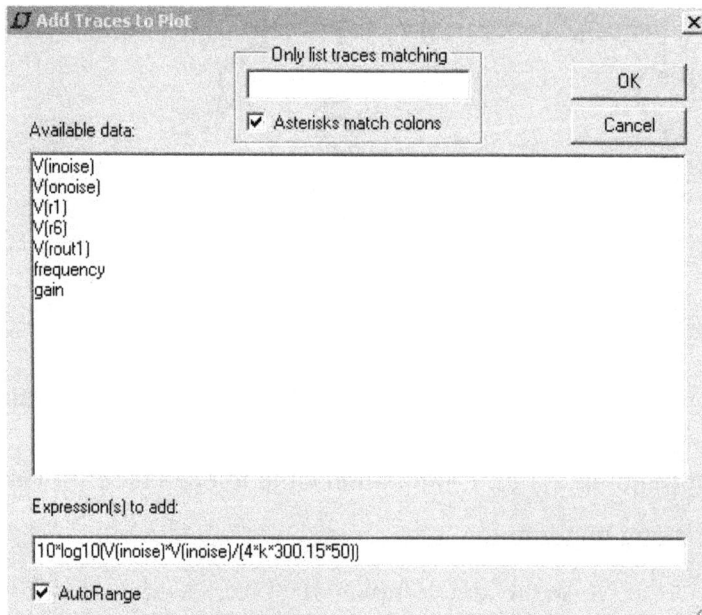

Figure 7-44. Expression for NF plot in dB.

7) Select OK to plot the noise figure versus frequency, as shown in Figure 7-45.

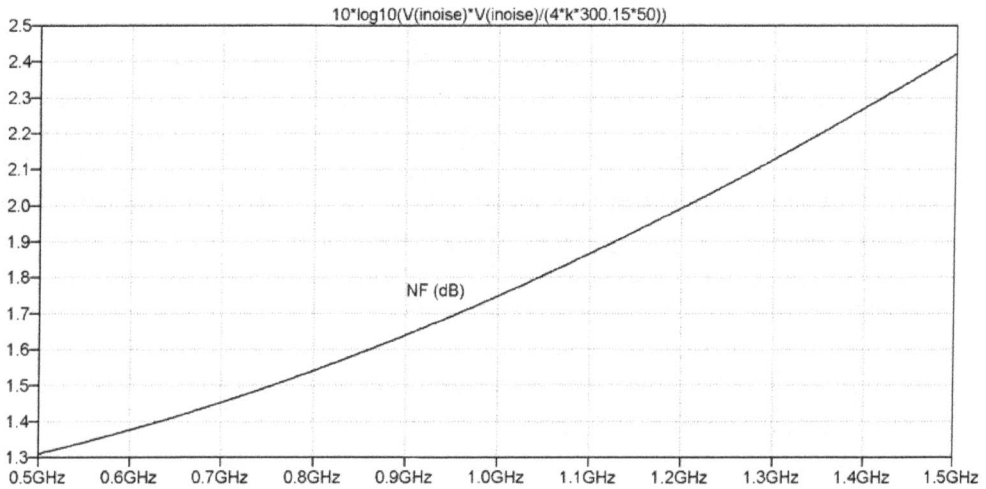

Figure 7-45. Noise Figure in dB as a function of frequency

To measure the noise figure at 1 GHz, move the cursor to the intersection of noise figure plot with the 1GHz line and read the value of noise figure.

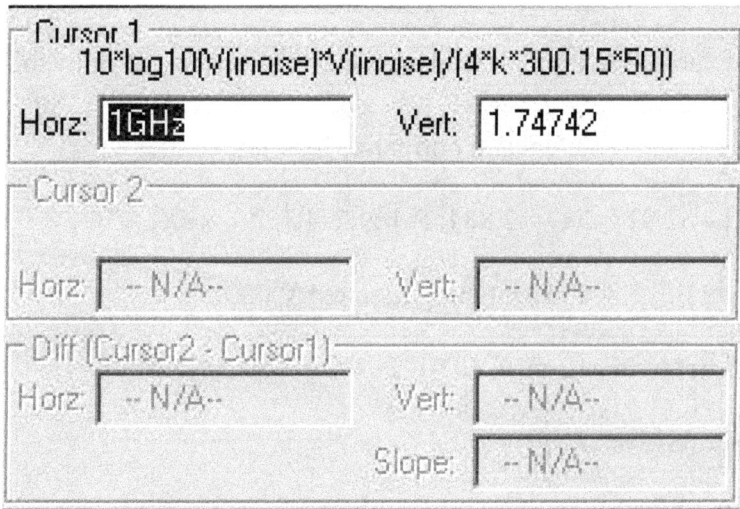

Figure 7-46. Noise Figure in dB at 1GHz.

Notice that the noise figure at 1 GHz is 1.74742 dB.

7.9 Design the LNA Input Matching Network

Example 7-16: Design the LNA input impedance matching network.

Solution: To design the LNA input impedance matching network and simultaneously achieve the optimum Noise Figure at 1 GHz, first we need to determine the optimum input impedance of the transistor, Zopt, at 1 GHz. According to the published data in Table 7-5 the optimum source reflection coefficient at 1 GHz is:

$$\Gamma_{opt} = 0.04 < 43°.$$

Converting Γ_{opt} to complex number we get:

$$\Gamma_{opt} = 0.0292 + 0.0272j \quad \text{at 1 GHz}$$

Using Equation $Z_{opt} = Z_0 (1 + \Gamma_{opt}) / (1 - \Gamma_{opt})$ we convert Γ_{opt} to Z_{opt}.

$$Z_{opt} = 52.927 + 2.884j$$

Design of the LNA input impedance matching network now follows.

1. Enter design parameters and normalize load impedance.

Z0=50; RL=52.927; XL= -2.884; f=1e9; r=RL/Z0; x=XL/Z0

2. Calculate B3, X3, and Matching Element Values

B3=(x+sqrt(r*(r^2+x^2-r)))/(Z0*(r^2+x^2)) = 3.653 x10^{-3}

X3=Z0*sqrt((r^2+x^2-r)/r) = 12.417

L2=X3/(2*pi*f)

C1=B3/(2*pi*f)

The MATLAB solutions are given in Table 7-7.

Name	Value
B3	3.652e-3
C1	581.3e-15
L2	1.976e-9
RL	52.927
X3	12.418
XL	-2.884
Z0	50
f	1e+9
r	1.059
x	-0.058

Table 7-7: MATLAB Solutions

The MATLAB solutions in Table 7-7 show that the parallel capacitor C1=0.581 pF and the series inductor L2 = 1.976 nH.

To verify the design of the LNA input matching network at 1 GHz; create a new schematic in LTspice and assemble the matching elements as shown in Figure 7-47.

C3

L4   OUT5

Rser=52.927   V4   55.2p   C2   1.976n   Rout

0.581p   50

AC 1

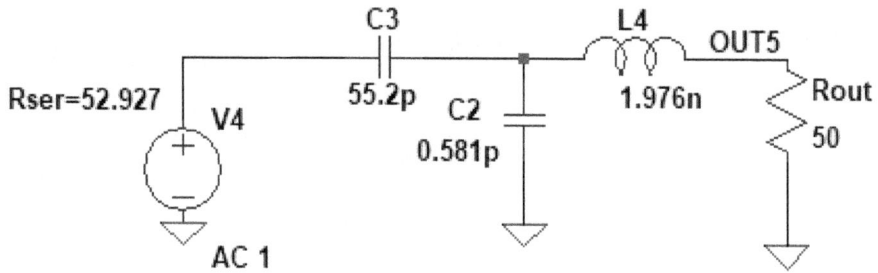

.ac lin 10000 0.5GHz 1.5GHz
.net I(Rout) V4

Figure 7-47: LNA input impedance matching network

Simulate the schematic and display the S11 response in Figure 7-48.

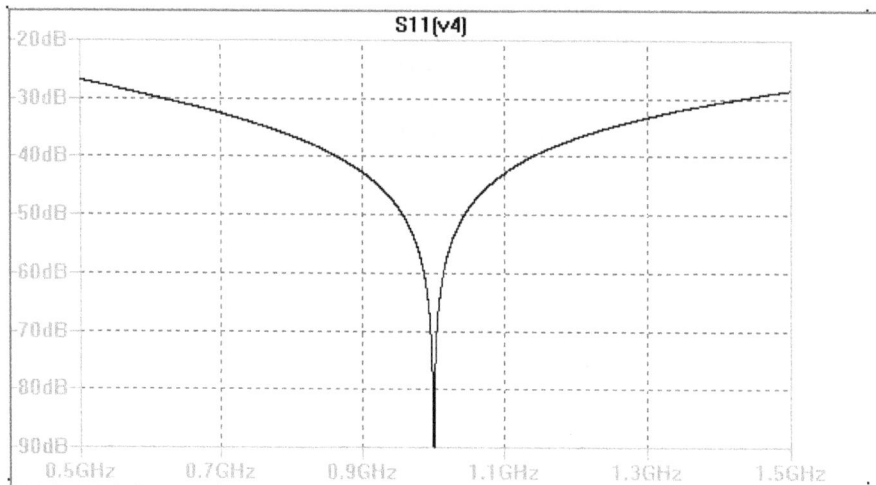

S11(v4)

-20dB
-30dB
-40dB
-50dB
-60dB
-70dB
-80dB
-90dB
0.5GHz   0.7GHz   0.9GHz   1.1GHz   1.3GHz   1.5GHz

Figure 7-48: LNA response of S11

The simulated response in Figure 7-48 shows that the input impedance of the LNA is perfectly matched to the source impedance at 1 GHz.

7.10 Design the LNA Output Matching Network

To design the output impedance matching network at 1 GHz we need to generate the transistor S parameters at 1 GHz and then calculate the LNA output impedance:

Example 7-17: Generate the transistor S-Parameters at 1GHz.

Solution: The following procedure shows how to generate the LNA S-Parameters at 1GHz.

1. From the stabilized transistor schematic in Figure 7-49, select Simulate > Edit Simulation Control.

2. Select AC Analysis and complete the simulation box as in Fig. 7-50.

3. Select OK > Simulate > Run.

4. From the waveform window select File > Export to open the Select Traces to Export shown in Figure 7-51.

5. Select the S-Parameter traces and Cartesian [re, im] format. Use Browse to save the text file in the desired directory.

6. Go to the directory and open the file to access tabulated S-parameters at 1GHz.

7. Organize the S-Parameter in Table 7-8.

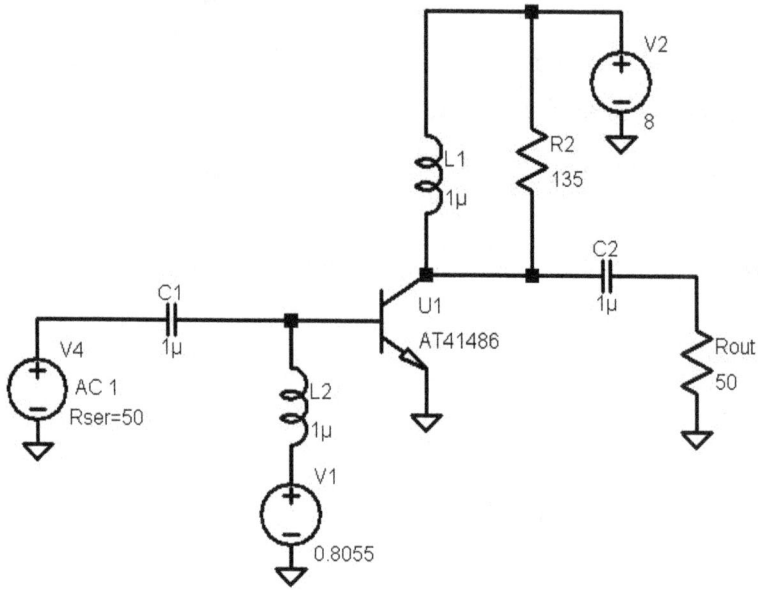

```
.ac dec 2 1G 2G
.net I(Rout) V4  ; Rin & Rout determined from V4 and Rout
```

Figure 7-49: Simulation of S parameter measurement

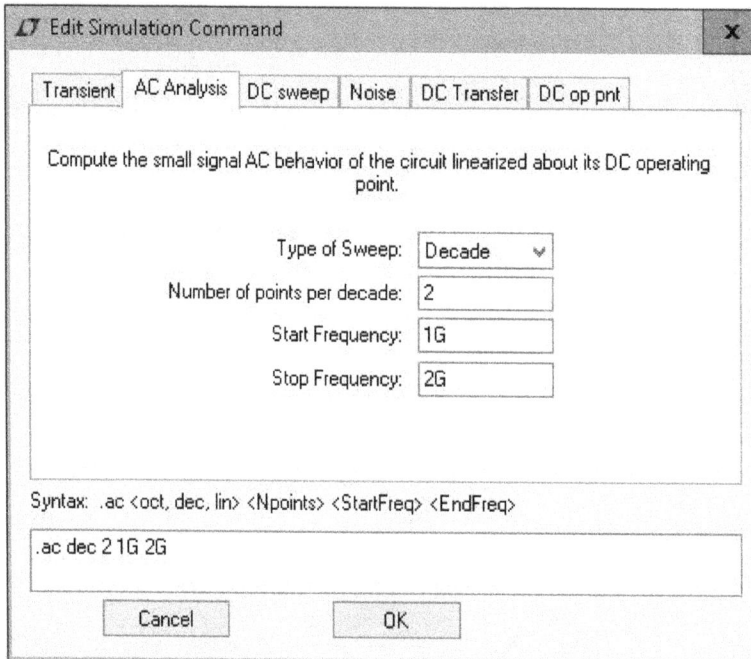

Figure 7-50: S-Parameter measurement at 1 and 2 GHz

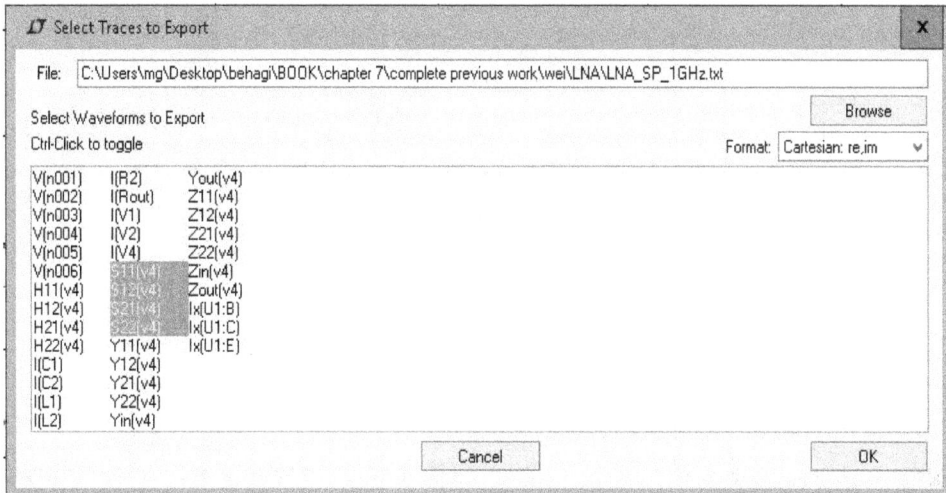

Figure 7-51: Window to Select Traces to Export

$$S11 = -0.5058 - j0.2099$$

$$S12 = +0.0198 + j0.0188$$

$$S21 = -0.4493 + j6.6770$$

$$S22 = +0.1739 - j0.1396$$

Table 7-8. Tabulated S parameters at 1 GHz

Next determine the LNA output reflection coefficient, Γ_L, and the load impedance, Z_L, at 1 GHz.

The LNA output reflection coefficient, Γ_L, is defined in terms of the source reflection coefficient, Γ_{opt}, and the transistor S parameters in Equation (7-28).

$$\Gamma_L = \left(S_{22} + \frac{S_{12} S_{21} \Gamma_{opt}}{1 - S_{11} \Gamma_{opt}} \right)^* \qquad (7\text{-}28)$$

To calculate the output impedance we first calculate the output reflection coefficient Γ_L in Equation (7-28) and then convert, Γ_L, to Z_L.

Using Equation $Z_L = Z_0 (1+ \Gamma_L)/(1- \Gamma_L$ and the given S-parameters at 1 GHz, the corresponding output impedance, Z_L, is :

$$Z_L = 66.73+19.541j \quad \text{at } 1GHz$$

Example 7-18: Design the LNA output impedance matching network.

Solution: designing the LNA output impedance matching network follows the procedure used in Example 7-16.

1. Enter design parameters and normalize load impedance.

Z0=50; RL=66.73; XL= -19.541; f=1e9; r=RL/Z0; x=XL/Z0

2. Calculate B3, X3, and Element Values

B3=(x+sqrt(r*(r^2+x^2-r)))/(Z0*(r^2+x^2)) = 3.653 x10^{-3}

X3=Z0*sqrt((r^2+x^2-r)/r) = 12.417

L2=X3/(2*pi*f)

C1=B3/(2*pi*f)

The MATLAB solutions are given in Table 7-9.

Name	Value
B3	5.207e-3
C1	828.7e-15
L2	5.333e-9
RL	66.73
X3	33.505
XL	-19.541
Z0	50
f	1e+9
r	1.335
x	-0.391

Table 7-9: MATLAB Solutions

The MATLAB solutions in Table 7-9 show that the parallel capacitor C1=0.828 pF and the series inductor L2 = 5.333 nH.

To verify the design of the LNA output matching network at 1 GHz; create a new schematic in LTspice and assemble the matching elements as shown in Figure 7-52.

Rser=66.73   C3   L4   OUT5
              8.15p   C2   Rout
    V4   5.333n   50
              0.828p

AC 1

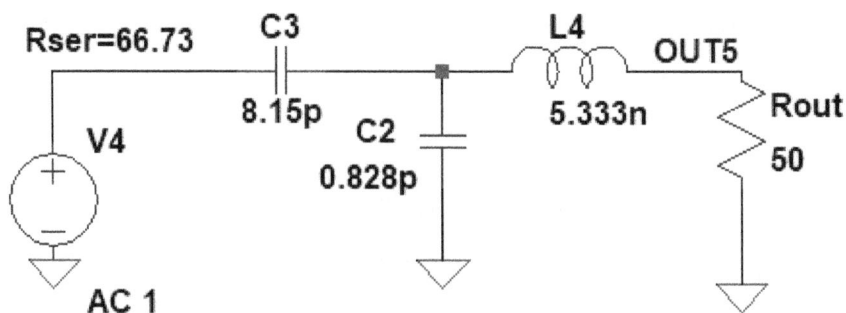

.ac lin 10000 0.5GHz 1.5GHz
.net I(Rout) V4

Figure 7-52: LNA input impedance matching network

Simulate the schematic and display the S11 response in Figure 7-53.

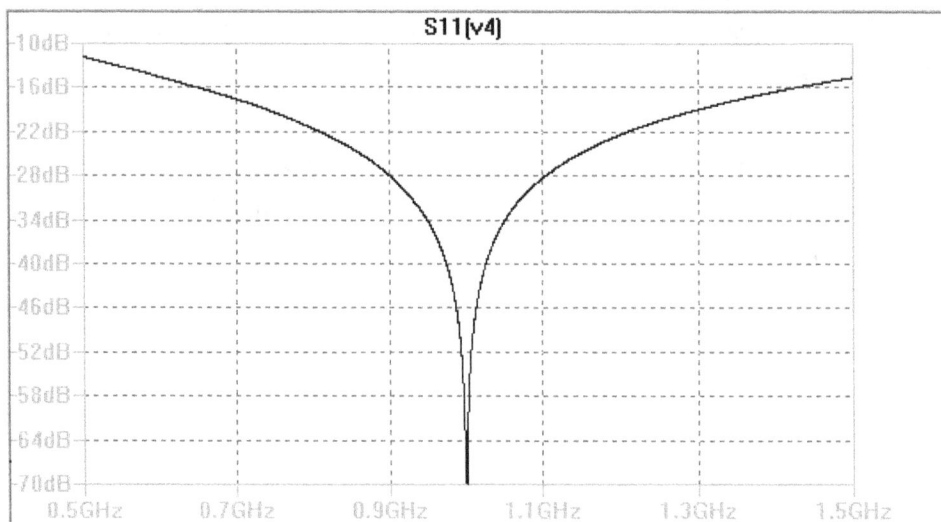

S11(v4)

-10dB
-16dB
-22dB
-28dB
-34dB
-40dB
-46dB
-52dB
-58dB
-64dB
-70dB
0.5GHz   0.7GHz   0.9GHz   1.1GHz   1.3GHz   1.5GHz

Figure 7-53: LNA response of S11

The simulated response in Figure 7-53 shows that the output impedance of the LNA is perfectly matched to the load impedance at 1 GHz.

7.11 Assemble the LNA and Measure NF$_{min}$ at 1 GHz.

Example 7-19: Assemble the low noise amplifier, simulate and measure the Noise Figure at 1 GHz.

Solution: The LNA schematic is assembled by adding the stabilizing elements and the input and output impedance matching networks. Connect a 50 Ohm AC source and a 50 Ohm load resistor, as shown in Figure 7-54.

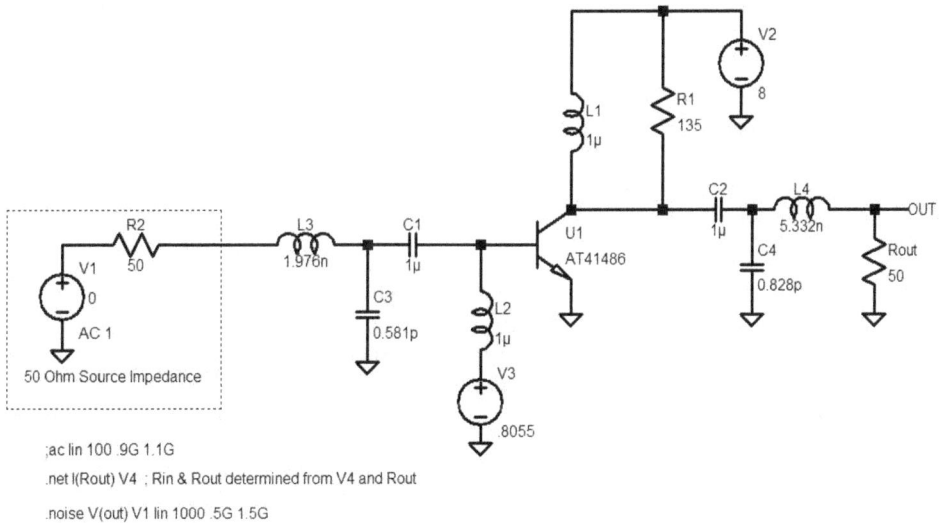

Figure 7-54: Schematic of the LNA to achieve NFmin at 1 GHz

Simulate the LNA schematic and display the LNA Noise Figure from 0.5 GHz to 1.5 GHz, as shown in Figure 7-55.

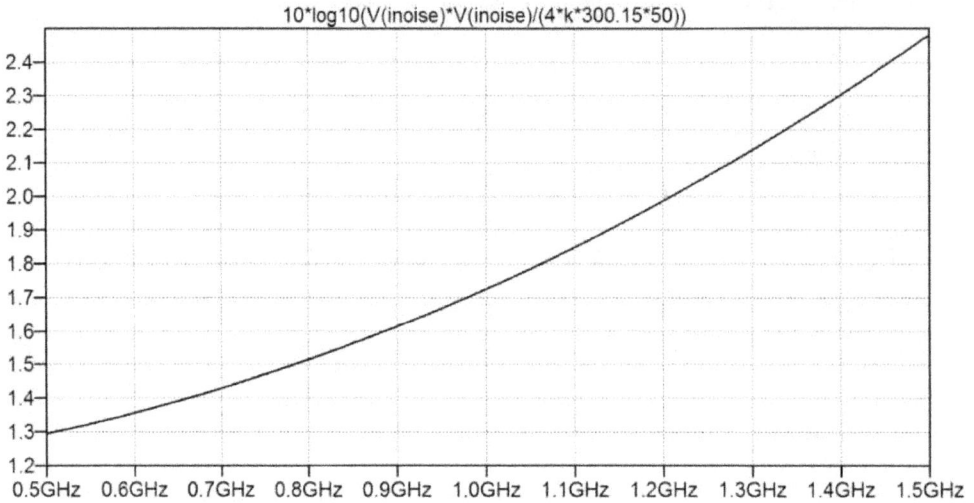

Figure 7-55: LNA Noise Figure Plot

Place the curser at 1 GHz and measure the Noise Figure, as shown in Figure 7-56..

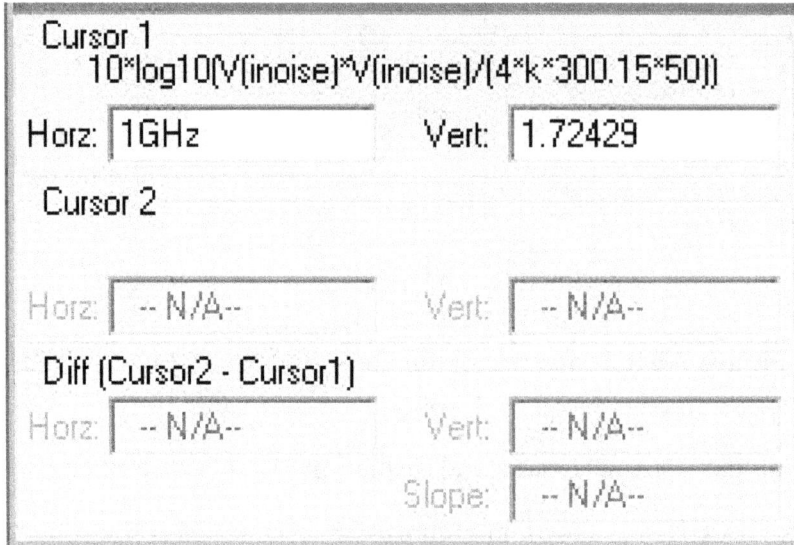

Figure 7-56: Captured NFmin of 1.72dB at 1 GHz

Figure 7-56 shows that the LNA Noise Figure at 1 GHz is 1.72 429.

This agrees closely with the measured Noise Figure in Figure 7-46.

References and Further Readings

[1]   Ali Behagi, *RF and Microwave Circuit Design*, A Design Approach Using (**ADS**), Techno Search, Ladera Ranch, CA 2017

[2]   Ali Behagi, *100 ADS Design Examples*, Based on the Textbook: *RF and Microwave Circuit Design,* Techno Search, Ladera Ranch, CA 2016

[3]   Ali Behagi and Manou Ghanevati, *Fundamentals of RF and Microwave Circuit Design,* Practical Analysis and Design Tools, Techno Search, Ladera Ranch, California 2017

[4]   Dale D. Henkes, FAST: *Fast Amplifier Synthesis Tool*, Artech House Publishers, Norwood, MA. 2004

[5]   Guillermo Gonzales, *Microwave Transistor Amplifiers – Analysis and Design*, Second Edition, Prentice Hall Inc., Upper Saddle River, NJ.

[6]   Randy Rhea, *The Yin-Yang of Matching: Part 1 – Basic Matching Concepts*, High Frequency Electronics, March 2006

[7]   Steve C. Cripps, *RF Power Amplifiers for Wireless Communications*, Artech House Publishers, Norwood, MA. 1999

[8]   David M. Pozar, *Microwave Engineering*, Third Edition, John Wiley a Sons, New York, 2005

[9]   R. Ludwig, P. Bretchko, *RF Circuit Design*, Theory and Applications, Prentice Hall, Upper Saddle River, NJ, 2000

[10]   Behzad Razavi, *RF Microelectronics, Second Edition, Prentice Hall*, New York, 2015

[11]   Ted Grosch, *Noise Concepts and Design, Noble Publishing*, Atlanta, GA, 2003

Problems

7-1. Design a Maximum Gain Amplifier

(a) For the Agilent HBFP0405 Transistor, create a symbol and bias the transistor for Vce and Ic given in S Parameter data file. Calculate the stability factor, K, and the stability measure, B1 as well as Γ_{MS} and Γ_{ML} and conjugate match impedances at 2 GHz.

(b) Sweep the frequency range of the device over the entire range of frequencies contained in the S parameter file.

(c) Plot of the stability parameters and GMAX for the Agilent HBFP0405 device from 100 to 4000 MHz.

(d) Employ a parallel RC circuit on the input of the device to stabilize the transistor. Make both the R and C values variable. Vary the resistance and capacitor values until K > 1 for frequencies above 500 MHz.

(e) Add an RL network in shunt with the input of the transistor for additional low frequency stability. Tune the resistance and inductor values until K > 1 for frequencies above 500 MHz.

(f) Use the analytical impedance matching techniques developed in Chapter 5 to design the input and output matching L-networks. When performing the analytical match we are; matching 50 Ω source impedance to the impedance looking into the device; and 50 Ω load impedance to the impedance looking into output of the device.

(g) Complete the amplifier design in LTspice by attaching the input and output matching circuits to the stabilized transistor. Use a linear sweep from 100 MHz to 4000 MHz to analyze the response of the amplifier, S11, S22, and S21 all in dB.

(h) Place a marker on S12 to measure the maximum gain of the amplifier at 2 GHz. Compare the maximum gain with the value originally calculated as G_{MAX}.

7-2. Design a Low Noise Amplifier

Design a single stage Low Noise Amplifier using Agilent AT-32011 device. The amplifier is intended to operate with a source and load impedance of 50 Ω. The design specifications are given as:

Center Frequency: 500 MHz
Gain: 18 dB minimum
Noise Figure: 1.2 dB maximum
Output Return Loss: Less than -10 dB

(a) Create a symbol for the device and bias it at Vce = 2.7 VDC and Ic = 5 mA. Create a new schematic in LTspice and setup a Linear Analysis and add a Table to display the stability parameters K and B1 and the noise parameters Γ_{opt} and Z_{OPT}.

(b) Utilize the "L" Network Impedance Matching equations developed in Chapter 5 to match the 50 Ω source impedance to Z_{opt} of the device at 500 MHz. In the impedance matching you must enter the conjugate of Z_{opt}.

(c) Use MATLAB script to calculate the load reflection coefficient, Γ_L. Convert Γ_L to output impedance Z_L. Z_{OUT} is the conjugate of Z_L. Check the location of Γ_L with respect to the output stability circle to make sure that the impedance does not lie in the region of instability.

Appendix A

Straight Wire Parameters for Solid Copper Wire

Wire Size (AWG)	Diameter in Mils	Resistance Ohms/1000 ft.	Area in circular Mils	Suggested Maximum Current Handling, Amperes[1]
0000	460.0	0.049	211600	1000
000	409.6	0.062	167800	839
00	364.8	0.078	133100	665
0	324.9	0.098	105500	527
1	289.3	0.124	83690	418
2	257.6	0.156	66360	332
3	229.4	0.197	52620	263
4	204.3	0.249	41740	208
5	181.9	0.313	33090	165
6	162.0	0.395	26240	131
7	144.3	0.498	20820	104
8	128.5	0.628	16510	83
9	114.4	0.793	13090	65
10	101.9	0.999	10380	52
11	90.7	1.26	8230	41
12	80.8	1.56	6530	32
13	72.0	2.00	5180	26
14	64.1	2.52	4110	20
15	57.1	3.18	3260	16
16	50.8	4.02	2580	13
17	45.3	5.05	2050	10
18	40.3	6.39	1620	8.0
19	35.9	8.05	1290	6.0
20	32.0	10.1	1020	5.0
21	28.5	12.8	812	4.0
22	25.3	16.2	640	3.0
23	22.6	20.3	511	2.5
24	20.1	25.7	404	2.0
25	17.9	32.4	320	1.6
26	15.9	41.0	253	1.2
27	14.2	51.4	202	1.0
28	12.6	65.3	159	0.80
29	11.3	81.2	123	0.61
30	10.0	104.0	100	0.50
31	8.9	131	79.2	0.40
32	8.0	162	64.0	0.32
33	7.1	206	50.4	0.25
34	6.3	261	39.7	0.19
35	5.6	331	31.4	0.16
36	5.0	415	25.0	0.12
37	4.5	512	20.2	0.10
38	4.0	648	16.0	0.08
39	3.5	847	12.2	0.06
40	3.1	1080	9.61	0.05
41	2.8	1320	7.84	0.04
42	2.5	1660	6.25	0.03
43	2.2	2140	4.84	0.024
44	2.0	2590	4.00	0.020
45	1.76	3350	3.10	0.016
46	1.57	4210	2.46	0.012
47	1.40	5290	1.96	0.010
48	1.24	6750	1.54	0.008
49	1.11	8420	1.23	0.006
50	0.99	10600	0.98	0.005

Current Handling based on 1 Amp/200 Circular Mils-no insulation and free air conditions.
Insulated and stranded Copper wire must be de-rated from the values in the Table

.

Appendix B: **Chapter 6 Data Point Figures**

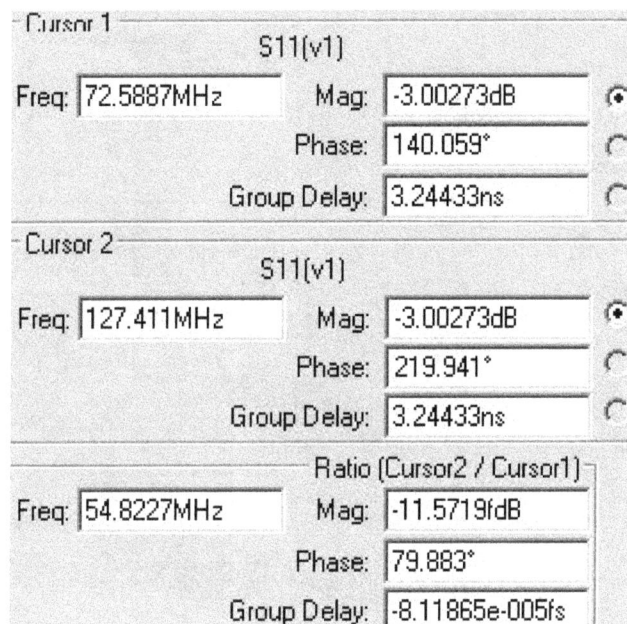

Figure 6-4A: 3 dB Data Points

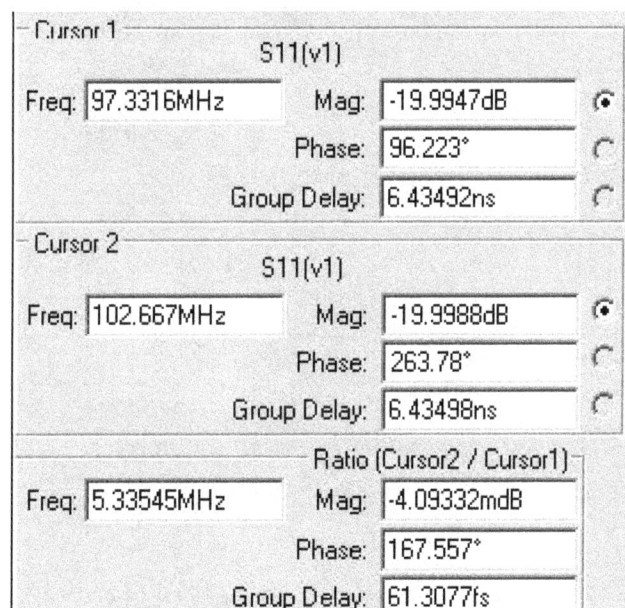

Figure 6-4B: 20 dB Data Points

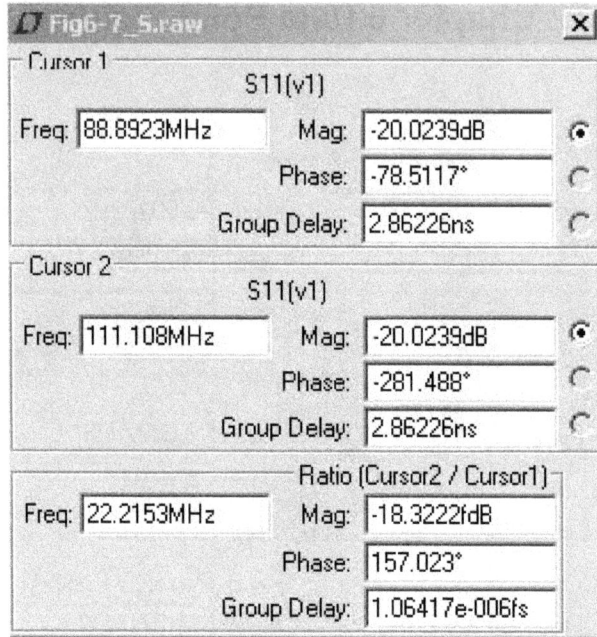

Figure 6-7A: 20 dB Data Points

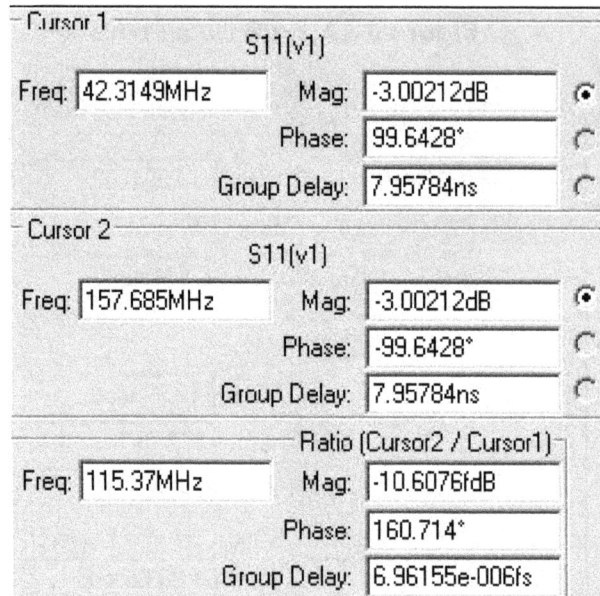

Figure 6-26A: 3 dB Data Points

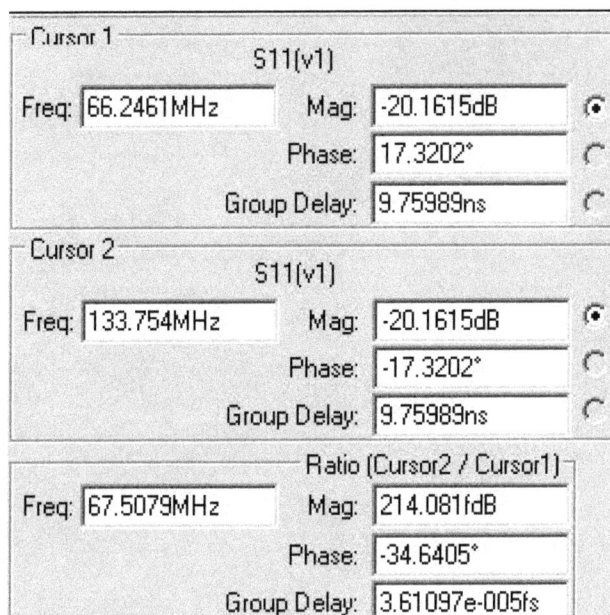

Figure 6-26B: 20 dB Data Points

Index

About the Author

Ali Behagi received the Ph.D. degree in electrical engineering from the University of Southern California and the MS degree in electrical engineering from the University of Michigan. He has several years of industrial experience with Hughes Aircraft and Beckman Instruments. Dr. Behagi joined Penn State University as an associate professor of electrical engineering in 1986. He has devoted over 20 years to teaching RF and microwave engineering courses and directing university research projects. While at Penn State he received the National Science Foundation equipment grant to establish a high frequency microwave lab, and the Keysight/Agilent software grant to use for teaching RF and microwave engineering courses. After retirement from Penn State he has been active in authoring and publishing RF and microwave circuit design textbooks. In authoring textbooks he is using various software including the Keysight ADS and Genesys software. Dr. Behagi is recognized as the Keysight Distinguished Author and the Keysight Certified Expert. He is a Life Member of the Institute of Electrical and Electronics Engineers (IEEE), and the Microwave Theory and Techniques Society.

www.ingramcontent.com/pod-product-compliance
Lightning Source LLC
Chambersburg PA
CBHW082002190326
41458CB00010B/3045